East African Vegetation

East African Vegetation

E. M. Lind and M. E. S. Morrison

with a contribution by
A. C. Hamilton

Longman

LONGMAN GROUP LIMITED
London

Associated companies, branches and representatives
throughout the world

First published 1974

ISBN 0 582 44149 8

Printed in Great Britain by
J. W. Arrowsmith Ltd., Bristol

Preface

In recent years there has been greatly increased interest in various aspects of East African vegetation. This is due, in part, to the development of botanical and ecological teaching in the universities, and also to the encouragement of research by the various government departments concerned with agriculture, forestry and wild life.

When they were teaching in the botany department of Makerere College, Uganda, the authors became increasingly aware of the need for a book which could serve for students and others as an introduction to the vegetation of East Africa as a whole. Its absence is not surprising in a country with such varying climatic conditions, and where plants are found from sea level to 6000 m, and vegetation ranges from semidesert to tropical forest.

An immense debt of gratitude is due to the early explorers and travellers for the descriptions they have left of plant and animal life: but, more recently, with the exception of notable books on the forests of Uganda and Kenya, much of the information about vegetation has appeared in reports of agricultural and forests officers and has become buried in government archives and is not easily available. In attempting to meet the need for a more comprehensive account, the authors have drawn on all these sources, adding their own observations and those of colleagues and they have tried to provide an account of the main vegetation types of Uganda, Kenya and Tanzania (excluding Zanzibar). Some apology is necessary for greater attention being paid to Uganda, which they knew well, rather than to Kenya and Tanzania, with which they were less well acquainted.

The approach is ecological, and considerable attention is paid to features of the environment which influence the distribution of vegetation. As the literature is so scattered, it seemed wise to include a detailed bibliography though the entries may not all be referred to in the text. Every effort has been made to keep the nomenclature up to date, and, where names have been recently changed, synonyms are given in the index.

When the book was begun, the authors were colleagues in the botany department at Makerere University College. The tragic and sudden death of Dr Michael Morrison in 1970 has left its completion to the senior author. Fortunately, the greater part of the work was finished; but there remained the sections dealing with climate and the chapter on the history of the vegetation. This was Dr Morrison's special research interest and had been left to the end. I have completed the sections on environment from Dr Morrison's notes with the help of Dr J. Kenworthy and Dr B. McLean who had both been postgraduate students at Makerere. The chapter on the history of the vegetation has been contributed by Dr A. Hamilton using material collected during three years' research with Dr Morrison.

It was Dr Morrison's wish that this book should serve to stimulate interest in the plant life of the beautiful country he so much enjoyed. I hope it may do this and also provide some memorial to a valued colleague whose death has been a sad loss to botanical studies in East Africa.

Acknowledgements

Publisher's Acknowledgements

This book could not have been written without the help of many friends who have contributed from their detailed knowledge of the vegetation. I am indebted to Dr P. J. Greenway whose knowledge of East African vegetation is unsurpassed, to Mr John Procter and Mr J. McCarthy for assistance with Tanzanian vegetation, and to Dr Patrick Denny who has contributed some of his own observations on aquatic vegetation. Professor and Mrs W. E. Isaac have helped me with the marine flora and Miss McClusky with the mangrove swamps, while to Mr J. Ford and Mr A. French I am grateful for information about the tsetse fly and termites respectively.

Acknowledgement is made in the text to those who have contributed figures or lists of plants from various habitats. In a few cases the lists were provided by Dr Morrison and I have not been able to trace their origin. I must therefore apologise for any omission of this kind. Although we are grateful for the help of these and other friends, the authors must take full responsibility for opinions expressed and for any mistakes they have inadvertently made in trying to cover such a vast subject.

Our thanks are due to the director and staff of the herbarium of the Royal Botanic Gardens, Kew, and of the East African Herbarium, Nairobi, both for identification of specimens and for help with the ever-changing nomenclature.*

Finally, we would thank those who have allowed us to use their figures and photographs, especially Professor Fergus Wilson, Mr H. A. Osmaston, Mr K. Lye and Commander Templer whose drawings of flowers help to relieve the manuscript. We must also thank our artists Miss Joanna Langhorne and Mrs J. Worthington; and our typist, Mrs Todd, whose experience of typing scientific material has been of such assistance.

We are grateful to the following for permission to reproduce copyright material:
Blackwell Scientific Publications Ltd extracts from the *Journal of Ecology*, Vol. 30, 1942, by B. O. Burtt and Vol. 46, 1958, by A. V. Bogdan; East African Natural History Society for an extract from *Wildflowers of the Nairobi National Park* by B. Verdcourt, and Weidenfeld and Nicolson Ltd for extracts from *The Life of Plants* by E. J. H. Corner.

* In a related connection it is appropriate here to mention the renaming of certain geographical features referred to in the text. The following changes have recently been announced by the Government of Uganda:
Lake Albert *is now* Lake Sessekou Mobutu,
Lake Edward *is now* Lake Idi Amin Dadi,
Queen Elizabeth National Park *is now* Ruwenzori National Park,
Murchison Falls National Park *is now* Kaberega National Park.

As we continually refer to published literature which uses the old names we have retained them in this book. The reader should, however, familiarise himself or herself with the new names as they will appear increasingly in the literature of this subject.

Contents

viii

List of Plates

38 *Dendrosenecio adnivalis, Lobelia wollastonii*, Ruwenzori, 4000 m. (Morrison)

39 *Dendrosenecio adnivalis*, Ruwenzori, 4000 m. (Morrison)

40 *Lobelia deckenii*. (Morrison)

41 Tussock grassland with *Dendrosenecio*'s, Mt Elgon. (Morrison)

42 *Dendrosenecio gardneri*, Mt Elgon. (Morrison)

43 Erosion gulley now used as hippopotamus track, Murchison Park. (Osmaston)

Introduction and Classification of Vegetation types

A glance at the contents will indicate what a wide variety of vegetation is to be found in East Africa. It is determined primarily by water availability with arid bushland at one extreme and closed rain forest at the other. Except where interference by fire or grazing gives a sharp boundary, the terrestrial vegetation changes gradually from one type to another along a gradient of increasing water availability, and sharp discontinuities are exceptional.

A good introduction to the vegetation changes can be obtained by taking a journey across Kenya and Uganda from the coast to the Zaire border and in this way reference can also be made to the cultivated crops in the settled areas which otherwise receive little attention.

The vegetation at the sea coast is confined to a narrow strip at sea level with an annual rainfall of about 1000 mm. The area is subject to winds from the Indian Ocean, is densely settled and characterised by coral vegetation and coconut palms (Plate 1). There is a rich covering of seaweeds and marine angiosperms on the shore, and gardens are gay with brightly coloured trees and shrubs. Mangrove swamps are a feature of some parts of the coastline.

Leaving the coast, the vegetation changes first to evergreen bushland and then to dry open country too dry for human settlement and inhabited largely by nomadic Masai tribesmen. Scattered trees and shrubs, many of them thorny or succulent, are interspersed with wide stretches of red soil, and it is here that some of Kenya's largest game reserves are found. Along the river-beds flat-topped *Acacias* are to be seen and palms and an occasional baobab.

This kind of country approaches quite near to Nairobi, but beyond the city it gives place to densely populated Kikuyu country where, among remnants of the original forest, are cultivated crops such as bananas, maize and coffee and plantations of the Australian *Eucalyptus*. Nairobi lies at 1680 m on the edge of the Kenya Highlands which are divided by the Eastern Rift Valley. Formerly this was a country of European farming with patches of African reserve; but now much of the land has been bought by the government and reallocated as smallholdings to Africans. The broad, flat floor of the Rift is still an area of extensive European farming, and to the east lie a series of steep terraces leading up to the 3660 m peaks of the Aberdares. Much of the original cedar (*Juniperus procera*) of the higher slopes has been replaced by plantations of cypress (*Cupressus lusitanica pinus patula*) and pine while the lower part is covered by dense bush in which the camphor bush (*Tarchonanthus camphoratus*) predominates. In the valley floor is wooded grassland with *Acokanthera*, *Euphorbia candelabra* and various species of *Acacia* including the fever tree (*Acacia xanthophloea*).

In the floor of the Rift is a series of beautiful lakes interspersed with extinct volcanoes. Naivasha is a freshwater lake surrounded by papyrus and Acacia woodland, while Lakes Nakuru and Elmenteita are saline with more sparse vegetation.

Climbing out of the Rift Valley to 2600 m we traverse the upland forest zone with much *Juniperus procera* and *Podocarpus* as well as *Ocotea usambarensis* (camphor wood) and associated species. Much of the bamboo (*Arundinaria alpina*) which used to grow at higher altitudes has been cleared to make way for farms or plantations, and round Eldoret are extensive plantations of wattle (*Acacia* spp.). Much of the country between here and Lake Victoria is devoted to tea growing.

Leaving Kenya, which receives a reliable 750 mm of rain in only 15 per cent of its territory, and on entering Uganda, which receives rainfall adequate for crop production over 75 per cent of its surface, the change in vegetation is immediately noticeable. The forests between the Nandi Escarpment and Lake Victoria resemble the high forest of Uganda rather than the cedar forests of Kenya, and large areas of grass and papyrus swamp become a feature of the

1   Coconut palms at the coast. (Fergus Wilson)

valleys. Passing through wooded grassland with mixed *Acacia* and broad-leaved trees we reach Jinja in the densely populated region of the lake basin. This area of high, well-distributed rainfall and fertile soil was formerly covered with forest, some of which remains, especially in the Mabira forest between Jinja and Kampala. Now it is broken by intensive cultivation interspersed with forest patches, swamps and plantations of *Eucalyptus*. The ubiquitous banana dominates the landscape, together with cotton, coffee, groundnuts, sweet potatoes, and mango trees.

It is a strange fact that the majority of subsistence crops raised by Africans are of foreign origin. Even the banana, of which there are at least fifty types in Uganda alone, with others in Kenya and Tanzania, which has been in East Africa for a very long time, is probably of Malaysian origin. It is among the most important tropical fruits and may have been one of the first to be cultivated. As it hardly ever has seed, it must have been propagated vegetatively ever since it originated. Besides the use of the fruit for food and for beer making, the plants provide shade, wrapping material, fibre and mulch (Plate 2). Of other edible fruits the jak fruit (*Artocarpus integrifolia*) comes from India, the orange from China, the paw-paw (*Carica papaya*) (Plate 3) and the pineapple (*Ananas comosa*) from tropical America, the last two probably introduced by the Portuguese.

Of the cereal grains, maize (*Zea mays*) is widely cultivated in all three territories. It is known to be of very ancient cultivation on the American continent, where it probably originated. The history of sorghum (*Sorghum bicoler*), finger millet (*Eleusine coracana*) and rice (*Oryza sativa*) may be somewhat different, for wild species and varieties of these genera still occur in East Africa, and from them cultivated forms may have originated. The groundnut (*Arachis hypogaea*) is known only in cultivation. Extensive sugar-cane plantations are a feature of this part of Uganda, as well as in parts of the other terri-

2 Buganda homestead with bananas. (Fergus Wilson)

3 Pawpaw trees; male and female. (Fergus Wilson)

tories, and the kind of climate and soil which formerly supported forest and is now heavily cultivated is also characterised by patches of elephant grass (*Pennisetum purpureum*) between the shambas.

Travelling westward, a clear pattern of vegetation begins to emerge and continues for some 240 km. It is associated with the regular topography of flat-topped hills about 1200 m high with swamp-filled valleys in between which resulted from the dissection by rivers of an ancient peneplain. The hilltops are usually eroded to rock, but are covered with short grass or sometimes by forest where soil patches remain. The soil which has been washed down the lower slopes is deep and fertile and here is to be found most of the cultivation with elephant grass and some trees. The valleys with clayey soil, often waterlogged, are filled with grass or papyrus swamp fringed by swamp forest. Intense settlement, together with the system of shifting cultivation practised by the Baganda, has resulted in the destruction of much of the high forest. Where patches remain, glimpses may be obtained of the tall, bare boles of the trees branching at the top to give a closed canopy, beneath which is a leafy, evergreen undergrowth with a tangle of creepers and woody lianes.

Further west, in a region of lower rainfall (750–950 mm), the Buganda pattern gives place to rounded, grassy hills with scattered trees and woodland. *Acacia* trees are common, as are *Euphorbia candelabrum* in the drier parts and, especially in the north, species of *Combretum* and *Terminalia*. Among the more showy trees are *Erythrina abyssinica*, bearing bright red flowers before the leaves, *Spathodea campanulata* (Uganda flame) and a tree which somewhat resembles it but has yellow flowers, *Markhamia platycalyx*.

On a clear day a first sight of the magnificent, snow-capped Ruwenzori Mountains may be obtained before descending through thorny and succulent vegetation into the Western Rift Valley. Stanley, in 1875, though he knew of the supposed existence of the Mountains of the Moon, failed to see them because the range was buried in mist. It was only on the return journey, in

1888, that he was able to describe this great range which rises to 5112 m between Uganda and Zaire. In the floor of the Rift lie lakes Edward,* Albert† and George and the wooded grassland of the Queen Elizabeth National Park.**

Had the journey begun at Dar-es-Salaam and traversed Tanzania, instead of Uganda and Kenya, we should again have passed through many miles of bushland with forest confined largely to the mountains. But Tanzania has a type of woodland almost unknown in the other territories. This is the open woodland known as 'miombo' and it is dominated by deciduous, compound-leaved trees belonging to the Leguminosae which, in December and January, burst into leaf with a beautiful flush of flame-coloured foliage followed by the flowering of *Brachystegia*, *Julbernardia* and other species. With the arrival of the long, dry period most of the trees shed their leaves and the country has very monotonous appearance. On the lower ground are patches of damp, grassy land with a few trees and known as 'mbugas'. A feature of some of the settled areas of Tanzania is the prevalence of both coconut and date palms (Plate 4) and countless mango

trees, the last two being a relic of the days of the African slave trade.

The main types of East African vegetation can be defined briefly as follows:

1 **Forests** consist of trees of columnar habit, often reaching a height of 50 m or more, with crowns touching and intermingling to form a continuous, deep canopy of complex structure. Lianes or woody vines are a characteristic component, often interlaced in the canopy. Epiphytes (bryophytes, ferns and orchids) are characteristic of the wetter forests. Most of the trees are evergreens, but in some—usually early seral stages of forests—deciduous trees may be prominent. The forest floor is usually incompletely covered with herbs and shrubs and the grasses are very broad-leafed.

2 **Woodlands** contain trees, usually more branched than columnar, often reaching a height of about 18 m with crowns which do not form a complex, deep canopy. Usually the trees are leafless for some part of

* *now renamed* Lake Idi Amin Dada
† *now renamed* Lake Sessekou Mobutu
** *now renamed* Ruwenzori National Park

4   Date palms near Tabora, Tanzania. (Fergus Wilson)

the year. Scattered evergreen shrubs may be present, but are not conspicuous. Epiphytes, with the exception of lichens, are rare. Grasses and herbs dominate the woodland floor.

3 **Bushland** is an assemblage of woody plants, mostly of shrub habit, having a height of less than 6m with occasional emergents and a cover of not more than 20 per cent. An emergent is a tree or bush growing to a considerably greater height than the surrounding vegetation.

4 **Grassland** is dominated by grasses, sometimes with widely scattered or grouped trees and shrubs, often on termite mounds. The canopy cover of the trees and shrubs does not exceed 2 per cent.

5 **Bushed grassland** is grassland with scattered or grouped shrubs, the shrubs always conspicuous, but having a cover of less than 20 per cent.

6 **Wooded grassland** is grassland with scattered or grouped trees, the trees always conspicuous, but having a cover of less than 20 per cent.

7 **Dwarf shrub grassland** is arid land sparsely covered by grasses and dwarf shrubs not exceeding 1m in height, sometimes with widely scattered large shrubs and stunted trees.

Barren land is naturally almost devoid of vegetation; in other words, semidesert and desert, where the vegetation is so thinly scattered that the landscape is dominated at all seasons by the colour of the soil.

Permanent swamp vegetation consists of grasses, sedges, reeds, herbs and ferns with their stems rising out of the free water accumulated on the surface of the soil. Forests growing on similarly swampy land are referred to as swamp or groundwater forests.

Some vegetation maps include a further type, namely, vegetation actively induced by man. This is important because large areas, especially in Uganda and some parts of Kenya, are used for intense cultivation. Further discussion of the vegetation types listed above with references to published descriptions and illustrations is available in a very useful mimeographed report prepared by Greenway (1943).

Special attention has been given by Pratt *et al.* (1966) to the classification of rangelands which are represented in the above types by numbers 2 to 7 inclusive. Rangeland includes all areas providing a habitat suitable for herds of wild or domestic animals, and they cover the greater part of East Africa. They are at present the subject of active research to discover their production potential.

The above classification is based largely on the broad types selected in 1940 by an International Pasture Conference in Nairobi. We have avoided terms of vernacular origin, such as 'savanna', 'steppe', 'veld', 'dambo'. In their correct usage they are too specialised, e.g. the term 'savanna' was originally used by the Carib Indians to describe certain treeless South American grasslands, while the term 'steppe' referred to treeless open grasslands in Russia and Central Asia.

The general distribution of the main types of vegetation is shown in the map (Fig. 0.1), though complete accuracy is impossible on such a small scale.

Bushland
and thicket

Grassland

Wooded
grassland

Woodland

Forest

Desert

Semi-desert

Afro-alpine heath
and moorland

Swamp

Mangrove

a – L. Rudolph
b – L.Albert (Sesekou Mobutu)
c – L. Edward (Idi Amin Dada)
d – L. George
e – L. Kyoga

f – L. Baringo
g – L.Victoria
h – L. Naivasha
i – L. Natron
j – L. Manyara

k – L. Eyasi
l – L. Tanganyika
m– L. Rukwa
n – L. Malawi

xvii

Fig. 0.1 General distribution of the main types of vegetation in East Africa.

Part One

Vegetation types

One

Forests

Introduction and classification

In any book dealing with plant life in East Africa, considerable space must be devoted to forests, partly because they are widespread and show great diversity of composition and also because the timber they produce is one of the area's most valuable assets.

During the colonial era forest departments were set up in Uganda, Kenya, and Tanzania whose duty it was to study the forests and produce working plans which would ensure their most economic use. Forest reserves were created in which the felling of timber was controlled; but in many cases a great deal of damage had been done by wholesale removal of some of the most valuable timber without replanting, or by gradual erosion of the forest edge by cultivation.

It soon became obvious that all too little was known about important questions, such as the growth rate of timber trees, the optimum conditions for regeneration and the effect on the more valuable trees of the associated vegetation. Research departments were set up and areas of forest were set apart in each country for experimental purposes. In addition, timber utilisation sections began a study of the wood of the commoner trees.

Every country has to decide the best and most productive use of its land; and one of the problems facing developing countries, such as those of East Africa, is to know whether to preserve and develop existing forests or to replace them with exotic conifers or *Eucalyptus* which may yield a quicker return. On the other hand, it may be more profitable to allow the forest to be cut and the soil used for cultivation of tea or other crops.

The East African forests present such a wide variety in relation to such factors of the environment as altitude, rainfall and soil that it is difficult to do justice to them in the space of a single chapter. We have there-fore tried to select some of the more important types and to give some idea of their distribution and floristic composition. More detailed descriptions will be found in the papers listed in the bibliography.

Each great forest in East Africa tends to be distinctive floristically. Dawe (1906a) first noticed this after his great tour of the medium-altitude forests of Uganda in search of latex-bearing trees and lianes. He remarked that almost every forest in Uganda possessed a character peculiarly its own given it by the predominance of one or two particular trees.

This same feature has been noted elsewhere in East Africa by Trapnell and Langdale-Brown. Such diversity poses an enormous problem in writing about the forests and naturally one searches for some way of assembling the data into a few major categories. The usual solution is along the lines adopted by Greenway (1943) and by Trapnell and Langdale-Brown (Russell, 1962) who make a primary division into lowland and highland forests. We follow Greenway in preferring the term 'highland' or 'upland' in place of 'montane' or 'mountain', because many of the elevated areas of East Africa are gently graded highlands rather than precipitous rocky mountains. We shall use the term 'upland'.

Greenway sets the boundary between lowland and upland forests at 1350 m above sea level; Trapnell and Langdale-Brown have settled for 2000 m; and Hedberg (1951) put the boundary between 1700 m and 2300 m. Since Hedberg's system of classifying highland vegetation is so widely used we have ourselves settled for approximately 2000 m.

These differences of opinion between the boundary of lowland and upland forest reflect the fact that the boundary is an *arbitrary* one and is not based on a well marked discontinuity in the distribution of the forest species. There is, however, a distinct impression of an abrupt change around 2000 m created by the discontinuity of the forest cover. Very frequently the land in the altitudinal range 1600–2500 m has been cleared of its natural vegetation and is covered by cultivation and post-cultivation vegetation.

In East Africa the forest, to judge from available evidence (Hamilton, 1970) changes *gradually* with increasing altitude, and this is true whether we consider floristic composition, the height of the trees, the occurrence of buttresses, the leaf-type (whether compound or simple), the leaf size or the percentage of trees with thorns. All these characters are commonly used to diagnose major forest categories, such as lowland and upland. However, although gradual change with increasing altitude is true for the characters mentioned, it is nonetheless clear that one or several species become dominant over certain parts of the altitudinal range and this allows us to recognize vegetation *zones* within the forest. One of the best-marked zones, certainly on the wetter face of any mountain, is the bamboo thicket or woodland which occurs mainly between 2400 m and 3000 m.

A feature which adds to the usefulness of the division into lowland and upland forests is the fact that many communities of the upland forests are common to all three countries in East Africa. For example, the structure and composition of *Juniperus* forest, bamboo thicket, and *Hagenia-Rapanea* woodland are essentially the same in Kenya, Tanzania and Uganda. The upland habitats obviously have much in common even when widely separated.

Within the upland forest belt, which extends upwards to between 3000 and 3300 m, there is a broad altitudinal differentiation into three zones. Thus, the lower part of the belt, from 1500–2400 m consists of broadleaved evergreen hardwoods and some conifers. In the middle part of the belt bamboo thicket is often dominant, and in the highest part there may be woodland with much *Hagenia* and *Rapanea*.

Inside each of these vegetation zones the distribution of forest types is closely determined by the availability of water. The direction of the moisture-bearing air streams creates wetter and drier faces on the mountains. On the Virungas, the Ruwenzori, and Mt Elgon the west side is distinctly wetter than the east. On the Aberdares, Mt Meru and Mt Kenya, the south side is wetter than the north, and on Mt Kilimanjaro the south-east side is wetter than the north-west. Steep hillsides and narrow valleys may, of course, alter the local distribution of wetness and dryness. However, it remains broadly true that the upland forest changes gradually in composition from the wettest side of the mountain to the drier side, and within this moisture availability continuum, several fairly distinct forest types are recognised. The arrangement of these forest types is shown very diagramatically in Fig. 1.1 which is also applicable

to a mountain in Kenya or Tanzania. The arrangement in Uganda differs slightly because of the absence of *Ocotea-Podocarpus* forest in the wetter parts of the upland forest belt. *Ocotea usambarensis* does occur, but rarely, in western Uganda, and it is recorded in eastern Zaire (Robyns, 1948). However, for reasons not yet known, it is not prominent even on the wettest forests of the western side of the Ruwenzori in Zaire between 1800 and 2300 m altitude. Robyns describes the forests there as comprising *Podocarpus milanjianus* associated with species such as *Albizia gummifera*, *Trichilia volkensii*, *Croton butaguensis*, *Macaranga kilimandscharica*, *Allophylus abyssinicus*, *Dombeya goetzenii*, *Symphonia gabonensis*, *Olinia usambarensis*, *Cassipourea gummiflua* var *ugandensis*, *Alangium chinense*, *Syzigium guineense*, *Polyscias fulva*, *Olea chrysophylla*, *Aningeria adolfii-friedericii*, and *Anthocleista orientalis*.

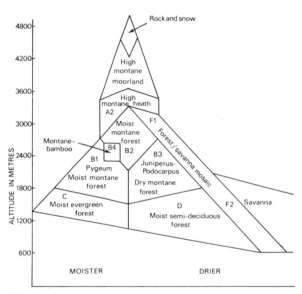

Fig. 1.1 Generalised altitude and moisture relationships of the montane forest vegetation of Uganda. (Langdale-Brown *et al.*, Department of Agriculture, Uganda, 1964).

Just as the forest types in each zone are controlled by availability of water, so too is the distribution of species *within* each of the forest types. The influence of topography and physical characters of the soil is very marked since both of these control water availability. Thus in each forest type it is usually possible to recognise distinct differences between the forest of the hill ridges and that of the valleys, and between that of the wide valleys open to the wind and forest in narrow enclosed valleys and in gulleys where there may be relatively little air movement. Other factors may accentuate these differences. For example, where elephants are abundant they will usually congregate on hill ridges and hilltops, do much damage, and give rise to a kind of secondary forest or even to grassland with a few trees and shrubs.

The naming of the forest types described in the following account is bound to be a source of dissatisfaction. As far as possible we have named each forest by the most abundant and/or characteristic tree of the hillslopes, avoiding species characteristic of 'extremes' such as valley, gulley, hilltop and ridges. This expedient is, of course, unsatisfactory and open to the same kind of criticism as a mapping which represents the soils of an area as red earths although extensive areas of grey soils may occur in the valleys. With forest vegetation, a further difficulty is created by the fact that a species may occur abundantly on hillslopes and even ridges in the wettest forests, but under drier conditions it may be limited to the lower hillslopes or even to valleys. Also the majority of descriptions of East African forests are very rudimentary; often they do not state clearly what type of site is being described, whether ridge, hillslope or valley, and species from all three habitats are indiscriminately placed together in one species list. Even if we do know that certain species characterise the hillslope we very rarely have information on relative abundance. Abundance will vary along moisture availability and temperature gradients which operate in a vertical and horizontal direction in each vegetation belt. The forests which are described represent only distinctive types in a three-dimensional vegetational continuum. It is very important to bear this in mind.

General features of East African forests

As East Africa has a large number of forest types it is best to begin by looking beyond the details of the individual forests and attempt to describe general features which they share. Once the common plan is established it will be easier to describe the principal communities and to appreciate that each, despite considerable floristic individuality, conforms to a general structure. As the norm for this introduction we have the forests at medium altitude (c. 1200 m) of Uganda. These forests occur in a belt around the north end of Lake Victoria and again along the shoulders of the Western Escarpment. In terms of general appearance and structure they are similar to some of the forests of Zaire and of West Africa and many of the genera found in these forests occur also in the west. But it is equally true that they lack many genera that are widely distributed in Zaire and in the west. A more widely based and more thorough treatment of many topics touched on in this introductory essay will be found in Richards (1952).

Diversity of species

The diversity of forest communities has its origin in many environmental and historical factors, but one major element is the large number of tree species which is a feature of tropical forests. These can be associated in many different ways.

In temperate places forests are dominated by a few and sometimes by one species and some of the richest, for example those of the Appalachian mountains in eastern USA, have only about 25 species of tree. Most tropical forests have not less than 50 and the number is often much greater. Richards (1952) obtained 109 species of 30 cm girth and over on a plot of 1·5 ha in the Cameroons. Cousens (1951) reported 197 and 227 species of 30 cm girth and over on two plots of 1·6 ha in the lowlands of western Malaya. Poore (1968) recorded 375 tree species belonging to the upper canopy in 23 ha of lowland forest in Malaya which was selected for its uniformity. Some of the Amazonian forests approach this degree of richness but the African rain forest is poorer in species. Dawkins has recorded nearly 200 species of flowering plants from Mpanga forest in the Lake Victoria belt, and in this total only 50 were tree species exceeding 30 cm in girth. This forest occurs at 1200 m under an annual rainfall of 1168 mm. The total number of canopy trees recorded from about 45 km^2 of Bugoma forest is certainly not more than 80. This is one of the forests of the Western Escarpment group, lying between 1200 and 1400 m and having a mean annual rainfall of about 1270 mm. The species lists in both these examples cover a variety of type site, such as ridge, hillside and valley, so that any uniform type site in either would be much poorer in species.

Genera containing a large number of closely allied species characterise the tropical forests of America and Malaysia, but in East Africa the number of species per forest genus is low (see Table 1.1). Indeed, chains of closely allied species are more characteristic of genera of wooded grassland.

Species richness is not confined to the tropical forests. For example, Kerfoot (1965) recorded 400 species of vascular plants in about 80 ha of vegetation in north-eastern Uganda. This was dry woodland and wooded grassland degraded in many places by over-

Table 1.1.

Number of species per genus	1	2	3	4	5	>5
Uganda *	59	18	6	4	5	8
Kenya†	55	17	10	3	4	11

*Data taken from Eggeling and Dale (1951).
†Data taken from Dale and Greenway (1961).

Genera with more than five species include the following: *Acacia, Albizia, Allophylus, Canthium, Capparis, Cassipourea, Celtis, Chrysophyllum, Clerodendron, Combretum, Croton, Dombeya, Euphorbia, Ficus, Grewia, Macaranga, Millettia, Ochna, Pavetta, Phyllanthus, Psychotria, Rhytigynia, Rinorea, Senecio, Strychnos, Teclea, Terminalia, Vernonia, Vitex.* With the exception of *Cassipourea, Celtis, Chrysophyllum, Macaranga, Ochna,* and *Teclea,* the majority of these genera having more than five species occur in the wooded grasslands and in thickets.

grazing to semi-arid thornbush and scrub. Richards (1969) suggested that sclerophyll scrub in the Cape Peninsula and some sclerophyll communities in Australia and New Caledonia may well be richer than most types of rain forest. However, the fact remains that moist tropical forest has more tree species per unit area than any other type of forest.

How can this richness be accounted for? In part, as Richards (1969) points out, the total richness of the rain forest depends on the large number of physiognomically and ecologically distinct synusiae present, e.g. trees, lianes, epiphytes and ground herbs. Federov (1966) believed that species richness and series of closely allied species in the forest could be accounted for by genetic drift. It was thought this would be favoured by the relatively low densities of most species, their spatial isolation, the irregularity and lack of coincidence in flowering, the small effective populations, and the domination of autogamy. However, Ashton (1969) in a detailed discussion, provoked in part by Federov's views, believes that our present knowledge of rain-forest ecology does not require modes of speciation essentially different from those within other terrestrial plant ecosystems. This is in line with Richards (1969) who asks whether it is really more difficult to explain the presence of fifteen *Eugenias* and twelve *Shoreas* in the same small area of moist tropical forest than the coexistence of numerous similar Cyperaceae in some temperate wetland communities.

Leaving aside the problem of *how* the forest species developed, Richards (1969) goes on to ask the interesting question of how many ecological niches they occupy in the forest. If there are, say, 100 species in 1·5 ha of forest are they *all* occupying different ecological niches? Or is it possible that the number of niches is much less than 100 and that whether species A, B, C . . . occupies a given niche when it becomes vacant is determined to a large extent by chance? This, according to Richards, is one of the most important questions of tropical ecology.

It is commonly supposed that the primary production of the tropical forests, with their large numbers of species, diverse ecotypes, and life forms, is very high. However, according to Dawkins (1964), in the close equatorial forest on lowland sites, outside the Malaysian region, no tree or community indigenous to the zone, planted or otherwise, appears capable of production exceeding 16 tonnes/ha/year and the common range is from 4 to 9 tonnes. On the exceptionally rich soils with high rainfall in the Malaysian region up to 27 tonnes have been measured but this may have been from higher rather than lower altitudes. Certainly from Malaya nothing over 12 tonnes can be hoped for from indigenous species in the lowlands. The primary production levels are very different when, within the same equatorial zone with no restriction on altitude, we consider the conifers or eucalypts, indigenous or introduced. Here the range obtained in forestry is from 13 to 40 tonnes with up to 50 from exceptional *Eucalyptus* sites. Dawkins considers that by no stretch of reason could this kind of production be achieved by the 'natural' forest communities which such crops have replaced.

These figures present a very attractive research problem, the answer to which perhaps will be in terms of the relative rates of photosynthesis and respiration. Muller (1967) compared the productivity of moist forest in the Ivory Coast with cultivated beech forest in Denmark. Gross production, loss through respiration and net production were measured. Although the gross production in the tropical forest (53 tonnes/ha dry matter) was considerably higher than in the beech forest (24 tonnes), the loss through respiration in the rain forest was very much higher than in the beech forest and the *net* production was approximately the same in both forests—about 9 tonnes.

Forest structure

Partly as a consequence of the large number of species, the forest is stratified so that the top layer or cover of the forest, referred to as the *canopy*, is not a simple structure with the tree crowns at one level. It is complex, has great vertical depth, and is made up of several more or less distinct layers or storeys. This tendency to layers in the canopy must be due largely, as Dawkins (1966) explains, to the fact that forest trees attain their ultimate or mature height relatively early in their lifespan and undergo a long period of subsequent growth in girth, during which time the overall height increases only slightly. Also, the crowns flatten with age, that is they tend to cease growing in depth while still growing laterally, and so increase their overall diameter. These tendencies, in combination, result in foliage being concentrated at distinct levels, where species of similar mature height are abundant. This layer effect is not dependent upon, though it is enhanced by a concentration of mature individuals. The fact that most of the tree's life is spent at mature height ensures coincidence of crowns, even in an all-aged stand. Therefore, stratification is inevitable where a single species is locally frequent, and this is quite common in East African forests. Dawkins (1966) is of the opinion that one is unlikely to find stratification in highly polyspecific natural forest, although he suggests that shade-enduring individuals, in a very densely populated community, may respond to competition for crown-space by accepting smaller crowns at different levels. This would tend to spread the canopy in vertical depth, and it is not easy to see why this, by itself, should lead to a stratification into *three* major layers. The precise structure of deep canopy has been much discussed and the problem requires further field investigation (Grubb *et al.*, 1963).

The literature on East African forests provides little evidence on canopy structure. One well-described canopy is that of Budongo forest, a mahogany-rich forest of the group on the Western Escarpment in Uganda. Eggeling (1940a and 1947) has described this in detail with transects (see Figs. 1.5 and 1.6) of the canopy in communities of seral stages. He adopted the three-layer concept for the canopy of the mixed community which is the midway stage in the development of some of these forests towards a climax of Uganda Ironwood (*Cynometra*); it is, floristically, the richest of the stages in the forest succession.

The three-layer canopy structure is probably applicable to most low- and medium-altitude forests in East Africa. At higher altitudes the structure may sometimes be less complex, with just two main tree layers.

According to Eggeling, the upper, or A layer in Richards' (1952) terminology, is about 21–36 m in height and the trees which project through it form the uppermost or emergent stratum, 36 m high and upwards. At Budongo the tallest tree was an *Entandro-*

phragma cylindricum measuring 55 m. The genus *Entandrophragma* includes some of the tallest of the tropical African forest trees. According to Bates (1961) the tallest tropical trees are somewhat less than 91 m. In the northern temperate forests, the average height for taller trees, in the least disturbed forests, is around 30 m with 46 m as exceptional. But tropical African trees do not reach the gigantic proportions of the California redwoods (*Sequoia*) or the Australian *Eucalyptus*; the tallest measured *Sequoia* reaches 110 m, the tallest *Eucalyptus* 107 m.

Taken together, the two top layers, that is layer A and the emergents, form a closed *upper* canopy, though individually they do not. The closing of this canopy is accomplished not alone by the crowns of trees, but also by the crowns of large woody vines or *lianes*, which are a characteristic component of the forest and are often abundant. It is easy to overlook this latter fact because, within the forest, they are sometimes not much in evidence. Being supported by the trees, they have dispensed with massive stems and have instead slender, often fluted ones which are largely conducting, and this *slenderness* belies the abundance of foliage and branches supported at canopy level. From a high vantage point, when the lianes flower, one realises, usually with surprise, their profuseness. They are especially profuse in the Lake Victoria belt forests and, in some, there may be as many species of liane in the canopy as species of trees.

Most medium-altitude forests in Uganda show some deciduousness in the top canopy throughout the year though this will be most marked during the dry seasons and there is little doubt that this indicates water stress. This is not surprising since at this altitude, at approximately 1200 m, the annual potential evaporation lies between 1500 and 2000 mm and the average annual rainfall in the forest belts ranges from about 1000 to 1400 mm. It is not certain that this deciduousness of the canopy would affect the net assimilation rate. Light passing downwards would be intercepted by lower strata and, unless these were suffering from severe water shortage, carbohydrate would be synthesised.

In Budongo the next layer, the B layer, was from 11 to 12 m high, usually with oblong crowns in lateral contact. The bottom or C layer of smaller trees up to 11 m usually had rounded or broadly pyramidal crowns often forming a closed canopy. In conjunction the B and C layers form a closed lower canopy.

The trees of the various strata in Budongo, as in other tropical rain forests, have characteristic features. Thus, the top-storey or A layer species have spreading, often umbrella-shaped crowns, supported by a candelabra-like system of branches in which the main axis is rarely well marked above the first fork. The trunks of the top storey and emergent trees are sometimes heavily buttressed or fluted. (Plate 5).

In contrast, the tall trees of the B layer of the middle storey have the main axis of the stem distinct almost to the top of the tree (Plate 6). Buttresses are not usual and, if present, do not extend far up the stem. The shorter trees of the middle storey have larger, much-branched crowns and the stems are unbuttressed. The under storey species are more variable, tending to have rather small, pyramidal crowns and slender unbuttressed stems.

Undergrowth and herbs—Microclimate

Below the lowest layer of trees there are tree saplings, shrubs, herbs, grasses and seedlings and all these, shielded by the deep canopy, experience a different climate from that in and around the crowns of the mature trees. The canopy reduces greatly the illumination and buffers the climate within the forest from the extremes of temperature and water-vapour saturation deficit experienced in the open. Also, the wall of lianes, secondary forest and thicket on the forest-edge form an effective wind-shield so that air-movement inside the forest is negligible. Figures 1.2 and 1.3 show quantitatively how conditions in the under storey contrast with conditions in the upper part of the canopy; further similar data for Uganda forests are given by Garnham, Harper and Highton (1946), Haddow (1945 a, b), Haddow and Corbet (1961), Haddow Gillet and Highton (1947), and Haddow, Corbet and Gillett (1961).

Haddow, working in various forests in Uganda found that near the forest floor the light intensity sometimes shows a strongly biphasic pattern, with peaks in the morning and afternoon and a period of lower intensity around midday, when the crowns of the tree cut out much of the sunshine. The biphasic effect is most pronounced where flat-topped trees such as *Newtonia*, *Piptadeniastrum* and *Albizia* prevail. He believes this has a bearing on mosquito biting patterns, as these are biphasic in many ground-haunting species.

The Haddow *et al.* (1947) study is of interest in showing that in *dry* weather some forests become stratified internally into two main microclimatic layers. The forest studied was Mongiro which lies at about 760 m on the eastern edge of the Semliki forest. It is described as very mixed, with *Mitragyna stipulosa* as the dominant tree. This would seem to indicate a swampy habitat, perhaps only locally or seasonally to judge from the associated trees which were *Alstonia congoensis*, *Celtis mildbraedii*, *Chlorophora excelsa*,

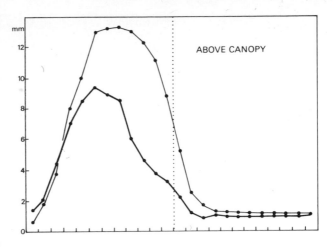

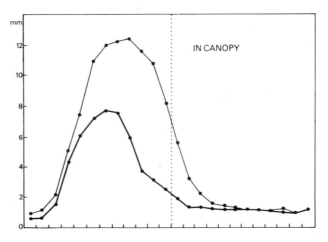

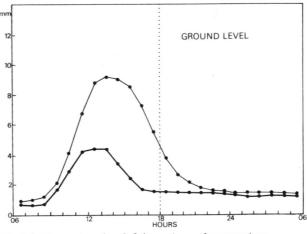

Fig. 1.2 Mean saturation deficiency (mm of mercury) on 18 days with afternoon rain (thick line) and 28 days with no afternoon rain (thin line). The levels shown are: above canopy (35 m), canopy (17 m), ground level (2 m). (Haddow and Corbet, 1961).

5 Buttressed tree and *Dracaena* sp. (Morrison)

Cola gigantea, Cynometra alexandri, Elaeis guineensis, Erythrina excelsa, Ficus capensis, Khaya anthotheca, and *Macaranga schweinfurthii.*

The main canopy lay between 15 and 24 m and nowhere exceeded 30 m; there were a few emergents rising clear of the main canopy with their first branches at about 24 m and their crowns reaching up to about 40 m. The main zone of the understorey was between 9 and 15 m and the crowns of the larger oil palms lay mainly in this stratum. Between 3 and 5 m there was a light, continuous zone of young trees including the screw pine *Pandanus chiliocarpus.* The mixed under-growth extended to heights of between 1 and 3 metres. Lianes were plentiful and epiphytic ferns and orchids were common on lower branches.

Figure 1.4 shows that the differences between the ground-level station and that at 5 m are relatively slight, both by day and by night. On the other hand, marked differences may be noted between the ground-level results and those obtained at 9 and 16 m. Further, it can be seen that the differences between the ground and the 16 m level are not much greater than those between the ground and 9 m. In other words, there are two main microclimatic strata in Mongiro forest. The lower, characterised by equability, coolness and low saturation deficit, extends from the ground level up to

Fig. 1.3 Mean hourly temperature (°C) on 18 days with afternoon rain (thick line) and on 28 days with no afternoon rain (thin line). The levels shown are: above canopy (35 m), canopy (17 m), ground level (2 m). (Haddow and Corbet, 1961).

6 Bugoma Forest, Bunyoro. (a) *Antiaris toxicaria.* (b) *Entandrophragma angolensis.* (c) *Entandrophragma cylindricum.* (d) *Trichilia splendida.* (e) *Alstonia boonei.* (f) *Entandrophragma utile.* (H. A. Osmaston)

at least five metres, that is, through the undergrowth to the discontinuous zone of small trees forming a lower understorey. The upper stratum extends from the lower limit of the understorey upwards to the main canopy and has a microclimate markedly different from that of the lower stratum. The temperature and saturation deficit traces from the 9 and 16 m stations show many of the characteristics of records made in the open air—a marked diurnal rise in temperature and saturation deficit. More particularly these traces show numerous sudden short-duration fluctuations, caused by puffs of wind, passing clouds, etc., which are conspicuously absent from traces made in the lower stratum. During the night, and particularly in the period before dawn, the differences between the two strata are much reduced, but they reappear soon after sunrise and persist until long after sunset.

Haddow et al. (1947) took care to emphasise that this internal stratification of forest into two main microclimatic layers is not necessarily widespread. On the other hand, there may well be parts of the Semliki forest which show even more sharply delimited strata than does Mongiro. In some rather dry areas, for example, the Uganda ironwood (*Cynometra alexandri*) occurs in almost pure stands forming an extremely light, though closed, canopy. The result is that the understorey of small trees—discontinuous at Mongiro—forms a dense belt of foliage between 4 and 6 m, below which the air is cool and very humid and the light intensity is low, while the conditions above

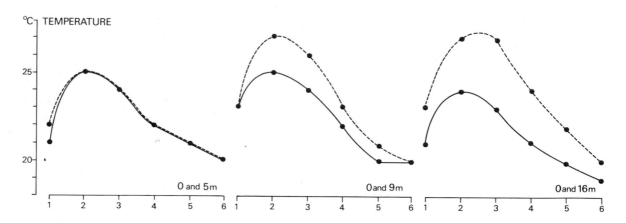

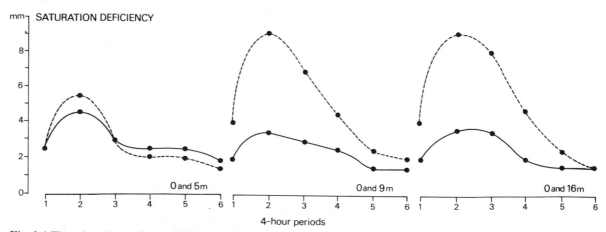

Fig. 1.4 The microclimate forest at Mongiro. Successive comparisons between the ground-level stations (0 m), and the three tree-platforms (5 m, 9 m and 16 m). The plotted points are four hour means calculated from hourly readings of thermograph and hydrograph traces (saturation deficiency also calculated from these traces). The numbers 1–6 represent the six 4-hour periods, namely, 06.10, 10.14, 14.18, 18.22, 22.02 and 02.06 hours. Continuous line: ground-level station; broken line: tree-stations. (From Haddow et al., 1947).

approximate to those of open air. But even in this type of forest the marked stratification into two microclimatic layers will tend to disappear during wet weather. Also in forests with a very dense canopy, which prevents the development of a well-defined undergrowth stratum, it may well be that climatic gradients rather than successive strata are to be expected. From South American forests Bates (1944) has described vertical gradients of temperature, saturation deficit and light increasing fairly regularly with increasing height.

Microclimatic features of the kind described above will have a substantial influence on the movement of insects within the forest and between the forest and its open surroundings. Certainly this is true for some important vector species of the mosquito (*Aëdes*). The effect on plant distribution may be fairly obvious for epiphytes (algae, bryophytes, orchids, etc.) but may not be so obvious where higher plants are concerned. However, certain effects come to mind. For example, the high humidity of the lower microclimatic layer will probably strongly favour the germination of trees with very large seeds like those of *Carapa grandiflora*. These four sided seeds are 4 to 5 cm in length and it is very unlikely that they would germinate successfully outside forest or even in dry forest. Their germination and establishment would probably be greatly favoured by a very humid, equable lower stratum in the forest.

On the other hand, the high saturation deficit in the upper layer will probably influence the size and sclerophylly of leaves. It is known that the rate of expansion of leaves, their ultimate laminar size, and their texture are affected by water stress which will be greatest in moving dry air. Small, sclerophyllous leaves are recorded as characteristic of some forest types in East Africa. Gilchrist (1952), for example, classified the *Cassipourea* forest of Tanzania as dry sclerophyll type and we could probably include in the same category the *Chrysophyllum* forest of south-western Uganda, and perhaps the *Brachylaena-Croton* forests of Kenya.

It might seem that these equable conditions in the understorey would be ideal for plant growth. However, although there is a rich microfauna and microflora, conditions are marginal for plants dependent on sunlight. This is demonstrated clearly by the small number of flowering species which live permanently on the forest floor. It is remarkable that so few flowering species from the profuse tropical flora have adapted themselves to the forest interior. Even saplings of the forest trees have difficulty in surviving; there are many East African forests with such a deficiency of saplings or recruits that it is apparent that regeneration will not be sufficient to retain the present floristic composition.

Many factors contribute to this situation. Root competition may be severe, there may be toxic products released in the breakdown of leaf litter, severe damage by browsing animals—especially elephants—climatic changes or changes in the surface soils. But it is widely believed that many higher plants are eliminated, and the survival of many rendered marginal, because of their inability to endure the very low light conditions which prevail; not only is the illumination feeble but its spectral richness has been reduced leaving it proportionally richer in red light. The intensity of the shade light is usually as low as one-hundredth of full sunlight, though for five to six hours in the middle of each day this is supplemented by sunflecks. These, which have the full spectrum of sunlight and range in duration from a few seconds up to a couple of minutes, more than treble the total energy received by plants on the forest floor, according to measurements made by Evans (1956) in lowland forest in Nigeria.

Under closed canopy it is often difficult to detect any increment in shrubs and small trees. The net production in the forest interior may often be very small on account of the poor illumination and the high rate of respiration relative to photosynthesis. Air temperature may be an important factor since the rate of photosynthesis is relatively constant between 10° and 30°C, but respiration is strongly dependent on temperature and an increase in temperature by 10°C often doubles the rate of respiration—at least in crop plants.

Many trees of the canopy, however, can tolerate low illumination, and their seeds germinate and become established as saplings. If the overlying canopy remains unbroken many of these saplings enter a stand-still phase during which they make scarcely any growth and they appear able to endure this for many years. Their leaves have a long life as indicated by the coating of lichens and liverworts which they acquire. Once, however, the canopy breaks, when an overmature tree topples, carries others with it and makes a large gap, then the saplings enter a phase of rapid growth and quickly reach mature height. At the same time, other trees not belonging to the canopy become established in the gaps. They include such trees as *Musanga*, *Polyscias*, which are fast-growing, with a relatively short life of thirty to forty years.

Apart from the tree saplings, there is a limited flora of shrubs and herbs which spend their entire life in the undergrowth. Many of these belong to either the Rubiaceae or Acanthacae. Usually several genera are present, but sometimes the shrub layer is dominated by a single species. A good example of this is the profusion of wild coffee, *Coffea canephora* or *C. eugenioides*, which occurs over large areas of the Kibale and Itwara forests in the Western Escarpment group in Uganda, and in much of the Zoka forest in West Nile. Elephants may be responsible for the distribution of wild coffee as they are very fond of the seeds.

Often the forest floor is covered, but not closely, with rhizomatous, broad-leaved forest grasses. The most common of these in the Uganda forests is *Leptaspis cochleata*; but others occur, namely, *Olyra latifolia*, *Oplismenus hirtellus*, *Streptogyne crinita*, *Setaria chevaleri*, *Commelinidium mayumbense*. *Leptaspis* has twisting petioles capable of orientating the leaf blade most efficiently to the incoming light. *Olyra* may grow much taller and, by sprawling through shrubs, may reach as much as three metres in height. *Oplismenus* is several centimetres in height and is common along the better lit pathways; it may be accompanied by tufts of tall *Setaria*. There are also, in Lake Victoria forests, rosettes of the broad-leaved liliaceous *Palisota schweinfurthii*, which has spikes of white flowers followed by bright red fruits. Similar in habit, but smaller, is *Pollia condensata*, which again has heads of white flowers, but followed by distinctive steel-blue fruits. *Costus* and *Aframomum* of the Zingiberaceae, are more common at the forest edge or in clearings.

Ferns of the forest floor are limited usually to specialised habitats, for example, *Marattia fraxinea* in wet places or ravines, and species of *Pteris*, *Gleichenia*, *Thelypteris*, along paths where the light is greater, and filmy ferns, *Hymenophyllum* and *Trichomanes*, on rocks and tree trunks, in the very shady and moist places.

Bulbous-rooted herbs are not common and there are very few ground orchids. Occasionally one may see the white, not very showy flowers of *Corymborkis*. Also on the forest floor one may occasionally find the red starlike flowers of the root parasite *Thonningia sanquinea*.

It is fortunate that the forest is not richer in grasses and herbs and that those which occur are not especially attractive to grazing animals, since greater abundance, especially of grasses, might serve only to entice more animals from the surrounding grasslands, which would perhaps feed on grass, herbs and tender tree seedlings, without much discrimination and with disastrous results to the last. The growing points of the tree seedlings are exposed, so that a single mouthful taken from them usually causes permanent damage. At present, with much of the game killed off or confined to parks, it is difficult to appreciate that a short time ago the countryside abounded in animals. Huxley (1960) emphasises how tremendously the past abundance of game has been reduced over most of East and South-east Africa. He refers to Meinertzhagen who records having seen a procession of some 700 elephants marching across what is now the township of Nyeri and who stated that the country swarmed with rhino and in one day in an area of 24 square miles, adjacent to Nairobi, he counted 684 zebra, 894 wildebeeste, 276 Cake's hartebeeste, 326 Grant's gazelle, 426 Thomson's gazelle, 184 impala and 46 eland, besides giraffe, rhino, warthog, wild dog, and over 4000 Masai cattle. This abundance of animals in the wooded grasslands surrounding the forests must, for immense periods of time, have been a major factor, apart from fire and cultivation, in restraining the spread of the forest.

In some forests, for example parts of the Uganda ironwood forests and the Tanzanian camphor forests, the ground-layer is almost devoid of higher plants. This is unlikely to be due wholly to low illumination but may be caused by unfavourable soil condition, excessive root-competition, or by toxic substances present in the surface litter and soil layers.

Microflora and microfauna

The general poverty of the ground layer in higher plants leads one to overlook the immensely rich and varied microflora and microfauna. (Dring and Rayner, 1967). Occasionally, flushes of larger fungi follow wet periods and myriads of termites on forest paths remind us of these inhabitants which have a vital role in breaking down the litter. Corner (1964) describes the ground layer as the goldmine of the forest and he outlines very well the activities there in the following passage:

First in tropical soils come the termites. In armies from earthern barracks they speedily organise on their six-legged carriages batteries of jaws which fracture and comminute the material; in their insides they can digest lignin. They generally work at night, when the munching can be heard. Thus, trunks may vanish before fungus or beetle has been able to work on them. Then, various insect grubs, woodlice, millipedes, snails and slugs rasp at the remains that the termites have left, earthworms tug pieces into their burrows where they are softened by bacterial action and are worked eventually through the worm. The fungi grow more slowly on all and sundry, but their chemistry is far more varied and they reduce to edibility such material as bark, which is not at first palatable. The animals excrete, defaecate, and leave slimy tracks. The fungus excretes and leaves tracks of threads. Bacteria multiply on these. Protozoa multiply on the bacteria. Rotifers, nematodes and other minute animals multiply in these soil colonies and add their excretions. Beetles, spiders, centipedes, and scorpions prey on the soil animals and add their quota of waste products, all of different chemistry. Fungus spores abound, germinating and un-germinated, and, on these microscopic eggs, many protozoa, rotifers and mites feed. A mouldy smell compounded mainly of the many minute fungus odours, marks the solid, liquid and gaseous decomposition of this microcosm of chemical engineering.

As well as breaking down and mineralising the litter some of the fungi may form beneficial associations with roots of trees in the form of mycorrhiza which may assist in transferring water and minerals to the tree. But, because of difficulties in excavating root systems very little is known about these associations, and tropical forest root systems have hardly been investigated (Singer and Morello, 1960).

At one time it was believed that the respiration of roots and micro-organisms in the surface layers of forest soils produced very high concentrations of carbon dioxide in the undergrowth layer, and this was believed substantially to increase the carbohydrate assimilation of the undergrowth flora. The term 'carbon dioxide flora' was used to describe plants of the undergrowth. Evans (1939) investigated carbon dioxide levels in the undergrowth of forests in southern Nigeria. He discovered that the carbon dioxide level was 'above normal' during the early hours of the morning up to about 10.00 am, but there was no evidence of very high concentrations, and earlier reports of these were probably based on inaccurate measurements. The effect of these higher carbon dioxide levels on the undergrowth flora is not known definitely and is difficult to predict. The rate of photosynthesis will be a function of the intensity of illumination or of the carbon dioxide concentration or of both, if we are dealing with the intermediate region of interaction of the two factors. In those forests having a high proportion of flat-topped trees in the canopy (*Albizia*, *Newtonia*, *Piptadeniastrum*, etc.), there will be a biphasic pattern of light intensity and high level of diffuse light together with sunflecks will reach the undergrowth in the morning period. In this period, therefore, leaves of undergrowth plants illuminated by sunflecks will be able to make use of the increased carbon dioxide concentration. On the other hand, in many forests, the most intense illumination will be in the early afternoon and, during the morning, light will be a limiting factor.

The soil stratum

The properties of soils and their influence on vegetation are described and discussed in chapter seven. Here it will be sufficient to say that it is widely believed that it is the *physical* rather than the chemical characteristics of soils which influence the distribution of forest types, determine their structure and their floristic composition. Physical soil properties provided the closest correlation in the ordination carried out by Ashton 1967) in the Dipterocarp forests of Brunei. Physical properties such as soil texture, predominant clay and soil depth, influence the availability of water and this is unquestionably the master factor in tropical forest ecology. This statement finds support, for example,

in the practice of Ugandan foresters in internally classifying their forests into sites—such as 'raised dry' and 'raised wet'—which depend on the average depth of the water table below the soil surface. However, we should take note of Webb's (1969) study of forests in eastern Australia. He was able to show that, provided soils were well-drained and the annual average rainfall exceeded about 1250 mm, the nature and distribution of the forest, its structure, and floristic composition, were broadly correlated with soil nutrient levels. He considered that his results supported Beadle's (1954, 1962) hypothesis that the distribution of vegetation types in Australia, e.g. vine frost, sclerophyll forest and heath, are correlated with soil phosphate levels, provided that soil aeration is not limiting and topographic factors are not extreme. He believed that, apart from phosphate, potassium and possibly magnesium and sodium are involved.

Root system

The amount of water available to any tree depends largely on the volume of soil exploited by its roots. In soils below field capacity the capillary movement of water is very slow and an extensive root system is necessary for absorption. Many foresters believe that, although the bulk of the roots of a tree lie close to the surface, there are roots which penetrate deeply, perhaps to a depth of as much as 10 m. These can be detected in deep pits if the soil face is gently washed away. Such deep-ranging roots may be important not only in absorbing water but in recruiting elements and ensuring the efficiency of the nutrient cycle. There is, however, very little exact information about forest root systems in East Africa. Kerfoot (1962a) found that *Albizia gummifera* had a lateral root spread of 10 m and a vertical extension in excess of 6 m. The roots of *Apodytes dimidiata* have a lateral extension of just over 10 m and a vertical extension of 8·2 m. Tea roots, often reputed to have a shallow root system, were found to have a lateral spread of 3·7 m and a vertical extension of 6·1 m. Pereira, Dagg and Hosegood's (1962) studies on soil moisture tensions in an upland forested catchment area in Kenya led them to believe that the bulk of active tree roots were in the first 3 m of the soil.

Lianes and climbing plants

Reference was made earlier to the abundance of climbing plants which are found inside tropical forests and on their margins. Some of them grow very tall and woody and are referred to as 'lianes', sometimes spelt 'lianas' and these are particularly characteristic of the tropical forest.

Contrary to popular belief the lianes are usually *not* active climbers; very few can climb up mature trees; normally they become attached with hooks or tendrils to tree saplings and grow with these to the canopy. Perhaps the only genuine climbing liane in East Africa is *Hippocratea* which lacks thorns and tendrils and ascends even branchless tree trunks by means of its sensitive, embracing, opposite branches.

Climbing plants can be classed in four groups according to the nature of the specialisation for attachment or climbing (Table 1.2).

With few exceptions lianes are a menace and an economic drain on forest management. Of the useful lianes *Landolphia landolphiodes* and *Clitandra cymulosa* contain a high percentage of pure latex caoutchouc which can be processed into good rubber, though at the present day there is not much demand for these sources which used to be adulterated in East Africa with latex from *Motandra altissima*. Locally *Mondia*

Table 1.2 Climbing plants in Uganda forests

1 TWINERS whose stem tips nutate or oscillate and wind around slender supports. Examples: Apocynaceae, Araliaceae, Aristolochiaceae, Bignoniaceae, Combretaceae, Connaraceae, Convolvulaceae (*Ipomoea*), Menispermaceae, *Arthropteris, Lygodium, Phaseolus, Thunbergia.*

2 CLIMBERS WITH SENSITIVE ORGANS
  (a) *Tendrils*, that is stems or leaves modified into threadlike organs which clasp slender objects after contacting them. Stem tendrils are *Landolphia, Passiflora, Vitis.* Leaf tendrils are Bignoniaceae, Cucurbitaceae, Leguminosae.
  (b) *Sensitive hooks* which clasp and become woody. Examples; *Landolphia, Paullinia, Strychnos, Uncaria, Uvaria.*
  (c) *Sensitive leaves.* Example; *Gloriosa.*
  (d) *Sensitive petioles.* Example; *Clematis.*
  (e) *Sensitive lateral branches.* Examples: *Hippocratea, Salacia.*
  (f) *Sensitive inflorescence branches.* Example; *Vitis.*

3 HOOK CLIMBERS; sprawling and attaching by hooks. Examples; *Caesalpinia, Entada, Acacia, Calamus,* Combretaceae, *Galium, Pterolobium, Smilax.*

4 ROOT CLIMBERS; with negatively heliotropic adventitious roots which adhere to supporting plants. Examples: Araceae (*Culcasia, Monstera*), Araliaceae, Bignoniaceae, Sapindaceae, *Ficus, Piper, Salacia, Stenochlaena.*

whiteii has been used for making fishing-lines because the fibres extracted from the stem are strong and durable. *Saba florida* is retained in many places because of its large edible acid fruits. The remaining lianes are troublesome weeds which retard natural regeneration and occupy space in the canopy which should be available to trees. In natural gaps in the forest, or gaps made by felling, they quickly form dense tangles because they are capable of branching from previously shaded stems and from severed stems. The tangle of vegetation then suppresses seedlings and retards the growth of young trees. Most of the East African forests have been combed for timber, and this has provided good opportunity for the spread of lianes so that some of the forests are, or have been before treatment, heavily infected; Dawkins and Philip (1962) suspect that there might be more lianes than trees and shrubs altogether in some of the Uganda forests in the Lake Victoria Belt.

Lianes are represented in many dicotyledonous families though they are most prominent in the Apocynaceae and Asclepiadaceae.

In these two families many of the genera are provided with highly plumed seeds which may be important for wind dispersal. The only important liane among the monocotyledonous families is the rattan cane, which is, in fact, a palm of the genus *Calamus*. Four or five species of the genus are located in tropical Africa and *C. deerratus* is common in many swampy valleys within some of the forests of the Western Escarpment group in Uganda.

These rattan canes have thin stems of immense length, often up to 185 m, and rattan cane walking sticks are made from the uppermost internodes, while strips of the stem are used for chair bottoms and baskets. The leaves are pinnate and the leaf rachis is prolonged into a long spiny hook or *cirrhus*. Each leaf when young, is erect at the top of the shoot but, when unfolding, it bends outwards and if in contact with another plant, the hook holds it firmly.

When overmature trees fall out of the canopy, the lianes which they supported may be brought down also and it is in this way that the liane stems become looped inside the forest. In any ordinary stem this looping would constrict the conducting elements and kill the plant, but the tissues within the liane stem are disposed in bands, furrows or strands so that great flexibility is achieved; very often the cross-section resembles a stranded cable.

The conducting elements themselves are usually of an advanced type. Certainly this is true of the water conducting vessels and may be true of the phloem sieve tubes, but not much is known about the latter. Chamberlain (1933) records sieve plates in *Tetracera*, which are large enough to be seen easily with a $\times 10$ pocket lens. Some species of this genus occur commonly on termite mounds and forest-edge thickets in East Africa, but we have not yet seen an East African species with such striking sieve plates. However, the water-conducting vessels in liane stems have a diameter which is so great that they are visible clearly to the naked eye, and they occupy most of the cross-sectional area. Large vessels are a necessity in these very long, relatively narrow stems so that they may supply water at a sufficient rate to the leafy crown of the liane in the canopy, a crown which may have as great a water demand as that of a tree. Increasing the diameter of the vessels greatly increases the volume of water which can move through the stem because—all other factors remaining steady—the volume conducted by the vessels is proportional to the fourth power of the radius (r^4); this relationship is fully stated in Poiseuille's Law which gives the rate of flow through narrow tubes. This striking relationship between volume conducted and the radius of the conducting element is shown in the following table:

Radius of conducting elements		Rate of flow per element per hour with pressure head of 1 atm in ml
cm	microns	
0.04	400	365
0.02	200	23
0.01	100	1.4
0.001	10	0.00014

Here, vessels with a radius of 400 microns would be representative of the vessels in many large lianes, while the smallest radius of 10 u is typical of the tracheids in conifers, where, of course, the flow could not be as great as that shown because the calculation makes no allowance for the resistance of the end walls; which occur only rarely in vessels.

Scholander et al. (1955, 1957) have done some interesting work on the conduction of water in the xylem of tropical American lianes. The fragility of the water columns moving under tension in the xylem vessels of long stems posed a problem for theories on the ascent of water (Dixon, 1964). It was difficult to understand how the water column would continue to move if gas breaks developed—as they seemed certain to—in the xylem vessels. Scholander was able to show that air in the vessels is arrested in its travel by occasional end walls in the vessels; in the plant he studied these occurred about every 60 cm. Moreover, the small dimensions of the connecting pits between vessel elements prevented the passage sideways of air from vessel to vessel. Thus gas breaks developing

in the vessels travel only a short distance and become 'hung up'. The movement of water of the transpiration stream is not seriously affected since it simply moves around these gas barriers. Thus, Scholander envisaged the water-conducting xylem in the liane as essentially a flooded, continuous, micropore system, scattered with elongate microcavities (vessels). The stem of a liane may be compared to a pipe filled with a sinter of fine sand, through which large longitudinal cavities are dispersed.

In contrast to the lianes inside the forest, the herbaceous and woody climbers on the forest margin are of immense value because they most effectively shield the forest from fires sweeping in from the surrounding grasslands and thickets. Along with shrubs and characteristic forest margin trees such as *Trema*, *Sapium*, and *Maesopsis*, these climbing plants of the forest margin create an unbroken protective wall of vegetation from ground level to canopy.

There are a great many herbaceous climbers of the forest edge and in thickets, many of them belonging to the families Vitaceae and Cucurbitaceae. For example, the genera *Ampelocissus*, *Cissus*, *Rhoicissus*, and *Vitis* are all of the Vitaceae, while the Cucurbitaceae are represented by *Diplocyclos*, *Luffa*, *Momordica* and *Shaerosicyos*.

The most widespread herbaceous climber inside the forest, and sometimes it is very abundant, is *Culcasia* which has white inflorescences followed by clusters of red berries. It may climb up the tree trunks by its clinging roots for perhaps 9 m but it does not reach the canopy. In fact, only two herbaceous climbers in East Africa are known to reach the canopy; they are *Smilax goetzeana* and *Smilax kraussiana*, both capable of very rapid growth. Climbing plants are often of much interest and it is fortunate that Darwin (1906) has provided such a good account of many aspects of their biology.

Epiphytes

Epiphytes are plants which grow on other plants but not parasitically. Their roots cling to the surface of the support, usually a tree, or penetrate cracks in its bark but, unlike parasites, they take no food directly from the support. This is strictly mechanical and never nutritional apart from substances available as bark exudates and from the decay of its outer layers. The epiphyte is a very characteristic life-form of the wetter tropical forest.

Epiphytes occur in nearly every plant group so that they range in size from bacteria to the herbaceous ferns and orchids; there are even a few tree epiphytes, such as *Schefflera*, one of the Araliaceae. Some begin as epiphytes and eventually become independent of the support, as for example, the figs. These, when growing epiphytically, produce immensely long aerial roots and after these reach the ground, the fig becomes a strangler and slowly kills the supporting tree. *Ficus brachypoda*, *F. capensis*, *F. dawei*, and *F. natalensis* develop frequently into large trees which strangle the host.

The wet forests of tropical America are the richest in epiphytes, while Africa's forests are relatively poor. The cause of this is not fully known, but it is due partly to the absence from Africa of the family Bromeliaceae. Willis (1966) lists about 1400 species of the Bromeliaceae, mostly epiphytes, in the tropical forests of the New World.

In East Africa vascular epiphytes—principally ferns and orchids—are characteristic of the lowland forests, and the upland forests are distinguished sometimes by vast amounts of non-vascular epiphytes, mosses and lichens. There is no general account of East African epiphytes, but Eggeling (1947) has given a list of those which occur in different communities of Budongo forest. In Budongo, excluding strangling figs, Eggeling found nearly 100 species of vascular epiphytes; including 30 species of fern, 60 species of orchids, and 10 species of other plants. The maximum number on a single tree was 26, but elsewhere in Uganda, on a single tree he was able to collect up to 45 species. He found that epiphytes occurred on trees in each of the canopy strata, but they were more frequent in the top storey and this reinforces the belief that many epiphytes are light-demanding plants. He observed also that they occurred chiefly on the largest individuals of any species of tree, showing that colonisation was successful late in the life of the tree; this partly explains the scarcity of epiphytes in the colonizing forests, although Eggeling believes that lower humidity in these forests is important. Probably, because of high humidity, epiphytes are common in some swamp forests. These species in the swamp forests are often different from those in dry land forests; for example, the orchid *Angraecum infundibulare* with large white flowers is very much a swamp or waterside species, and among the ferns *Stenochlaena mildbraedii*, which is a root climber with pinnate fronds up to 2 m in length, is rarely seen except on the stems of *Raphia* palms in swamps.

It is not known clearly why some plants have become epiphytic. It has been suggested, because many of the epiphytes are restricted to the high branches of trees and are never found inside the forest, that they require much light and are unable to survive the low light intensities of the forest undergrowth. Another explanation, perhaps more likely, is that they cannot compete successfully with other plants inside the forest and they escape competition by taking to the treetops. But this situation provides new problems of survival. It is a harsh habitat, being virtually a microdesert, and the exposure to drying winds is severe in the semideciduous forests. Water loss from the root-system is a major problem. This must be extensive and exposed in order to absorb water quickly during periods of rain, but an extensive and exposed root system becomes a liability in intervening dry periods. The roots of some epiphytic orchids and aroids have become modified internally so that water-uptake is enhanced and water loss prevented.

Some ferns have solved the problem by accumulating a nest of leaves and humus which provides soil for the roots. There are four of these humus-accumulating ferns in East Africa—*Asplenium africanum*, *Drynaria laurentii*, *Microsorium punctatum*, and *Platycerium elephantotis*. *Platycerium*, called commonly the elephant's ear fern, is well known because it is very large and occurs both inside and outside the forest, being found quite frequently on trees in wooded grasslands, and even on old mango trees. The fronds, as in many ferns, are dimorphic; there are erect vegetative fronds and pendent fertile fronds. The erect fronds when dying curl against the supporting tree or against previous fronds and so retain an accumulation of old leaves between them. In this way a mass of dead leaves and humus is built up. The dead fronds are spongy in texture so that the whole mass holds a lot of water and is ideal for the roots. The pendent or fertile fronds deserve special mention because their undersurface is one of the most delightful objects under a low-power binocular microscope or even a $\times 10$ lens. This undersurface is covered with a felt of golden or rusty brown, stellate hairs which conceal the large golden-coloured sporangia. The latter occur over the entire fertile surface and not in discrete *sori*, as in most ferns. There are about seventeen species of *Platycerium* in the Old World tropics and temperate Australia and one species in Peru; probably two species occur in tropical Africa.

Asplenium africanum is another common nest fern in many East African forests. It has a large number of long, tongue-shaped, succulent fronds arranged in whorls around a very short stem, so that they form a nest or basket in which old leaves are retained and rot to form humus. The plant resembles closely another common epiphytic fern *Microsorium punctatum*, which occupies the same habitat. When fruiting the two plants can be distinguished because the sori of *Asplenium* are linear and those of *Microsorium* are small, circular, slightly sunken in the undersurface of frond, and randomly arranged. Eventually the stem apex produces new fronds which at first grow vertically upwards but later bend backwards, holding the decaying debris between their bases and the bases of the preceding fronds. Roots then grow outwards from the stem and into the humus. The nest of humus is sometimes a habitat for other ferns, often *Davallia*.

Another well-known *Asplenium* is *A. dregeanum*—ordinarily called the walking fern. It gained this name because the tips of the fronds, where they rest against moist branches proliferate and give rise frequently to new plants. There are other species of *Asplenium* with buds on the ends of pinnae which proliferate in a similar way.

Our last example of a humus-accumulating fern, *Drynaria laurentii*, is another common epiphyte on high branches. It has dimorphic fronds borne on a stout scaly rhizome. The vegetative fronds of *Drynaria* are persistent and the fertile fronds are deciduous, although the midrib remains attached to the rhizome long after the pinnae have fallen. The persistent, vegetative unstalked fronds overlap each other by covering the rhizome and roots, and humus collects behind and between them. There are about twenty species of *Drynaria*; three in Africa and Madagascar; the rest in Asia and Australia.

An unusual looking fern which is epiphytic on forest trees is *Vittaria guineensis* which has pendulous succulent fronds about 30 cm in length and not more than 2 cm in width.

Not all the epiphytic ferns occur in the forest canopy; some are confined to lower levels of the forest. *Arthropteris palisoti* is not infrequently found and has dark rhizomes twined tightly around undergrowth shrubs. It is easily identified because the rachis of each frond is jointed distinctly or articulated at its base to the rhizome. Where old fronds have fallen, the rhizome is left with a small peg. Also the pinnae on the fronds are themselves jointed to the frond-rachis or midrib.

Of the epiphytic ferns of the undergrowth perhaps the most interesting, often the most beautiful, are the filmy ferns; so-called because their fronds consist of a *single* layer of translucent green cells. They grow often in deep shade, where the light intensity is only a hundredth, or less, of full sunlight. They absorb moisture directly through the surface of the frond and lose it very readily; in a dry atmosphere they shrivel quickly into dark green tufts, though they regain their delicate translucent beauty on rewetting. The filmy ferns include two genera, *Hymenophyllum* and *Trichomanes*; some of the species are small, the frond being not more than 15 cm in length while other East African species, such as *Trichomanes cupressoides* and *T. gigantioides*, have fronds up to 30 cm in length. The two genera contain about 300 species, the majority of which are tropical.

The number of epiphytic orchids is very great in Africa, although not as great as in Indo-Malaya and South America. There are keys for the identification of the East African orchids by Piers (1968) and by Copley *et al.* (1964), and Moreau and Moreau (1943) have produced a tentative field-key to the genera of epiphytic orchids in East Africa.

All orchids produce a colossal number of exceedingly light seeds, each ovary containing several million seeds, each weighing between 0.01 and 0.02 mg. Undoubtedly these are effectively dispersed by wind and are comparable in this respect to the spores of the ferns and mosses. On germination these seeds usually develop further only if infected by an appropriate fungus which forms an endotrophic mycorrhiza in the root. The fungus is, to some extent, parasitic on these outer cells, but in the deeper-lying cortical cells, the orchid dominates the penetrating fungus and brings its growth to a standstill. This mycorrhiza enables the orchid seedling to grow saprophytically, sometimes for many years, before relying on its own photosynthesis. Indeed, some orchids remain permanently dependent on the fungus, so that they are parasites!

The epiphytic orchid has two types of root. The primary need is for clinging roots which anchor the plant and these are insensitive to gravity and negatively *heliotropic*, so that the root tip moves away from light and into cracks of the bark or rock. The remaining roots, the aerial roots, hang in festoons and have a remarkable internal anatomy. In most plants the outermost layer of cells on the root develop delicate, water-absorbing root hairs. Obviously such structures are inappropriate for the microdesert of the orchids. Instead, this outer layer, the *periderm*, divides repeatedly and the resulting cells develop numerous interconnecting pores and pits, and their walls become reinforced with spiral or reticulate thickenings. Later the cells die, lose their contents, and remain on the root as a highly effective porous, water-absorbing sheath. These layers of dead cells, collectively termed the *velamen*, make the dry roots appear white or silvery. Water absorbed by the velamen passes through the passage cells of the exodermis to the interior water conducting xylem.

The form of the root growth in some epiphytic orchids probably enhances moisture uptake. Thus in certain of the leafless species of *Microcoelia* the roots, instead of spreading, form a densely interwoven mass. The most notable development of this kind has been perfected by *Ansellia*, the leopard orchid, which forms a root clump weighing several kilos, in situations that are usually hot and dry and subject to long droughts with desiccating winds. Moreau and Moreau (1943) describe how 'an *Ansellia* seedling first of all puts down roots that adhere closely to the bark of the host tree in the ordinary way. Once anchorage is secured, each season's growth consists of a forest of slender roots all thrusting vertically upwards. Each dies after a few inches growth, but the woody remains persist, so that in the course of years, a great fibrous mass is formed the constituents of which are so directed as to trap the maximum of rain.'

The orchid is further adapted to its drought-exposed habitat by leaf reduction and storage of food and water in the swollen fleshy part of the leaf known as the *pseudobulb*. The leaves of some orchids are small and narrow as in *Diaphananthe xanthopollinum* and *Ypsilopus graminifolia*, they are reduced to mere needles in some species of *Tridactyle*, while the species of *Microcoelia* appear to have no leaves at all, though a very close inspection of the stem reveals the presence of scales which are, in fact, degenerated leaves. The roots contain chlorophyll and have taken over the function of the leaves. The species of *Microcoelia* are among our most floriferous epiphytic orchids, and produce clouds of tiny, bell-shaped flowers on thin wiry racemes.

The effectiveness of the cuticle in preventing water loss is well illustrated in a note by Moreau and Moreau (1943). They record having kept a leafless stem of *Ansellia* for nearly two years, having only the end of the stem bound in moss and watered occasionally. At the end of this time the stem showed no sign of rooting and was as firm and sound as the day it came off the parent plant.

The orchids, like the ferns, are found mainly on the shaggy-bark trees, such as *Entandrophragma utile* and *Canarium schweinfurthii*, appearing rarely on smooth-barked trees, such as *Ficus*, *Morus*, and *Cynometra*, until these are so old that the branches have begun to decay; then these too may be heavily colonised. According to Eggeling (1947), however, two orchids *Ancistrochilis rothchildianus* and *Graphorkis lurida* prefer the boles of smooth-barked trees. Among other definite associations recorded are *Tridactyle wakefeldiana* limited to tidal mangrove swamps, *Ansellia gigantea* var. *nilotica* occurring frequently on the trunks of doum palms (*Hyphaene*) or, in western areas, on *Acacia lahai*, *Aerangis thomsoni* found mostly on the ragged trunks and branches of *Juniperus procera* in the drier upland forests, while the baobab seems to be the favourite support of *Angraecum dives*. One of the common habitats for epiphytic orchids is on trees along the water courses in the forest and in riverine forest where the high humidity favours them. The exception is where the riverine forests are composed of species of *Acacia* or *Ficus* which seem to be avoided by most epiphytic orchids.

The occurrence of several species of orchid on the same tree in close proximity and the apparent absence of hybrids between them raises an interesting problem in pollination biology. The study by Sanford (1968) of epiphytic orchids in West African forests showed that there was often a difference in flowering time between the species though sometimes overlapping flowering did occur on the same tree. It would seem therefore that some other barrier keeps closely allied species of tropical orchids apart.

Of the remaining vascular epiphytes, perhaps the most unexpected is *Rhipsalis baccifera* the sole African species of the New World family Cactaceae. It has green, dichotomously branching stems which reach a length 3 m hanging from the higher branches of trees. There can be little doubt that it was brought to Africa by migrating birds. It has a yellow-white berry with seeds embedded in a sticky pulp like the berry of the semiparasite *Viscum* which is bird distributed. *Medinilla*, a shrubby epiphyte of the Melastomataceae, has been recorded from the Sese Islands (Thomas, 1941) and from forest in Tanzania. Its small, bright reddish flowers might cause it to be confused with species of the semiparasitic genus *Loranthus* of which there are many examples in tropical and subtropical Africa mostly with showy tubular flowers in brilliant red, orange or yellow. The Piperaceae contribute species of the epiphyte *Peperomia* which occur commonly in patches of moss on trees and have circular, succulent and pellucid leaves and inconspicuous flowers.

We have left to the last, not because they are least important but because they are not well known, the non-vascular epiphytes, mosses, liverworts, lichens and bacteria. They are found in all forests and are especially prominent in the uplands where the wet woodlands of *Hagenia* and *Rapanea* have trunks and branches thickly swathed in mosses with the lichen *Usnea* pendent from the branches. *Usnea* is also sometimes very abundant in the dry upland forest of *Juniperus* and *Podocarpus* and the bark of the forest trees is often mottled in shades of grey and pink with patches of closely adherent thalloid lichens. Table 1.4a, p. 27, lists some of the commoner mosses and liverworts of lowland and upland forest. Very little has yet been published about the epiphytic mosses and lichens. But Jones (1952) has made extensive studies of the liverworts the majority of which belong to the Lejeuneaceae. Of particular value is his 'Provisional Key to the Genera of Tropical African Hepatics'.

Epiphyllous liverworts seem to be not very common in the East African forests and perhaps this reflects the relative severity of the dry season. Liverworts and mosses have no cuticular protection against water loss but the fine structure of their protoplasm is such that, even after prolonged desiccation, it returns to activity on wetting. Presumably the protoplasm of the filmy ferns, *Hymenophyllum* and *Trichomanes* is similarly adapted.

Very little is known of the tropical epiphytic algae but one of the green algae—*Trentepohlia*—forms orange-yellow cushions on tree trunks, leaves and rocks. The colour is due to the pigment haematochrome which occurs in fat globules that cluster around and obscure the green chloroplasts. *Trentepohlia* can withstand long periods of drying without appreciable change and some of its species constitute the algal component of lichens.

An account of the general features of East African forests is incomplete without reference to the animals. Haddow (1945 a, b) has written of the insects, Moreau (1935 a and b, 1966) of the birds, Mutere (1965) of the fruit bats, Schaller (1963) of the gorilla, Rowell (1966) of the baboon, Haddow (1952), Lumsden (1951) among others of the monkey, and Delany (1964, 1967) of the small mammals. Many of the large mammals besides the gorilla also frequent the forests. Unfortunately, all this information has not been brought together into a general account which might vividly illustrate the interrelationship of plants and animals in the way that Harrison (1962) has been able to do for the birds and mammals of the rain forests of Malaya and North Queensland.

The complexity of life in a tropical forest is well brought out in the following quotation from 'The life of plants' (Corner, 1964).

There is a giant tree, pre-eminent in a forest that stretches to the skyline. On its canopy birds and butterflies sip nectar. On its branches orchids, aroids, and parasitic mistletoes offer flowers to other birds and insects. Among them ferns creep, lichens encrust, and centipedes and scorpions lurk. In the rubble that falls among the epiphytic roots and stems, ants build nests and even earthworms and snails find homes. There is a minute munching of caterpillars and the silent sucking of plant bugs. On any of these things, plant or animal, funguses may be growing. Through the branches spread spiders' webs. Frogs wait for insects, and a snake glides. There are nests of birds, bees, and wasps. Along a limb pass wary monkeys, a halting squirrel, or a bear in search of honey; the shadow of an eagle startles them. Through dead snags fungus and beetle have attacked the wood. There are fungus brackets nibbled around the edge and bored by other beetles. A woodpecker taps. In a hole a hornbill broods. Where the main branches diverge, a straggling fig finds grip, a bushy epiphyte has temporary root, and hidden sleeps a leopard. In deeper shade black termites have built earthy turrets and smothered the tips of a young creeper. Hanging from the limbs are cables of lianes which have hoisted themselves through the undergrowth and are suspended by their grapnels. On their swinging stems grows an epiphytic ginger whose red seeds a bird is pecking. Where rain trickles down the trunk filmy ferns, mosses, and slender green algae maintain their delicate lives. Round the base are fragments of bark and coils of old lianes, on which other ferns are growing. Between the buttress-roots a tortoise is eating toadstools. An elephant has rubbed the bark and, in its deep footmarks, tadpoles, mosquito larvae, and threadworms swim. Pigs squeal and drum in search of fallen fruit, seeds and truffles. In the humus and undersoil, insects, fungi, bacteria and all sorts of 'animalculae' participate with the tree roots in decomposing everything that dies.

Distribution and floristic composition

Lowland Forest (below 2000 m)

Uganda;

Lake Victoria belt

Two major forest areas remain in Uganda, the forests round Lake Victoria and those of the Western Rift Escarpment.

It is in many ways remarkable that there are today any forests round Lake Victoria. On the fertile, well-watered crescent, the Baganda have for many generations practised a crop rotation of which the banana forms an important constituent. Possibly the relative permanence of this agricultural life to some extent explains the survival of some forest in an area which would certainly have been deforested if all the agriculture had been shifting and if the climate had not been so favourable for forest regeneration.

The forests begin in the east where the river Nile leaves the lake at Jinja and continue westwards along the northern shore, down the western shore, and across the Uganda-Tanzania border, fading out on the south-western shores. The southern and eastern shores of the lake have been deforested. A similar type of forest occurs near Kakamega in Kenya. Many of the islands in the lake including the Sese Islands (Thomas, 1941) are forest-covered, though some have sandy, shallow, and poor soils on which the climax vegetation appears to be a thin grassland of *Loudetia kagerensis*. The forests lie at an altitude of 1200 m, extend inland for about 56 km, and are of two types. Near Kakamega and from the Nile exit at Jinja, and as far as Dumu Point on the western lake shore they are angiosperm forests; but south of Dumu Point some way into Tanzania there are seasonal swamp forests dominated by the coniferous tree *Podocarpus*. The annual rainfall ranges from 875–1375 mm rising steeply towards the Sese Islands which receive about 2000 mm per year. There is no clearly marked dry season and conditions are favourable for forest growth and regeneration. The countryside consists of a succession of flat-topped hills, rising to heights of about 200 m above the lake with few areas of level ground. The general pattern of vegetation is of short grass on the hilltops and a complex of farms and elephant grass thickets on the hillsides, with forests and swamps in the valleys (Thomas, 1945–46). Most of the forests are very small, but some are large as, for example, the Mabira forest (c. 300 km²) described by Pitman (1934) and Webster (1961) between Jinja and Kampala, and the Jubiya forest established on sandy poor soils close to the lake shore, south-west of Masaka. There are brief accounts of the composition of most of these forests in the working plans set up by the Uganda Forest Department. The account of Mpanga forest 40 km west of Kampala (Dawkins and Philips, 1962) is particularly important since this is an area which was dedicated to research in 1951 with the prime object of studying the silviculture and productivity of the indigenous forests.

There is very little forest in Uganda to the east of the Nile; most of S.E. Uganda is farmland and elephant grass thicket, though it is dotted liberally with trees, particularly *Chlorophora excelsa* (mvule) and *Albizia* spp. This *Chlorophora–Albizia* wooded grassland mainly in Busoga, has been heavily depleted of timber, especially mvule, during the last few decades (Wood, 1960). Although this eastern part of the Lake Victoria belt does not now contain much closed high forest, there is little doubt that formerly it did.

The displacement of Angiosperm forest by coniferous forest, dominated by species of *Podocarpus*, on the western shore of Lake Victoria, is the more remarkable because elsewhere in Uganda and in East Africa species of *Podocarpus* are seen only in the altitudinal range of 1800–3000 m, while at Lake Victoria they flourish at about 1000 m. Probably this anomalous distribution is largely accounted for by the swampiness of this part of the Lake Victoria belt. These *Podocarpus* forests are, in fact, seasonal swamp forests and for much of the year they are unapproachable. They have been heavily exploited because of the excellent even-textured yellow wood of the slow-growing *Podocarpus*, a wood which compares very favourably with the finest coniferous woods of the northern hemisphere. At present, only small patches of these forests remain in the Lake Victoria belt and it is sad that our knowledge of their natural history is so slight. The main *Podocarpus* forest is the Maramagambo in Uganda which continues more or less into the Minziro forest in Tanzania, and finally dies out about 30 km south of Bukoba. The floristic composition is described under swamp forest.

Elephant grass thicket

Thicket of elephant grass in the Lake Victoria area is to be looked upon as the first stage of the succession back towards forest after this has been destroyed by shifting cultivation. It is therefore included here rather than under grassland. Elephant grass (*Pennisetum purpureum*) is an impressive grass reaching a height of about 5 m at flowering (Plate 7). The culms, 1–3 cm thick, bear long strap-shaped light green leaves 1–3 cm wide and sometimes as much as 1 m in length. Each flowering culm is topped by a hairy spike of densely set, golden or slightly purple-tinged spikelets; the entire inflorescence resembling a long, narrow bottle-brush. The bamboo-like shoots are used in house building to form a lattice which is plastered with mud.

Elephant grass is a desirable natural fallow. The tall growth suppresses most other weeds and grasses, the deep root system effectively 'mines' the soil profile for water and essential elements, there is a satisfactory maintenance of nitrogen levels, and the physical texture of the soil is improved. It will serve not only as a fallow but also as a fodder (Marshall and Bredon, 1963). Cattle may be allowed to graze enclosed paddocks or, if labour costs permit it is more effectively used if cut and fed to the stalled cattle.

The grass is frequent in western Uganda but less common in the northern and eastern provinces, though occasional patches may be found on the richer alluvial soils or on heaped up soils at the roadsides. Round the northern third of Lake Victoria it is abundant in a broad belt extending from the shore inland for 15–24 km. Though there is now much cultivation, including large plantations of sugar, tea or coffee, there is no doubt that this is an area which, in the absence of human disturbance, would be dominated by forest with occasional glades on the thinner soils of hilltops, on laterite platforms and in clearings made by large animals. Today only a few large forest patches remain.

The history of the large expanses of elephant grass east of Ruwenzori and the elephant grass thickets of Bunyoro is less clear. Eggeling (1947) considers the latter may indicate areas where well-wooded *Terminalia* grassland has been intensively farmed with the result that trees and short grass have been replaced by *Pennisetum*. Robyns (1948) describes how, in some of the valleys of the Ruwenzori foothills, elephant grass can be seen replacing *Cynometra* forest and sprinkled with trees of *Erythrina* and *Spathodea campanulata*. This is true also of the margins of the Budongo Forest in Uganda. Eggeling (1947) offers an interpretation of the *Pennisetum* thickets of western Uganda along the following lines. During the later Pleistocene period,

7 Elephant grass thicket in western Uganda. (Fergus Wilson)

peneplanation of the west, associated with major earth movements originating in the Rift Valley, may have caused extensive erosion of the existing lateritic soil blanket. Elephant grass then competed successfully with forest for the colonisation of the freshly formed soils and, with the aid of fire, has kept its hold ever since.

Scattered throughout the elephant grass around the lake are small groves and also some very large isolated trees, which because of their size and form have been claimed as relics from former forest. A few may be genuine relics left on account of their fruit, for shelter or because of ritual associations, or simply because they were too large to fell conveniently. One of the largest and commonest of these trees is the muwafu nut or incense tree (*Canarium schweinfurthii*). This has a trunk with rough flaking bark and a voluminous crown whose relatively stout branches end in pseudo-whorls or large pinnate leaves. The spikes of small yellow flowers are relatively conspicuous and the male and female flowers are usually carried on separate

trees. The nutty fruits are eaten and this, together with the size of most of the trees, probably accounts for their being so common. Other trees which grow to a considerable height in the elephant grass are *Albizia coriaria*, *A. zygia* and *Chlorophora excelsa* (mvule) (Plate 8).

Chlorophora (mvule), which yields a handsome mahogany-like timber, is most common in the Busoga area east of Jinja where there are many tall well-grown specimens. In some districts it is considered a sacred tree. Unfortunately, it is a difficult tree to establish in plantations and the early stages are frequently attacked by gall. (Butt, 1965; Jones, 1957). In Tanzania *Chlorophora* occurs naturally in the valley forests and up to 1000 m in the eastern Usumbaras where it is looked upon as a secondary tree following cultivation. The male and female flowers are on separate trees and the fruits seem to provide one of the main foods of the fruit bat (*Eidolon heavum*). According to Osmaston (1965) the bats have strong teeth and chew the fruits rather like a man chews a corn cob. The softer fruits are eaten completely, while the harder ones are passed to and fro in the mouth until most of the juicy pulp has been eaten together with many of the seeds, when the core is then discarded. He believes that a bat, in the course of a single night, consumes over 500 seeds and would scatter these over the countryside in its droppings. These animals may therefore be important agents in spreading the tree. It has been found that the seeds will germinate after passing through the gut, which is not surprising since they traverse it at remarkable speed. Osmaston fed bats with mvule fruits and observed seeds in their droppings at the end of two hours. For further observations on pollination by bats see Baker and Harris (1957, 1958).

Colonising forest

If left undisturbed, the elephant grass thicket is soon invaded by shrubs and bushes such as *Acanthus pubescens*, *Vernonia* spp, *Clausena* and *Acalypha* spp. and later, colonising trees appear—usually *Albizia* and *Maesopsis* together with some or all of the following—*Celtis africana*, *Dombeya mukole*, *Margaritaria dis-*

8   Mvule tree (*Chlorophora excelsa*) in eastern Uganda. (Fergus Wilson)

coidea, *Prunus africanus*, *Polyscias fulva*, *Sapium ellipitcum*, *Teclea nobilis*, *Blighia unijugata* and *Markhamia platycalyx*.

The poorer soils—those exhausted by over-cultivation—and the eroded soils of the upper hill slopes become dominated usually by lemon grass (*Cymbopogon afronardus*) with, on wetter parts, scattered trees of *Albizia* and other genera and, on drier plains, *Combretum* and its associates. In the course of time, both communities after a slow succession would seem to culminate in closed forest and there are certainly areas of closed forest even on thin soils.

Table 1.3 shows the actual composition of the colonising stage. This list was made in 1969 at Buto II forest reserve on the Fort Portal road about 24 km west of Kampala. The original forest was destroyed in 1964, but a few tall trees were untouched including *Aningeria altissima*, *Canarium schweinfurthii*, *Chrysophyllum albidum* and *Piptadeniastrum africanum*. In 1969, the ground below these was, apart from paths, entirely covered by a dense tangle or thicket reaching a height of 2 or 3 m with a few trees and shrubs up to 10 m. Buto forest is reported to be very rich in small mammals.

Intermediate stage and climax communities

Following the colonising forest, it is usual to recognise an intermediate stage in the forest succession before the climax. This contains a mixture of species and is characterised by some or all of the following trees. *Albizia zygia*, *A. grandibracteata*, *A. glaberrima*, *Antiaris*, *Blighia*, *Canarium schweinfurthii*, *Celtis africana*, *C. durandii*, *Entandrophragma*, *Fagara*, *Lovoa*, *Majidea* and *Pycnanthus*.

There are many different communities in the climax forests, but there is a tendency towards dominance by the genera *Celtis*, *Chrysophyllum*, *Aningeria*, *Piptadeniastrum* with lesser amounts of *Morus*, *Holoptelea*, *Antiaris* and *Alstonia*. The ultimate climax is sometimes mapped as (Pip-Albi-Celtis)[1] as on the Uganda atlas. The under storey is dominated mainly by *Trichilea*, *Teclea*, *Lychnodiscus*, *Lasiodiscus*, members of the Rubiaceae, and *Acalypha* spp. while the herb layer consists largely of broad-leaved grasses of which *Leptaspis cochleata* is the most abundant.

Details of the climax communities and their variation with soil depth and availability of water are given by Webster (1961). The details need not be repeated here but attention should be drawn to the site-type classification used by the Uganda Forest Department all over Uganda. It is given by Dawkins and Philip (1962) as follows:

[1] *Piptadeneastrum, Albizia, Celtis.*

Raised, never flooded or waterlogged: water table never within 1 m of the surface.

Raised dry: water table never or only briefly within 2 m of surface.

Raised wet: Water table always or frequently within 2 m of the surface.

Swamp: Always or sometimes flooded.

Seasonal swamp: Water table well below the surface for long periods, usually several months.

Permanent swamp: Water table never below the surface except for short periods.

Certain trees and herbs are indicators of the main site-types, although Dawkins emphasised that a particular ecological behaviour of these plants, say, in southern Uganda, is no guarantee of identical behaviour in other parts of Uganda or East Africa. For example, *Symphonia*, an indicator of swamp conditions in the Lake Victoria belt, is not confined to swamp forest on the Western Rift escarpment. Again, *Sapium ellipticum* is a swamp tree on some, but not all of the forests round Lake Victoria. These ecological differences may point to varieties within the species.

Probably the earliest major account of the lowland forest of Uganda, and one of great historical interest, is that of Dawe (1906a). His 4000 km safari was stimulated by his find, a few years before, of the rubber tree (*Futumia elastica*) in Mabira forest east of Kampala. Just prior to this discovery, young trees of *F. elastica* had been introduced from West Africa for experimental plantation. Its discovery as an indigenous species raised great hopes, later to be disappointed, that it would be a commercial success (Christy, 1911) as its indigenous occurrence established its suitability to the climate and soils of Uganda. The discovery was of such importance that Dawe continued his explorations and the results are contained in his report where the botany is enlivened by brief comments on the journey. Perhaps none is more indicative of his difficulties than the comment that, a short distance from Kampala, his mule became ill and had to be left behind while Dawe

Table 1.3   Buto II forest reserve, close to Kampala—Fort Portal road. An example of 5-year-old colonising forest in the Lake Victoria belt.

Trees and shrubs reaching a height of 6–9 m:	*Alchornea cordifolia*	Euphorbiaceae
	Albizia zygia	Leguminosae subfam. Mimosoideae
	Allophylus dummeri	Sapindaceae
	Tabernaemontana	Apocynaceae
	Ficus cf. *exasperata*	Moraceae
	Harungana madagascariensis	Hypericaceae
	Maesa lanceolata	Myrsinaceae
	Myrianthus holstii	Moraceae
	Parkia filicoidea	Leguminosae subfam.
	Polyscias fulva	Araliaceae
	Solanum cf. *indicum*	Solanaceae
	Tabernaemontana holstii	Solanaceae
	Trema orientalis	Ulmaceae
	Vernonia amygdalina	Compositae
Shrubs up to a height of about 2–3 m:	*Acalypha* sp.	Euphorbiaceae
	Acalypha bipartita	,,
	Acalypha ornata	,,
	Aframomum sanguineum	Zingiberaceae
	Costus afer	,,
	Cyathea deckenii	Cyatheaceae
	Dracaena sp.	Agavaceae
	Lantana camara	Verbenaceae
	Pennisetum purpureum	Gramineae
	Phyllanthus sp.	Euphorbiaceae
	Phytolacca cf. *dodecandra*	Phytolaccaceae
	Pteridium aquilinum	Dennstaedtiaceae

Herbs and shrubs up to about 1 m in height:	Aneilema beniense	Commelinaceae
	Carex manniana	Cyperaceae
	Dicrocephala integrifolia	Compositae
	Fleurya urophylla	Urticaceae
	Thelypteris cf. quadrangularis	Thelypteridaceae
	Achyranthes aspera	Amaranthaceae
	Mariscus sieberianus	Cyperaceae
	Microlepia speluncae	Dennstaedtiaceae
	Pteris dentata	Adiantaceae
	Pteris sp.	,,
	Pellaea viridis	,,
	Pellaea spp.	Adiantaceae
	Pseuderanthemum ludovicianum	Acanthaceae
	Solanum nigrum	Solanaceae
	Piper umbellatum	Piperaceae
	Palisota schweinfurthii	Commelinaceae
	Setaria chevaleri	Gramineae
	Tristemma incompletum	Melastomataceae
	Pseudoechinoclaena polystachya	Gramineae
Scandent herbs, scramblers, and lianes	Acacia brevispica	Leguminosae subfam. Mimosoideae
	Culcasia scandens	Araceae
	Hibiscus surattensis	Malvaceae
	Illigera pentandra	Hernandiaceae
	Mikania cordata	Compositae
	Olyra latifolia	Gramineae
	Panicum brevifolium	,,
	Rubus pennatus	Rosaceae
	Passiflora edulis	Passifloraceae
	Solanum terminale	Solanaceae
	Aneilema aequinoctale	Commelinaceae

continued on foot for the remainder of his journey.

Among the forests of the Lake Victoria belt, Mabira forest (Pitman, 1934; Webster, 1961) is one of the most interesting and possibly the richest. But the best known is undoubtedly Mpanga forest. A list of the flowering plants from Mpanga forest is given in Table 1.4. Or for a full account reference should be made to (Dawkins and Philip, 1962).

The forests of the Kakamega region of Kenya have been briefly described by Dawkins (1960) who claims that, despite the species they have in common with the Lake Victoria forests, they resemble more closely the higher level forests around 1500 m in Toro and Ankole in Uganda. He draws attention to the dominantly acanthaceous shrub layer and the frequency of the trees *Croton megalocarpus*, *Casearia*, *Drypetes*, *Fagara macrophylla*, and *Strombosia*.

Uganda;

Western Rift Escarpment

Distribution
The second major lowland forest region of Uganda is a more attenuated belt of large forests from about 500 to 1650 m, situated along the eastern escarpment and shoulders of the Western Rift. These great forests include Budongo, Siba, Bugoma, Itwara, Kibale, Kashaya-Kitomi, Maramgambo and Kalinzu and are all described in working plans of the Uganda Forest department. Of them, the Budongo forest is widely known through Eggeling's paper in the *Journal of Ecology* (1947) and references to it in Richards (1952). Moreover, it is distinguished by being, perhaps, the richest mahogany (*Entandophragma* and *Khaya*) forest

24

Table 1.4 Check list of flowering plants recorded from Mpanga forest, Uganda (1960)

Trees capable of exceeding 300 cm girth 100 cm diameter and commonly reaching the canopy

Albizia coriaria
A. ferruginea
A. glaberrima
A. grandibracteata
A. gummifera
A. zygia
Aningeria altissima
Antiaris toxicaria
Balanites wilsoniana
Blighia unijugata
B. welwitschii
Canarium schweinfurthii
Celtis mildbraedii
C. zenkeri
Chlorophora excelsa
Entandrophragma angolense
E. cylindricum
E. utile
Erythrina excelsa
Erythrophleum suaveolens
Fagara macrophylla
Ficus mucuso
Ficus (epiphytic spp.)
Holoptelea grandis
Klainedoxa gabonensis

Lannea welwitschii
Lovoa swynnertonii
L. trichiliodes
Maesopsis eminii
Majidea fosteri
Margaritaria discoidia
Mimusops bagshawei
Mitragyna stipulosa
Morus lactea
Newtonia buchananii
Parinari excelsa
Parkia filicoidea
Piptadeniastrum africanum
Premna angolensis
Prunus africana
Pseudospondias macrocarpa
Pycnanthus angolensis
Schrebera arborea
Strombosia scheffleri
Strychnos mitis
Symphonia globulifera
Syzygium guineense
Tetrapleura tetraptera
T. dregeana
Zanha golungensis

Trees rarely or never exceeding 300 cm girth 100 cm diameter, but commonly reaching 150 cm girth, 50 cm diameter

Afrosersalisia cerasifera
Antidesma spp.
Baikiaea insignis
Balsamocitrus dawei
Beilschmiedia ugandensis
Bosqueia phoberos
Celtis africana
C. durandii
Cleistanthus polystachyus
Dombeya mukole
Drypetes ugandensis
Ekebergia senegalensis
Fagara leprieurii
F. rubescens
Fagaropsis angolensis
Ficus capensis
F. exasperata

Funtumia africana
Linociera johnsonii
Macaranga schweinfurthii
Manilkara dawei
Markhamia platycalyx
Monodora myristica
Pachystela brevipes
Paropsia guineensis
Polyscias fulva
Sapium ellipticum
Spathodea campanulata
Spondianthus preusii
Staudtia stipitata
Sterculia dawei
Treculia africana
Trichilia prieureana

Small trees (by girth) or large shrubs, commonly exceeding 30 cm girth (10 cm diameter), but rarely or never 150 cm girth (50 cm diameter)

Aeglopsis eggelingii
Alangium chinense
Alchornea hirtella

Aphania senegalensis
Baphiopsis parviflora
Belanophora glomerata

Bequaertiodendron oblanceolatum
Bersama abyssinica
Bridelia micrantha
Canthium vulgare
Carpolobia alba
Cassipourea ruwensorensis
Chaetacme aristata
Clausena anisata
Croton macrostachyus
C. oxypetalus
Cussonia holstii
Diospyros abyssinica
Dicrostachys glomerata
Dictyandra aborescens
Drypetes sp.
Euphorbia teke
Flacourtia indica
Garcinia huillensis
Glyphaea brevis
Kigelia moosa
Lepidotrichilia volkensii
Leptonychia mildbraedii
Macaranga monandra
M. pynaertii
Maerua duchesnei
Maesa lanceolata

Memecylon jasminoides
Oncoba spinosa
Ouratea hiernii
Oxyanthus speciosus
O. uniloularis
Pancovia turbinata
Phoenix reclinata
Pittosporum mannii
Pleiocarpa pycnantha
Rauwolfia oxyphylla
Rinoria brachypetala
R. ilicifolia
R. oblongifolia
Rothmannia urcelliformis
Rungia grandis
Scolopia rhamniphylla
Tabernaemontana holstii
Teclea grandifolia
T. nobilis
Tetrorchidium didymostemon
Trichilia rubescens
Trema orientalis
Turraea vogelioides
Vangueria apiculata
Xymalos monospora

Small woody shrubs confined to the D and E strata, rarely or never producing stems over 30 cm girth (10 cm diameter)

Acalypha acrogyna
A. neptunica
A. ornata
A. racemosa
Allophylus africanus
A. dummeri
A. macrobotrys
Argomuellera macrophylla
Carpolobia conradsiana
Citropsis schweinfurthii
Cleistanthus sp.
Clerodendron myricoides
Coffea canephora
C. eugenioides

Dovyalis macrocalyx
Ficus urceolaris
Grewia flavescens
Maytenus gracilipes
M. undata
Lindackeria schweinfurthii
Mallotus oppositifolius
Ochna membranacea
Peddiea fischeri
Plectranthus decurrens
Securinega virosa
Suregada procera
Thecacoris lucida
Trichocladus ellipticus

Herbs confined to E and sometimes as high as D stratum

Aerangis rhodostricta
Aerva lanata
Afromomum spp.
Asystasia gangetica
A. vogeliana
Barleria brownii
Brillantasia madagascariensis
Celosia globosa
C. schweinfurthii
Commelina spp.
Corymborkis corymbosa

Costus afer
Cyathula prostrata
Cyrtorchis monteiroae
Desmodium repandum
Dracaena steudneri
D. laxissima
Elatostema orientale
Fleurya ovalifolia
Geophila uniflora
Haemanthus sp.
Hibiscus owariensis

Justicia betonica	*Piper capense*
Marantochloa leucantha	*P. guineense*
M. purpureum	*P. umbellatum*
Olyra latifolia	*Pollia condensata*
Oplismenus hirtellus	*Polyspatha paniculata*
Palisota schweinfurthii	*Scirpus fluitans*
Phrynium sp.	*Setaria kagerensis*

Table 1.4a Flowerless plants of lowland forest

LOWLAND FOREST
Bryophytes
From Dr. F. Rose

Epiphytic species		On soil and rocks	
Calymperes perrotetii	C	*Bryum argenteum* var. *majus*	
Dicranoloma billardieri		*Fissidens* spp.	C
Floribundaria cf. *cameruniae*		*Philonotis mniobryoides*	C
Hypopterygium viridissimum	A		
Papillaria cf. *africana*		*Anthoceros myriandraecius*	
Pilotrichella imbriculata	A	*Cyathodium* sp.	
Pinnatella le testui		*Fossombronia grandis*	
Rhacopilum crassicuspidatum	A	*Riccia stricta*	
R. macrocarpum			
Rhizogonium spiniforme			
Syrrhopodon mildbraedii	C		
Thuidium pycnangellium			
Vesicularia garelulata			
Frullania squarrosa	C		
Lophocolea cuspidata			
Plagiochila breviramea	C		
Plagiochila colorans	C		
Plagiochila cf. *subalpina*	C		
Porella hohneliana			
Ptychanthus striatus	A		
Lejeuneacean hepatics	C	A=abundant	C=common

Pteridophytes

The following are among some of the commoner ferns recorded from lowland forest.

Asplenium africanum	*Cyathaea deckenii*
A. barteri	*Doryopteris kirkii*
A. buettneri	*Marattia fraxinea*
A. dregianum	*Microlepia speluncee*
A. emarginatum	*Pellaea* spp.
A. erectum var. *usambarense*	*Pteridium aquilinum*
A. gemmascens	*Pteris acanthoneura*
A. cf. *kasserni*	*P. atrovirens*
A. laurentii	*P. burtoni*
A. unilaterale	*P. communata*
A. variabile var. *paucifugum*	*P. dentata*
A. warneckii	*P. intricata*
Blotiella currori	*P. pteridoides*
Bolbitis auriculata	*P. togoensis*
B. gemmifera	*Thelypteris dentata*

27

T. madagascariensis	*T. striata*
T. quadrangularis	*T. dentata*

Lichens of Lowland forest
(From Dr T. D. V. Swinscow)

Anaptychia boryi	*Leptogium coralloideum*
A. comosa	*Parmelia andina*
A. diademata	*P. austrosinensis*
Anaptychia spp.	*P. tinctorum*
Candelaria concolor	*P. wallichiana*
Coccocarpia parmelioides	*P. texana*
C. pellita	*Usnea articulata*
Collema mac-gregorii	*Usnea* spp.

in the Commonwealth, though many other Uganda forests are rich in fine hardwoods.

These forests lie in the rainfall range of 1250–1625 mm, and are isolated from each other by farmlands and various types of post-cultivation vegetation, notably elephant grass thicket and *Hyparrhenia* grassland. Formerly, many of them must have been connected with each other and with the Ituri forest in Zaire. The separation and contraction of the forests may be due partly to climatic changes, since the area is not densely settled today, and it is thought unlikely that human activities alone would account for the large areas of wooded grassland separating many of the forests. The most isolated is Zoka forest in West Nile Province. It is the only forest there, the remainder of the province being covered by two integrading wooded grassland communities of the *Combretum-Acacia* type and the *Combratum-Terminalia* type. It seems that the rainfall on West Nile in all except the north and north-west is sufficient to support forest and, according to Hamilton (personal communication), the two elements which prevent the spread of Zoka are fire and elephants. The latter are so numerous at some seasons in the forest that it is not only necessary to take a small army of game guards on a visit, but it is certain that the forest will cease to exist in a few years. They trample and eat the seedlings and debark many tree species—their favourites being *Albizia*, *Chrysophyllum*, *Cola*, *Cordia*, and *Khaya*. Hamilton suggests that it is unlikely that they would be so malicious if not concentrated by hunting and that therefore the forests, presumed to have covered much of West Nile Province, have been cleared largely by men.

By contrast with the Lake Victoria belt forests, those of the Western Escarpment show a distinct tendency to monospecific dominance, and it is not clear if this is due only to greater maturity.

Colonising forest

Eggeling (1947) classifies the colonising forest into two types: *Maesopsis* colonising forest on deep soils and a woodland forest on poor soils. *Maesopsis eminii* (musizi) is a fast-growing deciduous tree 15–27 m high, characterised by a straight bole with pale grey to almost white bark. The bark is longitudinally fissured and the bole is free of branches for perhaps 9–21 m. It produces an excellent light hardwood and it is under trial extensively as a plantation tree in East Africa.

The grassland develops to a woody thicket containing *Acanthus pubescens*, *Vernonia amygdalina*, *Milletia dura* and *Securinega virosa*, and in the shade of these shrubs, *Maesopsis* regenerates readily. It is probably distributed by hornbills and monkeys and the forest is often of almost pure *Maesopsis*. Other trees usually found are: *Sapium ellipticum*, *Polyscias fulva*, *Prunus africana*, *Albizia gummifera*, *Olea welwitschii*, *Markhamia platycalyx* and the shrubs *Margaritaria discoidea* and *Caloncoba schweinfurthii*.

This fringe of shrubbery and colonising forest, sometimes thickly laced with climbers and sprawlers, gives great fire protection, the fires which inevitably sweep through the grassland being prevented from entering the forest by this thick tangle of vegetation. The *Maesopsis* forest may be short-lived, perhaps not more than sixty or seventy years, and it gives way to mixed forest. The inability of *Maesopsis* to regenerate in dense shade may account for its occurring only as a relic in closed forest.

The second type of colonising forest (woodland forest) occurs on shallow soils such as are found on murram ridges. Here the initial shrubbery may be largely absent or it may begin on deeper soils of the termite mounds which also afford some protection from fire and grazing. If these shrubberies develop, they may become thickets and eventually coalesce to produce the woodland forest. Eggeling lists the following associated trees: *Albizia* spp., *Croton* spp., *Dombeya mukole*, *Olea welwitschii*, *Phyllanthus* spp., *Sapium ellipticum*, *Spathodea campanulata*.

Mixed forest

Both types of colonising forest develop to mixed forest (Fig. 1.5) and these of the Western Escarpment, like those of the Lake Victoria belt, show no single species dominance though certain species may be prominent. For instance, in some parts of the mixed forest at Budongo, three species of *Chrysophyllum* make up a high proportion of the canopy which may contain 43 or 50 tree species. Among other commoner trees are *Alstonia congoensis*, *Trichilia prieuriana*, *Khaya antotheca*, *Celtis mildbraedii*, *Cynometra alexandri*, but the composition of the mixed forest varies much from place to place.

Climax forest

Two main types of climax forest develop, the type appearing to depend on altitude. At lower levels, from 1000–1200 m *Cynometra alexandri* (Uganda ironwood) predominates, while from 1200–1700 m *Parinari*

excelsa (the grey plum) is most frequent. In Budongo and in parts of Kibale and Kashoya-Kitomi forests *Cynometra* seems to be the dominant species below 1200 m, but from 1200 to 1500 m a variety of species may predominate. In Kibale it is *Celtis* and *Chrysophyllum* while in Kashoya-Kitomi *Strombosia* and *Drypetes* are more important. In Kalinzu the latter two species may replace *Parinari* in parts where it is not regenerating and in the Impenetrable Forest above 1800–2100 m *Chrysophyllum fulvum* is an important dominant.

Cynometra forest (Fig. 1.6)

There are many species of *Cynometra* in East Africa, especially in Tanzania, but forest of *Cynometra alexandri* is confined to western Uganda, where it has been described by Eggeling (1947), Osmaston (1960) and Philip (1964). *Cynometra alexandri* is an unmistakable tree with smooth, grey-brown bark which flakes in a distinctive manner, very large, thin plank buttresses, a relatively short, often very gnarled bole, and a high spreading crown with dark green, compound leaves with small leaflets. It is overwhelmingly dominant forming 70–80 per cent of the single-layered

Fig. 1.5 Profile diagram of mixed forest, Budongo, Uganda. (Eggeling, 1947) Key on page 30

29

canopy. There are rarely any lianes, but there is an under storey sometimes dominated by *Lasiodiscus mildbraedii* and sometimes by *Celtis* spp. and *Strychnos mitis*. The wood of *Cynometra* is not unattractive but its weight precludes its use except for heavy construction and flooring.

In addition to *Cynometra*, the forest includes some very large trees of *Khaya* and *Entandrophragma*, but the stand lacks both the smaller exploitable trees and young regeneration of these valuable timber species.

Patches of characteristic colonising species, such as *Maesopsis*, mature alongside climax canopy species. This mosaic of climax and seral stages is believed to be due to the destruction caused by the fall of large trees, each of which may smash down a 4000 m^2 of forest. These patches provide conditions for seedling establishment closely resembling those of open colonising forest.

Parinari forest

The second type of climax community, the *Parinari* Forest, has been described by Dawe (1906) and by Osmaston (1960). *Parinari excelsa* forest is developed mainly in Uganda, but another species, *P. curatellifolia*, is recorded in wooded grassland in Kenya and is codominant with *P. excelsa* in parts of the East Usambaras. *Parinari excelsa* is a large-crowned tree with small leaves and, according to Osmaston (1960), in some valleys it forms almost pure stands, with *Carapa grandiflora* associated commonly as an under storey. The latter is remarkable for its enormous fruits and large oily seeds, so that they must limit

Key to symbols for Figs. 1.5 and 1.6

Ac	*Alstonia congoensis*	Fl	*Funtumia africana*
Cap	*Maerua duchesnei*	Kg	*Klainedoxa gabonensis*
Cb	*Celtis wightii*	Kh	*Khaya antotheca*
Cd	*Celtis durandii*	Ll	*Alchornea laxiflora*
Ceu	*Coffea eugenoides*	Lm	*Lasiodiscus mildbraedii*
Cyn	*Cynometra alexandri*	Ml	*Mildbraediodendron*
Chn	*Chrysophyllum* sp.		*excelsum*
Cs	*Celtis mildbraedii*	My	*Myrianthus arboreus*
Cz	*Celtis zenkeri*	Od	*Ouratea densiflora*
Dm	*Drypetes ugandensis*	Ra	*Rinorea ardisiaeflora*
Eu	*Euphorbis teke*	Tf	*Turraea floribunda*
Fu	*Funtumia elastica*	Tpr	*Trichilia prieuriana*
		Us	*Uvariopsis* sp.

Fig. 1.6 Profile diagram of *Cynometra* climax forest, Budongo, Uganda. (Eggeling, 1947)

30

its dispersal considerably. Other under storey trees include *Craterispermum laurinum*, *Trichilia prieuriana* and *Pleiocarpa pycnantha*. The *Parinari* itself may be assisted in its dispersal by the fruit bat *Eidolon helvum* which visits the Kalinzu forest in very large flocks (Osmaston, 1953). Unfortunately the wood of *Parinari* is of limited economic value. It is exceedingly hard and must be cut with specially-hardened steels since normal tools blunt rapidly. For some years, however, it has been successfully exploited at Kalinzu; the bulk of the product being sold for pit props to Kilembe Copper Mines.

Kenya lowland forest

Kenya has very little in the way of lowland forest and, if we exclude the mangrove forest, it is confined largely to patches on the coastal hills. Most of Kenya's forested country is above 1660 m on plateaux and upland slopes.

The lowland coastal region is unfavourable for forest because, being flat, it traps little rain from the south-west monsoon—the Kusi—which blows from April to August. Most of the rain is caught by the coastal hills, and thirty miles inland the countryside is almost arid, although the Teita Hills and Mt Kasiagu are high enough to be cloud centres and to receive water from rainfall and from mist.

Only small remnants of coastal forest remain, and they grow on the better-watered faces of the coastal hills, and riverine sites. At the lowest altitudes, that is below 330 m the moister places have mixed evergreen forest with much *Afzelia* and *Trachylobium*, while the drier places have *Brachylaena* woodland.

At higher altitudes in the coastal region, between 806 and 1500 m as on the Teita Hills, there is some woodland with a mixture of genera and above 1500 m evergreen forest with much camphor tree (Ocotea); Mt Kasiagu has a small area of forest near its summit, between 1300 and 1600 m and about half of it consists of *Newtonia buchanani*.

Tanzania lowland forest

Distribution

Much of Tanzania is arid and unfavourable for forest growth and recovery. The highest rainfall is on the mountains fringing the eastern edge of the great inland plateaux. These mountains, known to nineteenth century travellers as the African Ghats, receive moisture from the north-east trade winds coming inland from the Indian Ocean and some places have over 2000 mm of rainfall each year. On the northern mountains, the Usambaras, there are two rainy seasons; short rains from October to December and long rains from March to May. But on the central Uluguru Mountains and southwards there is one rainy season only, December to May.

Some of the mountains, such as the Pares, have been almost deforested, while, apart from gallery forest, the lower slopes and foothills of the others have been cleared and much of the present lower-level forest is secondary. According to Moreau (1933) vestiges of secondary forest extended north to the Shimba Hills, near Mombasa.

Between 660 and 1000 m this secondary forest, which is semideciduous, merges on the eastern faces of the mountains into evergreen rain forest with the usual dense canopy, numerous lianes, buttressed trees, and a luxuriant growth of shrubs with broad-leaved grasses in the ground layer. The total of this evergreen forest was estimated by Moreau (1933) as not more than about 800 km^2 and it occurs on the Usambaras, the Ngurus, and the Ulugurus. There can be no doubt that formerly it was more extensive and its potential area is thought to be indicated by large specimens of *Chlorophora* on farmland near the mountains. *Chlorophora* is not a tree of primary forest so that these isolated specimens are not relics from the earlier forest; they do, however, show where tree growth is possible.

Written evidence is scarce, and this makes it difficult to be clear as to what is truly representative of the low-altitude forests of Tanzania.

Gilchrist (1952) recorded a mixed broad-leaved forest occurring under not less than 1270 mm rain and below 1200 m in the Ulanga district. Its chief tree species were *Albizia*, *Antiaris usambarensis*, *Bombax schumannianum*, *Chlorophora excelsa*, *Newtonia pauci-juga*, *Zanha africana*, *Diospyros mespiliformis*, *Khaya nyasica*, *Pachystela* sp and *Terminalia kilimandscharica*.

At slightly higher altitudes, approaching 1200 m in the same district, Gilchrist found that *Allanblackia stuhlmanii* became dominant and was associated with *Bersama*, *Celtis durandii*, *Cephalosphaera usambarensis*, *Tabernaemontana holstii*, *Khaya nyasica*, *Macaranga kilimandscharica*, *Parinari excelsa* and *Newtonia buch-ananii*. Forest, apparently very similar to this was recorded at 914 m on steep topography on the East Usambaras and the Nguru Mountains by Willan (1965). It received a well-distributed rainfall of at least 1800 mm. It contained much *Cephalosphaera usambarensis*, and *Beilschmiedia kweo* and *Newtonia buchananii* were noted as valuable timber associates. According to Willan, there are fairly large tracts of this forest on the mountains mentioned and he regarded them as the finest and botanically richest forests of Tanzania. Brenan and Greenway (1949) record *Cephalosphaera* along with *Allanblackia* and *Polyalthia* on steep slopes between 900 and 1400 m on the East Usambaras and on the east side of the West Usambaras.

At altitudes above 1200 m and up to the edge of the upland forest belt at about 2000 m, it seems that *Ocotea-Podocarpus* forest occurs in the wetter parts and a mixed, dry sclerophyllous forest in the drier zones where the rainfall is less than 1500 mm. Gilchrist (1952) gives the range of this forest as 1200–2800 m, and he believes that it was once *very* much more extensive, probably extending into areas with a rainfall as low as 900 mm; but burning and grazing have caused these forests to disappear from the lower rainfall zones. He describes the forest canopy as reaching a height of about 15 m at lower altitudes, but its stature falls off rapidly with altitude. Lianes are frequent and epiphytes rare. Most of the trees are evergreen and many are sclerophyllous, especially at the lower altitudes. At Mufundi he recorded *Cassipourea gummiflua. C. malosana,* and *Rapanea rhododendroides* as the typical trees associated with *Albizia gummifera* and *A. schimperana* and for convenience we have described this *Cassipourea* forest under upland forest.

The above list from Mufundi forest is generally applicable to dry schlerophyll forest and is similar to the drier end of the *Ocotea-Podocarpus* forest, where, of course, one would add *Ocotea usambarensis* and *Podocarpus milanjianus.*

There is a high degree of floristic similarity between this *Cassipourea* forest of Tanzania and the Impenetrable Forest of south-western Uganda which is described in this account under the heading of mixed forest with *Chrysophyllum.*

In many parts of Tanzania large areas of former *Cassipourea* forest and of *Ocotea-Podocarpus* forest have been cleared away for cultivation and in some places the abandoned cultivation is marked by thickets of compact growth of woody shrubs and small trees growing to a height of about 7 m and more, with occasional emergent trees. There may be some lianes and scramblers, but practically no ground flora. This thicket or secondary forest occurs also around the upland forests. According to Gilchrist the bulk of the thicket consists of *Agauria salicifolia*, *Albizia* spp., *Apodytes dimidiata, Catha edulis, Cussonia* spp., *Ilex mitis, Kiggelaria* spp., *Maesa lanceolata, Myrica salicifolia, Nuxia* spp., *Olea* spp., *Crassocephalum mannii, Trichocladus ellipticus, Vernonia* spp., *Tecomaria shirensis.*

When the thicket is cleared again for cultivation, the first plants to come up are semi-woody genera such as *Hibiscus, Sparmannia* and *Triumfetta* followed by *Maesa lanceolata.*

In riverine forests the most important timber trees are *Chlorophora excelsa* (iroko or mvule), *Cephalosphaera usambarensis* (mtambara) and *Khaya nyassica, Khaya* is always, it seems, in valleys but not always confined to the riverine strips. McCarthy (personal communication) records huge *Khaya* up to 7 m in girth, in fair quantity though patchily distributed, in the foothills south of Ifakara in the Kilombero valley where riverine forest merges into lowland forest on the broad, alluvial flats.

These riverine and valley sites yield a variety of other timbers but few are available in large quantities. The following are some of the more important:

Antiaris toxicaria	Moraceae
Adina microcephala	Rubiaceae
Bombax rhodognaphalon	Bombacaceae
Erythrophleum suaveolens	Caesalpinoideae
Sterculia appendiculata	Sterculiaceae
Cordyla africana	Papilionoideae
Diospyros mespiliformis	Ebenaceae
Newtonia paucijuga	Mimosoideae

Upland forest (Above 2000 m)

Distribution

Upland forest in Uganda is usually separated from the medium-altitude forests by grassland and cultivation. Thus, the forests of the Western Escarpment reach a maximum altitude of about 1650 m and between this level and 2100 m much of the original forest has been removed. There is only one place in the south-west—the Impenetrable Forest—with a continuous forest cover from 1500 m to just over 2400 m. The lower part of this forest is in many places more or less dominated by the grey plum (*Parinari excelsa*), while the higher parts of the forest have large areas with *Chrysophyllum* and many associated species, and *Podocarpus* on the leached hill-ridges.

Hedberg (1951) regards the upland forest belt as extending to about 3150 m. It varies much in composition from mountain to mountain, but in Uganda four principal communities are found, *Juniperus* forest, *Chrysophyllum* forest, bamboo thicket, and *Hagenia-Rapanea* woodland.

The upland forest belt of Kenya begins with the plateau forests lying between 1300 and 2000 m and represented typically in the neighbourhoods of Nairobi, Ngong, Kiambu, and Nyeri, where the annual rainfall is 875–1000 mm and the climate is equable and cool. They are described as evergreen forests and they have a rich mixture of genera with *Brachylaena* and *Croton* predominating.

The largest areas of upland forest occur on the main mountains, Mt Kenya, Mt Elgon, the Aberdare Range, the Kikuya-Laikipia Escarpment, and the Mau-Elgeyo-Cherangani Mountain systems; they exceed in area the remaining forests of Kenya. They are evergreen and extend in altitude from about 2160 m to about 3500 m. There are four main types; *Ocotea* (camphorwood) forests, *Juniperus* forests, *Hagenia* woodlands, and bamboo thicket, though there are, of course, many intermediates.

Nearly all the forest of Tanzania is confined to the uplands which occur along a great figure 9. This begins in a cluster composed of the Crater Highlands, various small volcanoes, Mt Meru and Mt Kilimanjaro in the north-east. It continues via the Usambaras near the coast and curves inland along the Ngurus, Ulu-gurus, and the Usagaras towards the Southern High-lands, and volcanic Rungwe. From here the mountain line bifurcates; one branch continuing southwards into Malawi through the Nyika Plateau on the west of Lake Malawi, the other curving north-westwards to include some mountain forests of the Ufipa Highlands and ending at Kungwe on the east of Lake Tanganyika. Most of the valuable timber occurs in two regions of this mountain line; the larger covering Mts Meru and Kilimanjaro and the West Usambaras; the smaller on the East Usambaras.

The majority of the upland forests are isolated blocks and consist of forests similar to those elsewhere on the uplands of East Africa, namely, *Ocotea* forest, *Juniperus* forest, bamboo thicket, *Hagenia-Rapanea* woodland, and there is one additional type, namely, *Cassipourea* forest.

On small isolated mountains and outlying ridges of major ranges, and on mountains close to the coast, the upper limit of lowland forest is lower than on the main ridges. This is known as the Massenerhebung effect and the reason for it is obscure. Grubb (1970) offers an explanation in terms of availability of certain nutrients. The rate of mineralisation of organic matter decreases with either lower mean temperature or increase in soil water content. Nitrogen and phosphorus will be in lesser supply if either mean temperature decreases or frequency of fog increases as this raises the water content of the soil. The upper limit of lowland forest plants can be brought down below the temperature-controlled upper limit when the fog occurs at lower levels. The poorer supply of nitrogen and phosphorus is likely to have a strong effect when temperature is already beginning to be unfavourable. The lower limit of upland forest plants is likely to be governed by competition from plants of the next forest type downwards.

Floristic composition

In dealing with the floristics of upland forest, it is not necessary to consider each country separately. In all three countries the upland forest falls more or less into the same three zones:

1 A lower zone with mainly broad-leaved trees and, in some places, conifers.

2 A mid zone consisting mainly of bamboo thicket or woodland.

3 An upper zone of low stature forest or woodland containing usually much *Hagenia* and *Rapanea*.

1 The lower zone. A large number of forest types have been described from this zone. They can be arranged roughly according to their degree of wetness as follows:

Wetter types	*Ocotea—Podocarpus* forest
	Aningeria forest
	Ficalhoa—Afrocrania forest
	Cassipourea forest
	Mixed forest with *Chrysophyllum*
Intermediate types	*Juniperus* forest
	Podocarpus forest
Drier type	*Brachylaena—Croton* forest

2 The mid zone.   *Arundinaria* forest

3 The upper zone.   *Hagenia—Rapanea* forest
Afrocrania—Agauria forest (Zaire only).

Lower zone   Wetter type

Ocotea and Podocarpus milanjianus forest

The wettest climates with an annual rainfall in excess of 2226 mm in the altitudinal range 1700–2400 m often support forest dominated by *Ocotea usambarensis* and species of *Podocarpus*, mainly *P. milanjianus*. This type of forest has been recorded throughout East Africa, but it is important principally on the mountains of Kenya and Tanzania (Abraham, 1958). Forests dominated by *Ocotea* do not occur in Uganda though the tree occurs rarely in the west (Eggeling and Dale, 1951). Two species of *Ocotea* occur in East Africa, but only *Ocotea usambarensis* is important. It is a very large tree with a spreading crown and reaches a height of 45 m. The red-brown bark is shed in large flakes. The trunk may be as much as 3 m in diameter and, in girth, it is one of the largest trees in Kenya, being excelled only by the baobab (*Adansonia*). Its common name, East African camphor tree, is due to the camphor-scented leaves and wood which in fact contain no camphor, but a volatile oil known as cineol. The leaves are alternate, simple, and elliptic, 4–9 cm in length and 2·5 cm–4 cm wide, dark green above with a whitish pile on the undersurface. The margins are entire and recurved. Small specimens of *Ocotea* are easily confused, at first glance, with *Myrica salicifolia* which is also aromatic and has similarly shaped leaves, but leaves of *Myrica* can be identified by the golden-coloured glands which occur on both surfaces, especially the undersurface, and by the unequal base of the leaf. Also, *Myrica*, which is often associated with *Ocotea*, is usually only a shrub and, if a tree, never grows to more than 15 m in height.

The *Ocotea-Podocarpus* forests of Kenya occupy less area than their drier counterparts, the *Juniperus* forests, in the same altitudinal range, but their area is large enough to constitute a major asset in the country's indigenous forest estate. They are recorded from the eastern slopes of the Aberdares, and of Mt Kenya, these being the wetter slopes. *Ocotea usambarensis* is known to occur also on the hills in the Teita district and in the South Kamasia area of the Rift Valley, and is reputed to have occurred on the Machakos Hills. The ground covered by these forests is broken into a multitude of narrow ridges of fan-shaped stream and tributary systems, all with very steep valley sides, and it is on these ridge tops and valley sides that most of the camphor grows. It often grows at an angle to the slope and the trunk of some trees is so near the horizontal that one can walk up the trunk without holding on. The tension in the top buttress must be tremendous when the size of the trees is considered.

The lower parts of the valleys and ravines may contain *Syzygium* sp, *Ilex mitis*, *Ficalhoa laurifolia*, *Ensete vertricosum* (wild bananas) and tree ferns. These handsome plants have slender trunks, which may reach a height of about 7m and terminate in a single pseudo whorl of immense, delicately-incised fronds, each up to 2·5 m in length. Pl.9.

Ocotea-Podocarpus forests occur in Tanzania along the West Usambara Mountains and Pitt-Schenkel (1938), who has described these, states that on Mt Kilimanjaro—but not on the north slopes—there is a zone of *Ocotea* forests. Forests of similar composition —though lacking *Ocotea*—occur on Mt Meru, on the Mbulu Highlands, on the Ufiume-Mikiulu uplands, and on Ufiume and Hanang mountains. Pitt-Schenkel believes that the original forests along the Kiberege-Dagaba-Mufundi Escarpment in the Southern Highlands may have been similar to the West Usambara forests, but they are now all secondary having been cleared in the past by cultivators.

The principal species in the West Usambara *Ocotea-Podocarpus* forests are shown in Table 1.5. These trees occur together in different degrees of abundance so that Pitt-Schenkel (1938) was able to describe about eight communities in the area which he examined; the most widely distributed community,

and of greatest area, being the *Ocotea-Podocarpus-Lasianthus* forest. The structure of this forest seemed to be of the usual three-layered type with under-storey shrubs mainly of the Rubiaceae. There was no herbaceous ground flora and the soil, which was acid, was covered with a layer about 15 cm thick of organic matter. The most obvious differences between the various types of camphor wood forest on the West Usambaras was in the ground flora. There were three main types; those in which the ground flora was almost absent; those in which species of Acanthaceae predominated, and a third type—distinctly wet—in which the ground flora consisted of dense stands of a balsam (*Impatiens* sp.), growing to a height of between 60 and 150 cm and with large, pale mauve flowers.

Presentday conditions are no longer favourable for the *Ocotea* to maintain itself by regeneration, which implies that the present forests established themselves under different conditions from those obtaining now. Seed is produced plentifully in some years, but regeneration from seed is comparatively rare due to the predations of insects and larger creatures. *Ocotea* can, however, be regenerated from suckers when the trees are felled. Unfortunately elephants are fond of these suckers and in the Kenya forests normally very few survive to become big trees. In parts of Tanzania, with fewer elephants, the suckers survive, and this together with their very rapid growth makes the management of these forests somewhat easier (Willan, 1965).

The wood from *Ocotea usambarensis* seasons well, is strong, highly resistant to fungi and acids—though not to termites—and it is used for furniture, and joinery. Apart from timber exploitation the camphor forests have considerable value as protection forests, in preventing soil erosion and controlling the gradual release of water to the mountain streams.

Aningeria-Adolfi-friedericii forest

Aningeria-adolfi-friedericii is fairly widely distributed in wet montane forests in Uganda and Kenya within the altitudinal range 1524–2438 m. On some parts of Mt Elgon it becomes sufficiently common to give its name to the forest type (Dale, 1940). It is one of the commonest trees between the Sosia river in the south-east and the Sipi in the north-west. Towards its limits *Aningeria* and some of its associates are restricted to valleys. It is heavily buttressed and can reach a height of about 50 m. Dale (1940) has recorded it associated with *Alangium chinense, Albizia gummifera, Allophylus abyssinicus, Casearia battiscombei, Tabernaemontana* sp. *Croton macrostachyus, Neoboutonia macrocalyx, Pygeum africanum, Strombosia scheffleri, Syzygium guineense*, and, less commonly, with *Albizia*

**Table 1.5   Composition of the Camphorwood Forests (Ocotea Usambarensis),
west Usambara Mountains, Tanzania. (1700–2400 m)**

TALL TREE STRATUM

More important trees	*Entandrophragma excelsum*	Meliaceae
	Ocotea usambarensis	Lauraceae
	Podocarpus usambarensis	Podocarpaceae
	Prunus africana	Rosaceae
Less important trees	*Balthasaria schliebenii*	Theaceae
	Aningeria adolfi-friedericii	Sapotaceae
	Casearia battiscombei	Flacourtiaceae
	C. sp. nr. engleri	Samydaceae
	Chrysophyllum albidum	Sapotaceae
	C. gorungosanum	Sapotaceae
	Dombeya leucoderma	Sterculiaceae
	Ekebergia capensis	Meliaceae
	Ficalhoa laurifolia	Theaceae
	Ilex mitis	Aquifoliaceae
	Ochna holstii	Ochnaceae
	Mammea africana	Guttiferae
	Ocotea kenyensis	Lauraceae
	Polyscias kikuyuensis	Araliaceae
	Rapanea rhododendroides	Myrsinaceae
	R. usambarensis	Myrsinaceae

TREES IN THE STRATUM (9 m—21 m)

Afrocrania volkensii	Cornaceae
Albizia gummifera	Leguminosae subfam. Mimosoideae
Allophyllus abyssinicus	Sapindaceae
Aphloia theiformis	Flacourtiaceae
Apodytes dimidiata	Icacinaceae
Cassipourea malosana	Rhizophoraceae
Craibia brownii	Leguminosae subfam. Papilionoideae
Croton macrostachyus	Euphorbiaceae
Cussonia spicata	Araliaceae
Dasylepis leptophylla	Flacourtiaceae
Dracaena steudneri	Agavaceae
Faurea saligna	Proteaceae
Macaranga kilimandscharica	Euphorbiaceae
Myrica salicifolia	Myricaceae
Neoboutonia macrocalyx	Euphorbiaceae
Ochna stuhlmannii	Ochnaceae
Olea hochstetteri	Oleaceae
Pauridantha holstii	Rubiaceae
Podocarpus gracilior	Podocarpaceae
Strombosia scheffleri	Olacaceae
Syzigium guineense	Myrtaceae
Trichocladus ellipticus	Hamamelidaceae
Tabernaemontana holstii	Apocynaceae
Urophyllum holstii	Rubiaceae

Catha edulis	Celastraceae
Clausena anisata	Rutaceae
Croton sp. nr. *scheffleri*	Euphorbiaceae
Cyathea sp.	Cyatheaceae
Dodonoea viscosa	Sapindaceae
Dracaena deremensis	Agavaceae
Ensete ventricosum	Musaceae
Lasianthus kilimandscharicus	Rubiaceae
Memecylon cogniauxii	Melastomataceae
M. deminutum	Melastomataceae
Myrsine africana	Myrsinaceae
Pavetta hymenophylla	Rubiaceae
Pittosporum lanatum	Pittosporaceae
Piper capense	Piperaceae
Psychotria riparia	Rubiaceae
Syzygium cordatum	Myrtaceae
Teclea nobilis	Rutaceae
Xymalos monospora	Monimiaceae

BRYOPHYTES
(From Podocarpus forest on Mt. Elgon, Uganda at 2000 m)

Epiphytic species	*Pilotrichella cuspidata*
	Lophocolea moelleri
	Lejeunea cf. *lamacerina*
	Plagiochila breviramea
On soil	*Campylopus introflexus*
	Polytrichum piliferum
	Polytrichum juniperinum
	Ceratodon purpureus
	Pogonatum urnigerum
	Pohlia nutans
	Micromitrium abyssinicum
	Trachypodopsis sp.
	Tortula hildenbrandtii
	Cephalozia spp.
	Dumortiera cf. *nepalense*
	Nardia scalaris
	Pallavicinia lyellii
On rocks with soil filled crevices	*Asterella gibbosa*
	Bartramia pomiformis
	Fimbriaria sp.

---

grandibracteata and/or *A. zygia*, *Anthocleista* sp., *Bosqueia phoberos*, *Ekebergia capensis*, *Fagara macrophylla*, *Kigelia* sp., *Olea welwitschii*, and *Podocarpus milanjianus*.

Ficalhoa-Afrocrania forest

This type of forest is recorded by Gilchrist (1952) as occurring between 1200 and 2700 m under a rainfall of 2200 mm or more. Although its dominants occur throughout highland East Africa, they seem to appear in this particular combination only over large areas of the Mporoto Mts and Rungwe Mt in Tanzania. Gilchrist states that the forest is composed almost entirely of *Ficalhoa laurifolia* and *Afrocrania volkensii*.

Cassipourea malosana forests (Pillar wood)

The drier end of the camphor wood forest sometimes grades into a forest with much *Cassipourea malosana*. This is a common tree in the upland forest belt throughout East Africa from 2000 to 3000 m. It is an evergreen, small-crowned tree reaching about 20 m with a distinctly cylindrical or pillar-like trunk. The bark is greyish-white, smooth, with well-developed lenticels in horizontal lines. The slash is orange. The leaves are leathery in texture, xeromorphic in structure, elliptic, 3·5–6 cm long and 1·8–3 cm wide, with usually dentate margins. The flowers are small, yellowish green.

The wood gives a pale greyish or brownish timber with irregular darker markings. It is straight grained, fine textured, and moderately hard and heavy. The timber is easy to work and would be an excellent flooring timber, but it has a tendency to stain and is difficult to season. However, it is very suitable for construction work requiring great strength, and as a building pole for temporary buildings it is said to have no equal.

Cassipourea becomes plentiful in Uganda in some forest areas of Mt Kadam, Mt Elgon, and the Imatong Mts, and in Kenya it is common on south-west and north-east Mt Kenya and the north-eastern Aberdares. It would seem that it is best developed as a forest type in Tanzania, and perhaps the most important area commercially and ecologically is on Kilimanjaro. Gilchrist (1952) records blocks of this forest at Mufundi, Ukwama, Myumbanitu, Uzungwa, Dagaba, the west side of Rungwe, and Umalita. He believes that it is a forest type which, at one time, was more widespread and extended probably into rainfall areas of 875 mm per year, but fire and man's activities have caused the forest in the lower rainfall areas to disappear first.

The composition of the forest is shown in Table 1.6. It is a type which lies across our arbitrary division into lowland and upland forests and Gilchrist (1952) gives its range, in Tanzania, as 1220–2740 m. It is not certain that the forest extends much into the lower part of this altitudinal range in Kenya or Uganda.

The openness and dryness of the forest in Tanzania renders the forest susceptible to fire, but with sufficient rainfall it regenerates and stages of recovery are believed to be indicated by much *Myrica salicifolia* and *Macaranga kilimandscharica* in the upper canopy.

Mixed forest with Chrysophyllum

This type is well represented in the Impenetrable Forest (Table 1·7) which covers 29 800 ha of the Rukiga Mountains in south-western Uganda. In altitude the forest extends between 1400 and 2400 m, the higher range being dominated by *Chrysophyllum* while at the lower range it contains much *Parinari excelsa*. The Impenetrable Forest is of great intrinsic beauty and importance. It is one of the two habitats in Uganda of the rare mountain gorilla *Gorilla gorilla beringei*, and is rich in other animals including chimpanzee, red forest duiker, elephant and buffalo.

The rainfall is about 1160 mm with mist and cloud common in the early morning and undoubtedly this substantially reduces the evaporative demand. Usually the forest is too moist to burn extensively, although numerous gaps have been created by fires especially on ridges during dry seasons. The soils of the area (Harrop, 1960) are very acid and very poor in bases.

The forest was described by Langdale-Brown *et al.* (1964) as *Prunus* forest but, so far as we are aware, this tree is only locally common. The description which follows is based on that part of the forest close to Ruhiza forest station (Hamilton, 1969). All the forest in this part lies over steep ridges separated by deep valleys and gullies, and it thus provides an excellent opportunity for recording the effects of topography with its influence on water availability on the distribution of species.

On hilltops, which are uncommon, the trees are stunted, frequently multi-boled, and the canopy, which is closed and of even height, is only about 10 m tall. The most abundant tree is *Macaranga kilimandscharica*, with *Olea hochstetteri* also common. The shrub and herbaceous strata are open and poor in species, many of the shrubs belonging to the family Rubiaceae, while *Mimulopsis solmsii* and several species of ferns commonly occur.

On ridges, the canopy is taller, there are a greater number of species and a richer under storey. In places, *Faurea saligna* forms almost pure stands about 20 m tall. *Olinia usambarensis*, *Podocarpus milanjianus*, *Polyscias fulva*, and *Rapanea rhododendroides* are other common species found in the upper stratum. The understorey trees, such as *Macaranga kilimandscharica*, *Maesa lanceolata*, *Psychotria megistosticta* and *Rytigynia* sp. are spaced well apart. In places there is a tall shrub layer, 3 m tall with much *Cyathula* cf. *uncinulata* and *Mimulopsis solmsii* and, less commonly, *Cluytea* sp. and *Pteridium aquilinum*. The climber *Urera hypselodendron* is common.

On upper hillslopes, the forest is normally taller and denser than on ridges. In the canopy, which can be 30 m tall, *Chrysophyllum gorungosanum* is often the most abundant tree, and *Olea hochstetteri* is also common. Beneath the upper tree layer, there is a well-defined second tree stratum about 18 m tall, containing *Allophylus macrobotrys*, *Cassipourea ruwenzorensis*, and *Drypetes* aff. *gerrardii*. Rubiaceae are conspicuous in the under storey. *Cyathula* cf. *uncinulata* is a common shrub, forming a dense tangle about 2 m tall and scrambling up a tree trunk to a

Table 1.6 Composition of the Pillar wood, *Cassipourea malosana* forests in Tanzania (2500 m altitude). (After Gilchrist, 1952 and Rea, 1935)

TREES IN THE STRATUM (9—21 m)

Cassipourea malosana accompanied by—	Rhizophoraceae
Albizia gummifera	Leguminosae subfam. Mimosoideae
A. schimperiana	Leguminosae subfam. II Mimosoideae
Apodytes dimidiata	Icacinaceae
Calodendrum capense	Rutaceae
Casearia battiscombei	Flacourtiaceae
Dombeya spp.	Sterculiaceae
Ekebergia capensis	Meliaceae
Fagara amaniensis	Rutaceae
Hagenia abyssinica	Rosaceae
Ilex mitis	Aquifoliaceae
Macaranga kilimandscharica	Euphorbiaceae
M. conglomerata	,,
Myrica salicifolia	Myricaceae
Nuxia spp.	Loganiaceae
Ocotea usambarensis	Lauraceae
Parinari excelsa	Rosaceae
Podocarpus milanjianus	Podocarpaceae
Prunus africana	Rosaceae
Rapanea rhododendroides	Myrsinaceae
Strombosia scheffleri	Olacaceae
Syzygium guineense	Myrtaceae

TREES AND SHRUBS IN THE STRATUM (0—9 m)

Bridelia brideliifolia	Euphorbiaceae
B. scleroneuroides	,,
Mystroxylon aethiopicum	Celastraceae
Cassipourea gummiflua	Rhizophoraceae
Cordia africana	Boraginaceae
Flacourtia spp.	Flacourtiaceae
Gnidia glauca	Thymeliaceae
Psychotria spp.	Rubiaceae
Maytaenus spp.	Celastraceae
Olea africana	Oleaceae
Pittosporum viridiflorum	Pittosporaceae
Teclea spp.	Rutaceae
Syzygium guineense	Myrtaceae
Tetrorchidium didymostemon	Euphorbiaceae

TREES USUALLY IN THE STRATUM (0—9 m)

Acalypha sp.	Euphorbiaceae
Allophylus macrobotrys	Sapindaceae
Glyphea brevis	Tiliaceae
Mimulopsis sp.	Acanthaceae
Myrianthus holstii	Moraceae
Teclea nobilis	Rutaceae
Xymalos monospora	Monimiaceae

Table 1.7   Composition of the Impenetrable Forest, Uganda 1400 m—2600 m

TALL TREE STRATUM (21 m—37 m)

More important trees	*Chrysophyllum gorungosanum*	Sapotaceae
	Entandrophragma excelsum	Meliaceae
	Newtonia buchanani	Leguminosae subfam. Mimosoideae
	Parinari excelsa subsp. *holstii*	Rosaceae
	Podocarpus milanjianus	Podocarpaceae
	Prunus africana	Rosaceae
	Symphonia globulifera	Guttiferae
Less important trees	*Balthasaria schliebenii*	Theaceae
	Allanblackia floribunda	Guttiferae
	Ekebergia capensis	Meliaceae
	Fagara macrophylla	Rutaceae
	Ocotea usambarensis	Lauraceae

TREES IN THE STRATUM (9 m—21 m)

Albizia gummifera	Leguminosae subfam. Mimosoideae
Allophylus abyssinicus	Sapindaceae
Beilschmiedia ugandensis	Lauraceae
Carapa grandiflora	Meliaceae
Cassipourea ruwensorensis	Rhizophoraceae
Croton macrostachys	Euphorbiaceae
Dombeya goetzenii	Sterculiaceae
Drypetes gerrardii	Sterculiaceae
Faurea saligna	Proteaceae
Ficalhoa laurifolia	Ericaceae
Guarea mayombensis	Meliaceae
Harungana madagascarensis	Hypericaceae
Ilex mitis	Aquifoliaceae
Macaranga kilimandscharica	Euphorbiaceae
Neoboutonia macrocalyx	Oleaceae
Olea hochstetteri	Oleaceae
Olinia usambarensis	Oliniaceae
Polyscias fulva	Araliaceae
Dichaetanthera corymbosa	Melastomaceae
Strombosia scheffleri	Olacaceae
Syzigium guineense	Myrtaceae
Tetrorchidium didymostemon	Euphorbiaceae

TREES USUALLY IN THE STRATUM (0—9 m)

Allophylus macrobotrys	Sapindaceae
Cyathea deckenii	Cyatheaceae
Psychotria megistosticta	Rubiaceae
Lobelia gibberoa	Campanulaceae
Myrianthus holstii	Moraceae
Rytigynia sp.	Rubiaceae
Tabernaemontana holstii	Apocynaceae
Teclea nobilis	Rutaceae
Xymalos monospora	Monimiaceae

9   Tree ferns (*Cyathea deckenii*) in Uganda forest. (Uganda Ministry of Information).

height of 6 m. Ferns are abundant on the forest floor.

On lower hillslopes and in valleys, the upper storey is often over 30 m tall, and the under storey and shrub layers are well developed. Among the large trees, *Chrysophyllum gorungosanum*, *Newtonia buchananii*, and *Symphonia globulifera* are particularly common. *Croton megalocarpus* is said to form almost pure stands in some parts (J. Ball, personal communications). Though *Parinari excelsa* also occurs, it is probably not as common as is sometimes supposed, being mistaken for other species.

Gully forests vary in species composition and in structure from their upper to their lower parts. They are usually very open, this being partly attributable to the instability of the soils and partly to elephants which can in places cause considerable damage by uprooting trees and digging for their roots. The main species concerned appear to be *Macaranga kilimand-scharica* and *Polyscias fulva*. The usual appearance of the upper parts of these gullies is of a dense tangle of shrubs and climbers, standing about 2 m tall, from which scattered trees emerge. Common tangle species are *Cyathula* cf. *uncinulata*, *Impatiens* sp., *Pteridium aquilinum*, *Rubus* sp. and *Urera hypselodendron*. Common trees are *Dombeya goetzenii*, *Neoboutonia macrocalyx* and *Polyscias fulva*, and common climbers on these trees, *Urera* and species of *Clematis*.

Lower zone: Intermediate type

Coniferous, dry upland forest

There are five conifers on the uplands, namely, *Juniperus procera* and four species of *Podocarpus*. *Podocarpus milanjianus* occurs in the wetter parts and the remaining species together with the *Juniperus* and other hardwoods occur together in the drier parts, generally in the annual rainfall range 875–1375 mm. Although these coniferous forests occur in all three countries of East Africa, they have been most fully developed—and exploited—in Kenya where *Juniperus* is often referred to as 'cedar'.

Juniperus forests

Juniperus is a genus with about forty species which occupy a continuous broad belt round the northern hemisphere with some species extending southwards through Arabia. The distribution of *Juniperus procera* begins in western Arabia and extends southwards across the Ethiopian highlands and throughout eastern Africa as far as southern Tanzania.

The tree grows to 47 m in height and achieves a bole 3 m in girth at breast height. Its foliage when young is prickly and needle-like, but the mature foliage consists of minute, xeromorphic closely overlapping scale-like

leaves. The female cones are extremely modified, contain about three seeds, and have the general appearance of a glaucous blue berry; the seeds are almost certainly distributed by birds. The heartwood is reddish brown, strong, and very resistant when seasoned, to insect, fungal and termite attack. It is easy to work, but is apt to be brittle and splits easily. It is used for joinery, flooring, shingles for roofing, and for the wood casings for lead pencils. The yield is not being sustained and the available supplies are not expected to last more than about twenty years.

Juniperus forests develop in two principal ways. The tree will be found growing up under the shelter of bushes such as *Philippia* on the edge of forests. Here the resultant trees are usually comparatively short and branching and, where the process has been allowed to continue for a long time, progressively older trees are found further into the forest. The other principal method is natural regeneration, usually very dense, after a fire. This gives rise to tall, even-aged stands of timber with long, branch-free boles. The nearly pure stands resulting from fire naturally last longer than the type arising from forest spread.

In Uganda there are no extensive areas of *Juniperus* forest, but only patches occurring on the mountains of the north-east and on Mt Elgon. These patches are presumably relics of once somewhat larger communities.

Kenya is the main home of *Juniperus* in East Africa and it is widely distributed in the drier highland areas, ranging from 1067 m to 2896 m, but attaining its maximum development between 1829 m and 2896 m within an annual rainfall of 1016–1156 mm. Some years ago it was regarded (Gardner, 1932) as almost the commonest *timber* tree in Kenya.

In Tanzania *Juniperus* occurs in some quantity on the drier north-west slopes of Mt Kilimanjaro, Mt Meru, and the West Usambaras and it is recorded also from Mt Hanang and the Livingstone Mountains. Sayers (1930) believes that in Tanzania *Juniperus* forest was—perhaps not so long ago—much more extensive. He records logs, possibly some centuries old, lying half-buried on the floor of mixed forest and of *Podocarpus* forest, suggesting that, in the course of time, competition may have ousted the essentially light-demanding *Juniperus* to drier, less competitive habitats. It may seem strange, in tropical Africa, to attribute such a great age to logs on the forest floor, but the dead timber of *Juniperus*, because of its high resistance to attack and decay, is extremely durable in or out of the ground. This contrasts with the living wood which is susceptible to damage by the cedar fungus, *Fomes juniperinus*. In dry forest (Gardner, 1932) claimed that nearly every tree was infected, but in the moister forest infection may amount to not more than 50 per cent. The fungus creates cavities in the wood, from the size of an orange to that of a large football and, in bad attacks, a large tree may be reduced to a mere shell.

There are very few accounts of the structure and composition of *Juniperus* forest. Katende and Lye (personal communication) observed *Juniperus* forest on the eastern side of Mt Meru. In places the forest was almost pure *Juniperus* with very little growing with it. But in many places the *Juniperus* was accompanied by large trees of *Olea africana* (mainly in the higher parts) and *Olea welwitschii* (on the lower parts), and smaller trees including *Buddleya polystachya*, *Diospyros abyssinica*, *Dombeya* sp., *Euclea microcarpa*, *Halleria lucida*, *Myrsine africana* (on rocky slopes), *Peddiea fischeri*, and *Teclea simplicifolia*. Epiphytes on the trees included *Elaphoglossum deckenii*, *Hymenophyllum capillare*, *Lycopodium ophioglossoides*, and *Sedum meyeri-johannis*. Gardner (1932) recorded that the branches of *Juniperus* are often infested with the hemi-parasite *Loranthus*. The shrubs in the Meru forest included *Afrocrania volkensii* and *Clutia robusta*. Smaller shrubs included *Micrococca holstii*, *Polygala persicariifolia*, *Rubus pennatus*, *Struthiola thomsonii*. Among the herbs recorded were *Aristea alata*, *Aneilema pendunculosum*, *Bidens kilimandscharica*, *Carex fischeri*, *C. johnstonii*, *Linum volkensi*, and the beautiful blue-flowered *Streptocarpus glandulosissimus*, probably mainly in the wetter parts. The more riverine parts of the forest contained the fern *Blechnum attenuatum*.

In Kenya there are only fragmentary records of the species associated with *Juniperus*. On hills in the northern frontier region with an annual rainfall in the range 559–635 mm. Gillett (1951) recorded *Juniperus* with *Olea africana*, *Podocarpus gracilior*, and *Croton megalocarpus*. Gardner (1932) was of the opinion that *Olea africana* occurs with *Juniperus* in the drier areas while *O. hochstetteri* occurs in the more moist regions. The forests on the hills and mountains in northern Kenya (Leroghi, Matthews Range, Kulal, Marsabit, the Ndotos, etc.) are valuable protection forests and their conservation is vital for water supplies in the area. Unfortunately, many of them are heavily infested with game, such as elephant, buffalo, and rhino which have caused much damage, as has fire. At one time it was believed that the forest on Marsabit contained about one elephant to every 80–100 acres, as well as buffalo, rhino and giraffe. To maintain these forests the game density should not be allowed to rise too high, and some planting is required to make good past damage.

From the generally more moist areas of Kenya,

Battiscombe (1936) recorded that *Juniperus* was generally associated with *P. gracilior*, *milanjianus*, *Olea africana*, *O. hochstetteri*, and other hardwoods noted were: *Allophylus abyssinicus Celtis africana*, *Cassipourea malosana*, *Dombeya goetzenii*, *Ekebergia capensis*, *Faurea saligna*, *Hypericum revolutum*. *Myrsine africana*, *Nuxia congesta*, *Olinia usambarensis*, *Teclea* spp., and *Warburgia ugandensis*. *Acanthus eminens*, a blue-flowered shrub up to about 2·5 m in height has been recorded as covering large areas inside what is probably a moist variant of *Juniperus* forest on the Mau Elgeyo, Elgon and Cherangani regions, and it occurs on Mt Kenya. *Clutia abyssinica*, a shrub which grows up to about 2 m, is common in *Juniperus* forest above Naivasha, and the solanaceous *Discopodium penninervium* is recorded as an undershrub of *Juniperus* and *Podocarpus* forests of North Tindaret and the Aberdares.

The Podocarpus forests

Podocarpus with 110 species, is the largest genus of all present day conifers. The majority are all tall trees some of which reach to 60 m, but many are shrubs. In the western hemisphere the genus extends from Patagonia to the West Indies, and in the eastern from New Zealand to Japan, with extensions westwards to Burma and West Africa. There are only four species in East Africa; one of these, *P. usambarensis* var. *dawei*, is confined to seasonal swamp forest on the western side of Lake Victoria; one species and one variety are endemic in Tanzania; and the remainder occur quite widely in the upland forest belt. *P. ensiculus* is known only from the Usambara and Uluguru Mountains, though it may be more widespread as, when not in fruit, it is easily confused with *P. milanjianus*, while *P. usambarenis* var. *usambarenis* occurs on the Usambara Mountains and on Mt Hanang, some 480 km west of the Usambaras and on other mountains. The remaining two widespread species are thought to differ slightly in their moisture preferences; *P. milanjianus* occurring in the wetter and *P. gracilior*, which is more local, in the drier upland forests. In the Cheranganis *P. milanjianus* is regenerating under *P. gracilior* (personal communication, 1968). However, not too much should be made of this distinction because, in the past, there has been much confusion in identifying the various species. Vegetatively they are similar and the very gradual change from juvenile to fully adult foliage—extending over many years—leads to confusion of the leaf forms of one stage of a species with those of a different stage of another species. Melville's (1958) revision of the genus should help to eliminate future errors of identification. In Kenya and Uganda, where there are only *P. gracilior* and *P. milanjianus* on the uplands, no difficulty should be found in separating these two because their fruits are very characteristic, as indeed are the fruits of all E. African species. Both have highly modified female strobili which resemble green olives. That of *P. gracilior* is very hard-shelled, but covered with a yellow, fleshy pericarp and nearly half an inch in diameter; that of *P. milanjianus* has, in addition, a fleshy aril-like body which is bright red when ripe. The fruits of both species are greedily eaten by monkeys, chimpanzees, and hornbills.

The two mature trees differ in general appearance, the most noticeable feature being the trunk which is often strongly fluted in *P. milanjianus*, but usually clean and cylindrical in *P. gracilior*. The latter is a stately, handsome tree with a total height of up to 33 m and a diameter, at breast height, of up to 2 m. It commonly reaches a considerably greater diameter on Mt Elgon where some of the large *P. gracilior* forest is set aside as a nature reserve. It is found in pure stands and also mixed with *Juniperus* and *Olea* and is often localized in occurrence.

P. milanjianus is generally a much smaller tree though it may reach a height of 27 m and a diameter, at breast of $1 \cdot 5$ m. It has a shaggier bark than *P. gracilior*, a bark resembling that of *Juniperus*—thin, pale to dark brown, fibrous, and peeling in long narrow strips.

The timber of both trees is of high quality. It is creamy or pale yellow in colour, occasionally showing reddish zones which may be due to compression wood. It is straight grained and very even textured being composed of uniform sized tracheids; there are no well-defined growth rings. It is light, but unfortunately not resistant to insect or termite attack though it can be treated. It is easy to work and takes an excellent finish, but it is inclined to be brittle so that it needs care in working; some of the difficulty is because it is sold green. Basically, it is a high-quality joinery timber suitable for internal uses, but in East Africa it is used for a variety of constructional purposes from roof trusses to dock piling.

There are very few descriptions of *Podocarpus* forest and this is because it relatively rarely occurs in pure stands and is often associated with *Juniperus*, *Ocotea*, *Olea* and other upland forest belt hardwoods. Pitt-Schenkel (1938) has described a *Podocarpus*-rich forest from the Shume-Magamba reserve on the West Usambara Mts of Tanzania.

In Uganda, a forest with much *Podocarpus gracilior* and *Juniperus procera* has been described from the north side of Elgon by Langdale-Brown *et al.* (1964) and by Dale (1940). Langdale-Brown *et al.* regard this type of forest as the natural climax between 1660 and 3000 m on the mountains of Karamoja and on the drier northern slopes of Elgon. But it has often been much reduced in area by fire, and only relic patches remain with *Protea* or mixed wooded grassland in between. The upper margin of the forest sometimes passes into a dry version of *Hagenia* woodland.

On the Kimothon track on the eastern side of Mt Elgon in Kenya between 2743 and 2819 m Hamilton (1970) notes that the trees of *Podocarpus gracilior* were widely spaced, growing to a height of about 30 m and accompanied by a few large trees of *Olea hochstetteri* growing to about 20 m. Smaller trees included *Bersama abyssinica*, *Dombeya goetzenii*, *Hagenia abyssinica*, *Ilex mitis*, *Nuxia congesta*, *Olea africana*, *Rapanea rhododendroides*, *Schefflera* sp., and *Xymalos monospora*. Among the shrubs recorded were *Buddleya polystachya*, *Crassocephalum* spp., *Sambucus africana*, *Clutea abyssinica*, and *Vernonia* spp. A *Podocarpus gracilior* forest, very similar in structure, was seen on the North Elgon track at 3000 m.

Lower zone: drier type

Brachylaena-Croton forest

These are semideciduous forests occurring within the annual rainfall range of 875–1000 mm. They are found mainly on plateaux in Kenya so that they are known usually as 'plateau forests'. They are represented typically in the neighbourhoods of Nairobi, Ngong, Kiambu, and Nyeri, and something of their type is seen also in the Kilimanjaro area. These forests have been described by Troup (1922 a and b), Battiscombe (1936), and by Verdcourt (1962). They tend to be dominated by *Brachylaena hutchinsii*, and *Croton megalocarpus* accompanied by a modest number of species (see Table 1.8).

Brachylaena is the only species of the East African Compositae to develop to timber size; the tree, which is deciduous, reaches 27 m in height with a diameter of $0 \cdot 6$ m, but it has a fluted and curved bole so that it is difficult to use for timber. The oblanceolate leaves are about $12 \cdot 5$ cm in length, and have a grey felt on the undersurface. The white flowers are in small clusters in the leaf axils. The timber, which is known in the trade as 'muhugu', is extremely durable and very suitable for heavy-duty flooring. *Croton megalocarpus* also yields a durable timber suitable for flooring. It grows to a height of 40 m or more and it is not too difficult to recognise by its leaves; these are pale silvery on the underside with petioles with several large, yellow, sessile or subsessile glands near the junction with the lamina. The yellow flowers grow in racemes up to 25 cm in length.

Although *Juniperus procera* and *Olea hochstetteri* grow in these plateau forests, this is the lower end of

Table 1.8 Composition of the plateau forests (Brachylaena—Croton 1500 m–2000 m)

TALL TREE STRATUM 21—37 m

More important trees	*Brachylaena hutchinsii*	Compositeae
	Croton megalocarpus	Euphorbiaceae
Less important trees	*Apodytes dimidiata*	Icacinaceae
	Canthium schimperianum	Rubiaceae
	Elaodendron buchananii	Celastraceae
	Celtis africana	Ulmaceae
	Diospyros abyssinica	Ebenaceae
	Newtonia buchananii	Leguminosae subfam. Mimosoideae
	Schrebera alata	Oleaceae
	Warburgia ugandensis	Canellaceae

TREES IN THE STRATUM 9—21 m

Acokanthera schimperi	Apocynaceae
Calodendrum capense	Rutaceae
Dombeya burgessiae	Sterculiaceae
Drypetes gerrardii	Euphorbiaceae
Ficus eriocarpa	Moraceae
F. thonningii	Moraceae
Juniperus procera	Cupressaceae
Markhamia hildebrandtii	Bignoniaceae
Ochna ovata	Ochnaceae
Margaritaria discoidea	Euphorbiaceae
Rawsonia lucida	Flacourticaceae
Teclea nobilis	Rutaceae
T. simplicifolia	Rutaceae

TREES AND SHRUBS IN THE STRATUM 0—9 m

Acokanthera longiflora	Apocynaceae
Canthium keniense	Rubiaceae
Erythrococca bongensis	Euphorbiaceae
Olea africana	Oleaceae
O. hochstetteri	Oleaceae
Strychnos henningsii	Loganiaceae
Tarrena graveolens	Rubiaceae
Teclea trichocarpa	Rutaceae
T. villosa	Rutaceae

HERBS

Anagallis arvensis	Primulaceae
Diclis ovata	Scrophulariaceae
Isoglossa lactea	Acanthaceae
Vernonia holstii	Compositae

44

their altitudinal range and they are usually small, and stunted.

Verdcourt (1962) records the occurrence of the plateau type forests in the City Park Nairobi, around the Coryndon Museum, and to the north-west of the city. Large remnants are preserved at Karura, Langata and Ngong. Since many Nairobi people visit the forest inside the city parks we reprint his description:

The semi-evergreen forest within the park boundary, in effect the southern fringe of the Langata forest, is of a dry type (e.g. *Newtonia* and *Sapotaceae* do not occur there) and in places scarcely deserves more than the name of woodland. It is thickest and most luxuriant towards the west and the canopy varies from about 12–14 m. There are really very few tall trees but several specimens of *Margaritaria discoidea* reach about double this height on the forest margins. *Croton megalocarpus* and *Schrebera alata* tend to dominate in the western part, whereas in the east *Brachylaena hutchinsii* and *Olea africana* are common. The actual number of tree species involved is quite limited and the following are the commonest —*Apodytes dimidiata*, *Brachylaena hutchinsii*, *Canthium schimperianum*, *Elaodendron buchannii*, *Croton megalocarpus*, *Diospyros abyssinica*, *Ficus thonningii*, *Phyllanthus discoideus*, *Schrebera alata*, *Teclea simplicifolia*. Tall shrubs or small trees in the understorey include *Strychnos henningsii*, *Acokanthera schimperi*, *Ochna ovata*, *Dombeya burgessiae*, *Erythrococca bongensis*, *Teclea villosa*. There are very few lianes and the herbs are much less numerous than on the surrounding plains. They include such characteristic species as—*Clitandra somaliensis*, *Isoglossa lactea*, and *Vernonia holstii*; minute herbs such as *Anagallis arvensis* and *Diclis ovata* also occur.

Mid zone

Arundinaria alpina forest or thicket (Bamboo)
The mountain bamboo *Arundinaria alpina*, a member of the Gramineae or grass family, occurs in the altitudinal range 1800–3330 m being more common towards the upper part of this range. It appears to grow most vigorously and to form continuous stands where the average annual rainfall exceeds 1250 mm, where the soils are relatively deep and the slopes not very steep. These conclusions are supported by the fact that it is absent on Mt Kilimanjaro which is regarded as one of the driest of the East African mountains; that it forms continuous stretches on the moist western side of Mt Elgon, but occurs more patchily on the drier eastern side; that on the wet Ruwenzori it is poorly developed on very steep slopes in the Mubuku and Bujuku valleys, but elsewhere on the Ruwenzori, in the range 2200–3200 m, it dominates hillslopes and rounded ridges.

In Uganda the largest stretches of mountain bamboo occur on the Ruwenzori, the Virunga volcanoes, and the Echuya reserve on the Ruchiga mountains in the south-west. In Kenya it covers large stretches on the Mau and the Aberdares, and it occurs patchily on the

eastern side of Mt Elgon. In Tanzania, Rea (1935) and Gilchrist (1952) report it on Mts Meru and Rungwe, Brenan and Greenway (1949) record it from the Ulugurus and the Livingstone Mountains (see also Gilchrist, 1952) and they think it may be in the Iringa Highlands and the Mporotos. Gilchrist (1952) notes scattered patches in forest areas of southern Tanzania, notably Mufindi, Dagaba, Uzungwa, and Ukwama, and states that there are over 7932 ha of almost pure *Arundinaria alpina* on the Livingstone Escarpment.

The tallest stems of *Arundinaria* grow to a height of 10–12 m or more and are topped by whorls of slender branches covered with pale green leaves. The plant does not grow in clumps, but in dense even stands, often the stems are so close together that it is difficult to pass through (Plates 10 and 11). On deep soils the stems reach a diameter of 7·5 cm and on shallow soils, such as those of the lava beds of Mt Muhavura in western Uganda, they may be not more than 3 cm in diameter, and the whole plant is stunted and appears not unlike a strongly-growing elephant grass.

Perhaps the best descriptions of bamboo forest are from Uganda where it has been described by Dale (1940), Snowden (1953), Langdale-Brown et al. (1964) and Watt (1956).

The slender branches of the tops of the bamboo stems form a fairly dense canopy so that vigorously growing bamboo has little in the way of undershrubs. Herbs and grasses, and mosses and ferns form a large part of the ground-cover and Snowden (1953) notes that in clearings and along pathways there are usually shrubs and herbs and grasses which are the same as those in the nearby mixed upland forest, for example, *Afrocrania*, *Phyllanthus*, *Clutia*, *Dipsacus*, *Cineraria*, *Mimulopsis*, *Pycnostachys* and *Lobelia gibberoa*. The most commonly recorded grasses, certainly in Uganda, are *Agrostis kilimandscharica*, *Acritochaete volkensii*, *Hyparrhenia mobukensis*, *Panicum adenophorum*, *P. calvum*, *P. hochstetteri*.

Of the species of shrubby *Mimulopsis*, which occur in the bamboo thicket, *M. elliotii*, and *M. solmsii* flower and die gregariously at intervals of three to ten years (Tweedie, 1965). This gregarious flowering is also a feature of the bamboo itself but the interval is about fifteen years. Unlike some of the Indian bamboos, the entire community does not flower and die at the same time, but it dies in patches which vary in size from a few to several hectares, large enough as Dale (1940) says, to raise the anger of the uninitiated who are apt to assign the cause to careless use of fire. In some areas the flowering interval is longer than fifteen years and may be as much as thirty years in south-western and western Uganda. Flowering has never been observed in the bamboo of Echuya reserve where it occurs in the rather low altitudinal range 2286–2500 m.

The dead patches, according to Dale, usually become covered with a strong growth of brambles (*Rubus* sp.) mixed with *Sambucus*, *Hypericum* sp. *Lobelia gibberoa* and *Impatiens*. It is possible that the other forest trees, often localised in the bamboo, seed themselves in this regeneration period. Trees commonly recorded are *Afrocrania volkensii*, *Cassipourea* sp., *Dombeya* spp., *Faurea saligna*, *Hagenia abyssinica*, *Hypericum* spp., *Myrica*, *Nuxia*, *Rapanea* sp., *Schefflera volkensii*, *Trichilia volkensii*, *Xymalos monospora*, *Podocarpus gracilior*, *P. milanjianus*.

Some of the plants of the herbaceous layer in the bamboo forest may owe their survival to their twining or scrambling habit which enables them to reach openings in the canopy. This group includes—*Stachys aculeata* var. *afromontana*, *Thunbergia* sp., *Plectran-*

10 Bamboo forest, Kigezi, Uganda. (Morrison)

thus ramossisimus, Mimulopsis solmsii. Also recorded among the herbs are *Droguetia iners* and *Urtica massaica.*

Cushions of epiphytic mosses grow in the axils of the branch whorls on the culm and these small cushions provide a place for delicate aphyllous orchids.

Bamboo has been considered as a source of paper pulp, but this is not thought to be a commercial proposition. It is often claimed as a first class conserver of water, but catchment area experiments in the Aberdare Mountains show that its evapotranspiration is as great as an ordinary forest stand. Nevertheless, it is an extremely valuable forest crop and is locally much used for house building; indeed, it is so much in demand for this that upland bamboo would have disappeared from many areas during the last few decades had it not been for the creation of forest reserves. In some areas people eat the young bamboo shoots after removal of the larger leathery bracts, and elephants and buffalo also feed on the young shoots and leaves.

Upper zone

The upper zone of the upland forest belt usually consists of low stature woodland ranging in height from 12 to 18 m. Under very wet conditions *Hagenia-Rapanea* woodland is common and it can extend into fairly reduced conditions. It seems, however, that sometimes reduced availability of water and/or very thin soils over lava favour its replacement by *Afrocrania-Agauria* woodland. So far as we know, this type of woodland has not been described from East Africa, but it is well developed nearby at the Zaire end of the chain of Virunga volcanoes. Lebrun (1942) gives a full description of it and maps it on Mts Nyiragongo and Shaeru in the altitudinal belt between 2300 and 2800 m at the very top of the upland forest belt (Fig. 1.7).

11   Inside bamboo forest (*Arundinaria alpina*). (Morrison)

Hagenia woodland

Hagenia abyssinica is an untidy-looking tree reaching a height of about 18 m. It has reddish-brown bark, large stipulate imparipinnate leaves at the ends of the gnarled branches and pendulous bunches of male or female flowers, the female flowers being reddish, the male flowers pale orange, buff or white.

It has a wide altitudinal range. It has been recorded by Dale (1940) as low as 1800 m on the western slopes of Mt Elgon, and it is often gregarious, forming woodland in the upland forest belt around 3200 m. But the upper boundary of this woodland has, in many places, been lowered through burning and grazing when it has been replaced in part by tussock grassland. This is clearly observable on Mt Elgon where pastoralists live between 2700 and 3300 m (Weatherby, 1962). Under favourable conditions of minimum disturbance *Hagenia* woodland would probably extend upwards towards 3400–3500 m. In this highest range *Hagenia*, *Faurea saligna*, *Prunus africana* and *Rapanea rhododendroides* are the only trees which occur, often as isolated and dwarfed specimens in the ericaceous belt in sheltered places among tall bushes of *Erica* and/or *Stoebe*.

The factors which control the occurrence of this woodland are not quite clear. There is no doubt that *Hagenia* woodland often does coincide with the wettest zone of the upland forest belt, the rainfall decreasing above and below it. Often the trees are clothed from the base of the trunk right to the ends of the branches in a thick layer of epiphytes, including enormous cushions of bryophytes (Plate 12). Ferns and orchids, many of which are pendent, are rooted in these cushions. The ends of the branches, where the bark is exposed, are occupied sometimes by crassulaceous epiphytes such as *Sedum* and *Cotyledon* and orchids. The branch-ends are often hidden under a veil of grey lichens of the *Usnea* type, though *Usnea* seems to occur more commonly on *Rapanea* than on *Hagenia*. On the other hand *Hagenia* woodland can be found under what are believed to be drier conditions. For example, it occurs *throughout* the upland forest belt on the southern slopes of Mt Meru, though it is interesting to note that it is weakly represented a few miles away on Mt Kilimanjaro. It is said to be common in colonising forest on the Imatongs. Where it occurs in the lower part of the upland forest it is generally on the edge of forest and in clearings; it is found only very rarely within the lower altitudinal forests.

All these observations indicate that *Hagenia* forms climax forests at the top of the upland forest belt, where low night temperatures exclude many other trees and competition is low. At lower altitudes it occurs serally again in habitats—forest margins and clearings—where competition from other trees is minimal.

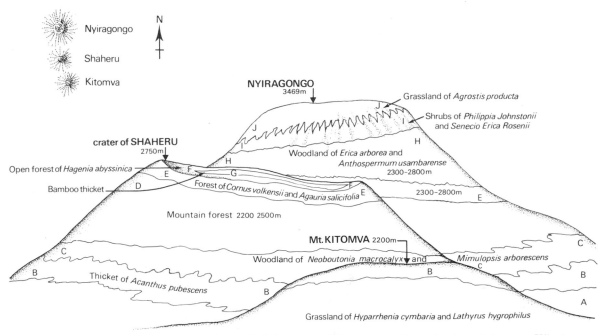

Fig. 1.7 Distribution of the vegetation on the south facing slope of Nyaragongo, after a sketch made between Kibati and Kitomva. (Lebrun, 1942)

In climax woodland *Hagenia* is commonly accompanied by *Rapanea rhododendroides*, a relatively straight-boled tree sometimes growing as tall as the *Hagenia*, but often just slightly shorter so that it forms something of an under storey. In some places the under storey is formed by *Hypericum* and perhaps more rarely by *Sambucus*. On parts of the Virunga volcanoes the *Hagenia* woodland gives way in fact to *Hypericum* thicket. On account of these variations this uppermost zone of the upland forest belt is variously designated as *Hagenia* woodland, *Hagenia-Rapanea* woodland, *Hagenia-Hypericum* woodland.

The composition of *Hagenia* woodland is shown in Table 1.9 and it is important to notice the richness in

Table 1.9   Composition of Hagenia woodland on Mount Elgon

TREES IN THE STRATUM 9 m

Afrocrania volkensii	Cornaceae
Dombeya goezenii	Sterculiaceae
Hagenia abyssinica	Rosaceae
Hypericum lanceolatum	Hypericaceae
Prunus africana	Rosaceae
Rapanea rhododendroides	Myrsinaceae

TREES AND SHRUBS IN THE STRATUM 0—9 m

Lobelia gibberoa	Campanulaceae
Sambucus africana	Sambucaceae
Senecio amblyphyllus	Compositeae

HERBS

Agrostis quinquiseta	Gramineae
Anthriscus sylvestris	Umbelliferae
Bromus leptocladus	Gramineae
Caucalis melanantha	Umbelliferae
Carex erythrorrhiza	Cyperaceae
Cerastium afromontanum	Caryophyllaceae
Clutia abyssinica	Euphorbiaceae
Crassocephalum sp.	Compositeae
Deschampsia flexuosa	Gramineae
Dipsacus pinnatifidus	Dipsacaceae
Erlangea sp.	Compositeae
Echinops hoehnelii	Compositeae II
Euphorbia spp.	Euphorbiaceae
Galium spp.	Rubiaceae
Geranium aculeolatum	Geraniaceae
Helichrysum odoratissimum	Compositeae
Hydrocotyle mannii	Umbelliferae
Impatiens eminii	Balsaminaceae
Microglossa pyrifolia	Compositeae
Myosotis vestergrenii	Boraginaceae
Peucedanum kerstenii	Umeelliferae
Plectranthus sylvestris	Labiateae
Poa leptoclada	Gramineae
Pseudobromus engleri	Gramineae II
Rumex spp.	Polygalaceae
Stachys aculeata	Labiateae
Sanicula elata	Umbelliferae
Vernonia abyssinica	Compositae
Viola eminii	Violaceae

Antitrichia curtipendula
Breutelia stuhlmannii
Ceratodon purpureus
Grimmia spp.
Hookeriopsis mittenii
Leptodontium squarrosum
Macromitrium abyssinicum
Sphagnum davidii
Anastrepta cf. *orcadensis*
Anastrophyllum gambaragarae
Anthoceros laevis
Chandonanthus cf. *giganteus*
Dumortiera spp.
Herberta dicrana

In bamboo forest

Brachymenium elgonense
Campylopus introflexus
Erythrodontium subjulaceum
Hypopterygium spp.
Macromitrium abyssinicum
Porothamnium cf. *hildebrandtii*
Prionodon vehmanii
Rhodobryum keniae
Rhynchostegium sp.
Syrrhopodon sp.
Also many epiphytic Hepaticeae, mainly Lejeuneaceae

species of the herbaceous layer. It seems likely that this richness depends on a high availability of moisture since less rich herbaceous layers occur in the drier types of *Hagenia* forest. However, the structure of the canopy may afford very favourable light conditions for growth. Openings in the canopy are often accompanied by clumps of *Arundinaria*, *Lobelia gibberoa* or *Senecio amblyphyllus*. All these grow to a height of between 6 and 10 metres.

The timber of *Hagenia* is dark red in colour, but it is not sufficiently durable for outside work, though its handsome appearance makes it suitable for furniture and flooring. Not much of the timber is available since many of the *Hagenia* woodlands are on steep slopes where they are preserved from exploitation and kept strictly as protection forests to prevent soil erosion and to control the release of water to streams. On Mt Meru flat terraces in the *Hagenia* woodland have been cleared and found suitable for the cultivation of *Pyrethrum*.

Swamp forest

Although they occupy only a small proportion of the land surface, swamp forests are of considerable interest partly because they are the main habitat of certain characteristic trees including the palms. We shall give an account of three swamp forest types in Uganda and describe more fully some of the trees, which commonly occur in this forest type in East Africa.

The three Uganda types are: mixed swamp forest of the Lake Victoria belt; Podocarpus—Baikeaea forest; Myrica—Rapanea—Syzigium forest.

Swamp forest of the Lake Victoria belt

In the Uganda forest department, the height of the water table is used to define three arbitrary categories of wet or swampy forest, namely, raised wet, seasonal swamp and permanent swamp. In raised wet forest the water table is frequently or always 1 or 2 m below the surface; in seasonal swamp forest the water table varies from being at or above the soil surface to more than 1 m depth for prolonged periods; in permanent swamp forest the water table is never below the surface except for short periods and may be permanently at the surface. It is unwise to try to define these sites too closely floristically, as trees confined to permanent swamp in areas of low rainfall may extend to seasonal swamp or even raised wet sites where the effective rainfall is higher. Eggeling (1947) in his description of the swamp forests in Budongo forest in west Uganda claimed that, of the 80 tree species recorded, the following 15 were confined to swamps and found only very exceptionally on dry ground sites. *Bombax reflexum, Cathormium altissimum, Cleistanthus* cf. *polystachyus, Cleitopholis patens, Erythrina exelsa, Euphorbia teke, Macaranga schweinfurthii, Mitragyna stipulosa, Parkia filicoidea, Phoenix reclinata, Pseudospondias microcarpa, Spondianthus preussii, Treculia africana, Beilschmiedia ugandensis, Voacanga obtusa.*

The general composition of swamp forests of the Lake Victoria belt is given in Table 1.10. The canopy seldom exceeds 24 m and seems to be of only two layers, of which the lower consists mainly of *Raphia, Tabernaemontana* and *Voacanga*, sometimes with *Phoenix reclinata*, though this is more commonly seen at the forest edge. The floor of the forest is slightly uneven, dry and firm in parts and breaking into swampy patches and pools. Below the litter layer the soil profile continues downwards for about 30 cm through a pale-brown, humus-rich, clayey layer which changes to a pale grey, sandy horizon, and this, at a depth of about a metre, passes into a grey, tough, sandy clay. The high humidity of the swamp forest undergrowth is probably responsible for the growth of aerial and stilt roots in such species as *Uapaca* and *Musanga*.

Podocarpus—Baikiaea seasonal swamp forest

This is confined to alluvial deposits on the western shore of Lake Victoria where former lake and river deposits were elevated to their present altitude of around 1200 m during the latter part of the Pleistocene period. It is probably the largest area of swamp forest in Uganda and this type is not found elsewhere in tropical Africa. The alluvial sands are covered mainly by very tussocky *Cymbopogon—Heteropogon* grassland and the forest is on the more clayey soils. Because of the surrounding swamps which sometimes contain

12   Inside *Hagenia-Rapanea* woodland, Mt Elgon, 3000 m. *Lobelia wollastonii* and *Senecio* sp. (Morrison)

Table 1.10   Composition of permanent swamp forest in the Lake Victoria forest belt

TREES

Erythrina excelsa	Leguminosae subfam. Papilionoideae
Macaranga monandra	Euphorbiaceae
M. pynaertii	,,
M. schweinfurthiii	,,
Mitragyna stipulosa	Moraceae
Musanga cecropoides	Euphorbiaceae
Parkia filicoidea	Leguminosae subfam. Mimosoideae
Pseudospondias microcarpa	Anacardiaceae
Phoenix reclinata	Palmae
Raphia farinifera	,,
Spondianthus preussii	Euphorbiaceae
Uapaca quineensis	,,
Voacanga obtusa	Apocynaceae

SHRUBS AND SMALLER TREES

Dracaena laxissima	Agavaceae
Euphorbia teke	Euphorbiaceae
Ficus spp. (epiphytic)	Moraceae
Myrianthus holstii	,,
Schefflera barteri	Araliaceae
Ataenidia conferta	Marantaceae
Costus afer	Zingiberaceae
Marantochloa mannii	Marantaceae
M. lucantha	,,
M. purpurea	,,
Piper capense	Piperaceae

EPIPHYTIC PLANTS

Asplenium africanum	Aspleniaceae
A. barteri	,,
A. dregeanum	,,
A. laurentii	,,
Davallia chaerophylloides	Davalliaceae
Microsorium punctatum	Polypodiaceae
Arthropteris palisoti (scandent fern)	Davalliaceae
Culcasia scandens (scandent liane)	Araceae

GROUND LAYER HERBS

Asplenium buettneri	Aspleniaceae
A. subaequilaterale	,,
A. cf. *uhligii*	,,
A. variabile var. *paucifugum*	,,
A. warnecki	,,
Thelypteris dentata	Thelypteridaceae
T. madagascariensis	,,
T. quadrangularis	,,
T. striata	,,

Doryopteris kirkii	Adiantaceae
Impatiens niemniamensis	,,
Leptaspis cochleata	Gramineae
Blotiella cf. *currori*	Dennstaedtiaceae
B. glabra	,,
Microlepia speluncae	,,
Palisota schweinfurthii	Commelinaceae
Phymatodes scolopendrium	Polypodiaceae
Pollia condensata	Commelinaceae
Pteris burtoni	Pteridaceae
P. togoensis	,,
P. communata	,,

several feet of water the forests can only be safely studied in the dry season. They were first described by Dawe (1906 a, b) and further accounts by Holtz (1906), Fyffe, Wright Hill, Philip and Swabey are in the records of the Uganda Forest Department.

The forest is evergreen with a two-layered canopy; the upper storey reaches 30 m and the lower or under storey is seldom more than 15 m in height, often lower (see Fig. 1.8 and Table 1.11). In relatively undisturbed forest *Podocarpus usambarensis* var. *dawei* and *Baikiaea eminii* tend to be the dominant trees of the upper canopy. It is difficult to visualise the structure and composition of the forest under natural conditions since the *Podocarpus* has been heavily exploited. In 1906 Dawe recorded:

Its insect resistant qualities are well known to the natives; instances have come to my notice where the poles have been taken inland, a distance of nearly 15 miles, for building purposes. A chief, whose father had died and left a house built six years ago, is said to have rebuilt it with the same poles that were as sound as ever; this is unusual in these parts where white ants occur. The young poles are very largely used for building purposes and the result is that forests in thickly populated districts, like Sango and Kanabulemu, do not contain many large trees.

P. usambarensis var. *dawei* is claimed to be endemic in all except two of these forests—Tero and Namalala —where it is replaced by *P. milanjianus*. Its closest relative is *P. usambarensis* var. *usambarensis* which is endemic on the coastal rim mountains of Tanzania and inland on Mt Hanang about 576 km south-east of the *Podocarpus-Baikiaea* forests. *P. usambarensis* var. *dawei* has, in the past, often been misidentified as *P. gracilior* which is now believed to be confined in Uganda to the mountains of the east, but *P. milanjianus*, which occurs in the swamp forests, is found on the mountains of western Uganda. (Melville, 1958).

Baikiaea insignis is a caesalpiniaceous tree reaching a height of about 30 m and bearing handsome large, white flowers followed by dark velvety pods about 50 cm in length. *Baikiaea* extends south along the western lake shore belt as far as Bukoba district; it

occurs in the swamp forests in the northern Lake Victoria forest belt but, according to Webster (1961), it is very rare there except in Mabira forest where occasionally it is the climax of seasonal swamp forest.

Although *Podocarpus* produces an abundance of seed which germinates readily it is difficult to find adequate regeneration. Some of the seeds are attacked by borers and others are carried underground by some small mammal. Of those which germinate many are killed by browsing buck, while others are killed during the twice yearly inundation of March–May and September–December. The forest floor is very uneven and the *Podocarpus* seedlings which have been buried—perhaps by elephants' feet—in the bottom of the hollows, are submerged for long periods and eventually die. As a consequence of the low density of establishment of *Podocarpus*, and, perhaps, the opening of the canopy, most of the exploited areas have an under storey of *Lasiodiscus* (see Fig. 1.8) which is a relatively worthless timber tree. Large areas of these former *Podocarpus-Baikiaea* forests are degenerating from valuable closed forest to almost pure stands of low-growing *Lasiodiscus*. *Lasiodiscus* is also a common understorey tree in the Budongo forest in western Uganda where it occurs under *Cynometra*.

A sustained yield of *Podocarpus* from the seasonal swamp forests may be difficult to achieve although seedlings can be established on the more elevated parts of the forest floor. But, this is a relatively expensive operation and, coupled with the slow growth of *Podocarpus*, there is a balance of evidence in favour of replacing some of these forests with plantations of *Maesopsis*.

Myrica—Rapanea—Syzygium forest

This appears at a late stage of the herbaceous swamp succession and it is often preceded by *Miscanthidium-Sphagnum* or papyrus swamp. Though trees of *Anthocleista zambesiaca*, *Syzygium cordatum*, *Mitragyna rubrostipulata*, *Ficus* spp. and *Myrica kandtiana* occur sometimes in *Miscanthidum* and papyrus swamp, it would appear that conditions rarely favour the development of *Myrica—Rapanea—Syzygium* swamp forest. This is, at present, known only from a few places in the uplands of south-western Uganda around 1800 m and neighbouring Rwanda (Deuse, 1959) although something similar occurs on swampy ground in the Bukoba district. *Syzygium cordatum* has, of course, a widespread distribution in wet places from sea level to about 2000 m; Brenan and Greenway (1949) record it as locally common along permanent streams in *Brachystegia* woodland.

Myrica kandtiana, Rapanea rhododendroides and *Syzygium cordatum* form the canopy of the forest at a height of about 10 m; the relative abundance of the species varying from place to place. A species of *Psychotria* occurs slightly below the canopy, and various woody and herbaceous scramblers help to close its gaps. *Culcasia scandens* is often abundant on the trunks—running out on to the forest floor—and thin stems of pink-flowered *Begonia* hang from branches which are usually loaded with an abundance of epiphytes. A red-flowered semi-parasitic *Loranthus* is often common.

The forest floor is permanently moist and is sometimes underlain by deep peat. Here and there the floor breaks into marshy pools. The drier surface is usually completely covered with short grasses and herbs and in some places by a species of *Selaginella*.

The forest is economically unimportant except as a site for beehives, and possibly these insects collect nectar from the *Syzygium*. At present, many of these swamp forests are being cut down for firewood or included in the draining programme in the uplands of south-western Uganda, where many peaty swamps are being successfully drained and taken over for growing vegetables and other crops for local use and for export to other parts of Uganda.

Important trees of swamp forest

Mitragyna stipulosa

Mitragyna stipulosa is a common constituent of the swamp forest flora especially in some of the forests bordering the lake. The tree can reach a height of 45 m and is without buttresses. It produces a good general-utility timber and is probably the highest volume producer of wood in the freshwater swamp forests. For this reason there has been considerable interest in its exploitation and regeneration, though the relatively slow girth increment is somewhat against

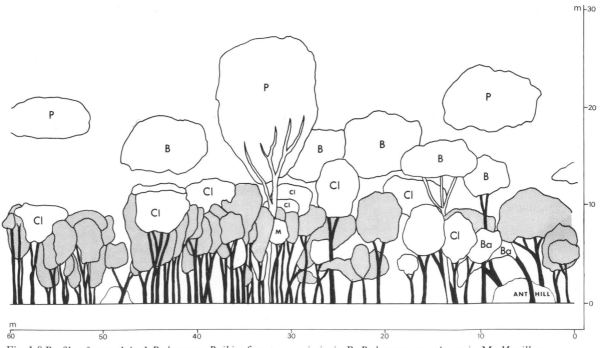

Fig. 1.8 Profile of unexploited *Podocarpus-Baikiea* forest. (After Farrer)
Key to Symbols: Cl. *Cleistanthus polystachus*; B. *Baikiea insignis*; P. *Podocarpus usambarensis*; M. *Manilkara obovata*; Ba. *Baphiopsis parviflora*; Trees shaded; *Lasiodiscus mildbraedii*.

Table 1.11 Composition of the Podocarpus—Baikiaea seasonal swamp forest (1200 m altitude)

TALL TREE STRATUM 21–37 m

More important trees	*Podocarpus milanjianus*	Podocarpaceae
	P. usambarensis var. *dawei*	
	Baikiaea insignis subsp. *minor*	Leguminosae subfam. Caesalpinoideae
Less important trees	*Apodytes dimidiata*	Icacinaceae
	Canarium schweinfurthii	Burseraceae
	Cassipourea ruwensorensis	Rhizophoraceae
	Croton megalocarpus	Euphorbiaceae
	Diospyros abyssinica	Ebenaceae
	Heywoodia lucens	Euphorbiaceae
	Klainedoxa gabonensis	Simaroubaceae
	Maesopsis eminii	Rhamnaceae
	Manilkara obovata	Sapotaceae
	Parinari excelsa subsp. *holstii*	Rosaceae
	Pseudospondias microcarpa	Anacardiaceae
	Pycnanthus angolensis	Myristicaceae
	Suregada procera	Euphorbiaceae
	Symphonia globulifera	Guttiferae
	Tetrapleura tetraptera	Leguminoseae subfam. Mimosoideae
	Trichilia splendida	Meliaceae
	Uncaria africana	,,
	Warburgia ugandensis subsp. *ugandensis*	Canellaceae

TREES IN THE STRATUM 9–21 m

	Pericopsis angolensis	Leguminosae subfam. Papilionoideae
	Albizia grandibracteata	Leguminosae subfam. Mimosoideae
	Baphiopsis parviflora	Leguminosae subfam. Papilionoideae
	Bersama abyssinica	Melianthaceae
	Chaetacme arisata	Ulmaceae
	Croton macrostachyus	Euphorbiaceae
	Fagara melanacantha	Rutaceae
	Ficus spp.	Moraceae
	Garcinia buchananii	Guttiferae
	Linociera johnsonii	Oleaceae
	Macaranga monandra	Euphorbiaceae
	Mitragyna rubrostipulata	Rubiaceae
	Parkia filicoidea	Leguminosae subfam. Mimosoideae
	Sapium ellipticum	Euphorbiaceae
	Spondianthus preussii var. *glaber*	,,
	Syzygium guineense	,,
	Teclea nobilis	Rutaceae
	Trema orientalis	Ulmaceae

Acalypha neptunica	Euphorbiaceae
Aframomum sanguineum	Zingiberaceae
Barteria fistulosa	Passifloraceae
Clausena anisata	Rutaceae
Cleistanthus polystachyus	Euphorbiaceae
Coffea canephora	Rubiaceae
C. eugeniodes	,,
Craibia brownii	Leguminosae subfam. Papilionoideae
Crassocephalum mannii	Compositae
Dovyalis macrocalyx	Flacourtiaceae
Lasiodiscus mildbraedii	Rhamnaceae
Marantochloa leucantha	Marantaceae
M. purpurea	Marantaceae II
Memecylon myrianthum	Melastomaceae
Mussaenda erythrophylla	Rubiaceae
Peddiea fischeri	Thymeleaceae
Phoenix reclinata	Palmae
Scolopia rhamniphylla	Flacourtiaceae
Syzygium cordatum	Myrtaceae
Trichocladus ellipticus	Hamamelidaceae

its commercial use. McCarthy's (1961) investigations suggest the need for clear felling to ensure the necessarily high light intensity for subsequent good seed regeneration. The breathing roots of *Mitragyna* are formed from negatively geotropic roots which grow for about 5 cm above the swamp surface and then curve over and enter the soil. They keep pace with water-level fluctuations by renewed vertical growth and by the production of further knees. They are well supplied with lenticels at the water surface and are in direct gas communication with more profundal roots.

Raphia farinifera (Plate 13).

This tree, which is usually scattered through the swamp forests but is sometimes gregarious, is claimed by Corner (1966) to be Africa's most distinguished contribution to the palms. The *Raphia* palms are massive plants not adequately investigated and our knowledge of their taxonomy is unsatisfactory. There are about nineteen species on the African mainland and one species in tropical America. Of the African species, *R. farinifera* occurs in Uganda, Kenya and Tanzania. Russell's study (1965) raises the possibility that all East African *Raphia* palms should be included under *R. ruffia* (Gaertn.) Hylander. The Uganda plant has a completely upright trunk, reaching a height of 12 m, and crowned with a massive cluster of once-pinnate, spreading leaves often as much

as 15 m long. They are some of the largest leaves in the plant kingdom and Corner (1966) describes their intricate development. The leaf bases remain attached to the trunk for a very long time and this rough armour supports various epiphytic plants including the large scandent fern *Stenochlaena mildbraedii*, which is so far confined to this habitat.

This handsome fern with once-pinnate fronds up to 2 m in length, has the pinnae grading from rose pink to green. The lamina of the pinnae on the fertile fronds is much reduced and replaced with sporangia so that the fertile fronds resemble long, rusty-brown plumes. Towards the lower part of the *Raphia* trunk the palm leaf bases may be pushed off by a vigorous growth of many stout adventitious roots which form a closely adherent sheath over the lower metre. The underground root system sends up numerous peg-like pneumatothodes which project a few centimetres above the soil surface.

Raphia appears to flower once and die. There is some doubt, however, whether the production of the massive, branched inflorescences actually consumes the apical bud or causes its abortion. The inflorescences, when mature, are heavy, about two metres long, and contain large numbers of attractive, cone-shaped fruits covered in smooth, overlapping scales. It is not known how these fruits are dispersed, but the ground around the palms is usually disturbed as if some forest animal, perhaps wild pig, had been rooting there. Whatever the animal, it removes fruits so effectively from the immediate neighbourhood of the palm that much time may be spent in locating a single fruit at the base of a plant which is producing fruits by the hundreds and thousands.

The trunks of *Raphia* palms are often used for building and there is extensive use of the leaflet fibres for raphia twine which is obtained by stripping the fibrous tissues from the uppermost side of young leaflets. The demand for it in East Africa is great enough to threaten the extinction of the species. Palm wine can be obtained from the sap by severing the inflorescence, but in East Africa the palms mostly used for this purpose are *Borassus*, *Cocos* and *Elaeis*. Eggeling and Dale (1951) record that *Phoenix* was tapped for wine in southern Uganda, but the practice has almost died out.

Raphia occurs in swamp forests beyond the limits of the Lake Victoria belt, though it is rare in western Uganda. It extends eastwards through Kenya, south via Mafia and Pemba Islands, then inland along the Pangani River in Tanzania. In western Uganda it appears to be replaced in swamp forest similar to that of the Lake Victoria belt by the African oil palm *Elaeis guineensis*.

Elaeis guineensis (oil palm)

Elaeis is of similar stature to *Raphia*. It grows to a height of about 30 m and is crowned with a dense head of shining leaves about 5 m in length with more or less drooping leaflets. This plant is apparently centred on W. Africa but it is indigenous in some of the forests of the Semliki Valley in western Uganda, is recorded by Williams (1949) as abundant in the wild state in the south of Zanzibar and on Pemba Island and is reported from Tanzania. Williams claims that in

13   Raphia palm (*Raphia farinifera*). (Fergus Wilson)

Zanzibar the fruit of the wild tree is of little or no economic value, most of the oil being obtained from a cultivator. According to Corner, the African oil palm is second only to the coconut in the world's supply of palm oil.

The brief description by Haddow, Gillett and Highton (1947) of the Mongiro forest appears to be the only account of the plant growing in natural conditions in East Africa. The forest lies at 762 m on the eastern side of the Semliki forest in West Uganda and *Elaeis* is dominant throughout. Other trees include *Mitragyna stipulosa*, *Cynometra alexandri*, *Chlorophora excelsa*, *Khaya anthotheca*, *Celtis mildbraedii*, *Antiaris toxicaria*, *Alstonia congoensis*. Lianes, epiphytic ferns and orchids are common on the trees. The under storey includes *Pandanus chilocarpus* and in the herb layer there is a dense growth of *Marantochloa*.

Calamus deeratus (rattan palm)

Calamus is a slender, thorny, scrambling plant which can grow to immense lengths and reach the forest canopy. It occurs, though probably rarely, in swampy forests of the Lake Victoria belt (Dawe, 1906) but Dawe found that it was common in the forests of the Western Escarpment and Osmaston (1906) has reported it from Bugoma and Maragambo. There are three other rattan palms in tropical Africa, but only *Calamus* has been recorded from East Africa and, so far, only from Uganda.

Phoenix reclinata (wild date palm) (Plate 14)

This common palm occurs in clumps mainly on the edge of swamp forests and along lines of seepage but it is not limited to swamps and is found in a variety of sites from sea level to 3000 m. It is closely allied to the edible date palm (*Phoenix dactylifera*) which was introduced into East Africa by the Arabs and is scattered throughout Tanzania wherever there have been Arab settlements. The fruit of *P. reclinata* is edible, though it is rarely eaten and must first be soaked in water. Many people have difficulty in distinguishing between *Phoenix*, *Raphia* and *Elaeis*, though *Phoenix* is a much more slender plant usually with a curved trunk. Its pinnate leaves are rarely more than 3 m long and are turned so that the leaflets lie more or less in a vertical plane, the lowermost leaflets being reduced to spines. Closer inspection shows that the *Phoenix* leaflets are induplicate, that

is, V-shaped, in section while those of *Raphia* and *Elaeis* have the form of an inverted V in section and are reduplicate. *Phoenix* is the only pinnate-leaved palm with induplicate leaflets.

Bogdan (1958) describes swamp forest of *Phoenix reclinata* at 900 m in Kenya on waterlogged flats and depressions close to the Kiboko River on the Nairobi–Mombasa road. He states that the general habit of the *Phoenix* swamp seems similar to that occupied in the same region by wooded grassland with *Acacia xanthophloea* and there are sometimes mixed stands with *Acacia* trees rising above a dense growth of palms. Under a dense, pure stand the ground may be bare, but there is usually a well-developed ground cover and herbaceous climbers including *Commicarpus plumbagineus* and *Melothria microsperma*. On the basis of herbs, Bogdan distinguishes several types within the palm forest but points out that they are not well defined. Examples of such types are: forest with bare ground; forest with *Setaria chevalieri*, *Panicum maximum* and other tall herbs; forest with *Acacia xanthophloea* forming a thin upper storey; palm forest, more swampy, with *Imperata cylindrica*; *Cyperus flabelliformis* and other tall species of *Cyperus*; forest on more alkaline soil with *Pluchea sordida*.

14 Wild date palm (*Phoenix reclinata*). (K. Lye)

Two

Rangelands

Rangelands are defined in this book as 'all areas providing a habitat for wild or domestic animals'. They cover the greater part of East Africa and include a wide variety of vegetation types from grassland through bushland, bushed and wooded grassland to woodland. In the scope of a single chapter it is only possible to give examples of some of these and then to discuss the utilisation of rangelands now and in the future.

Bushland and thicket

Bushland is very widespread especially in Kenya and Tanzania where it forms a good deal of the country included in the game parks. It is familiar to tourists as the kind of landscape often forming the background to photographs of wild life. It occupies the more arid parts of the country and varies in appearance and plant composition largely according to the rainfall it receives. In some parts of northern Kenya, for example, there is a semidesert landscape with widely spaced bushes and bare earth in between. Here there is an erratic rainfall of about 250 mm. Rather different is the bushland of the Tsavo National Park where there is grass and occasional trees among the shrubs and the rainfall here may be up to 750 mm per annum. This, in a temperate climate, would support good grazing land, but, in tropical Africa, the evaporation rate per month may equal the rainfall and little water is available for the vegetation.

Semi-arid bushland in Kenya

Edwards (1940) describes the dry desert bushland of North Kenya at an altitude of 600–1200 m, where extensive areas are devoid of trees and support only isolated tufts of grass and occasional low, thorny bushes with bare, red soil in between. *Commiphora* spp (Fig. 2.1) occur with the low-growing *Acacia mellifera*, while two noticeable plants in these very dry parts are *Sericomopsis hildebrandtii* and *S. pallida* which have dry, almost papery but quite attractive compound spikes of white flowers. Edwards reckons that semi-arid bushland (Plate 15) occupies not less than one-third of the total area of Kenya and, together with the dry bushed grassland, constitutes two-thirds of the total. This is not surprising when we realise that 72 per cent of the total area of Kenya can expect less than 500 mm of rainfall per year and is not suitable for crop production. Nearly 66 per cent of East Africa suffers from drought for six months of the year, while only 2 per cent of the land receives rainfall of over 50 mm per month.

Bushland of Tsavo National Park

The appearance of the bushland depends largely on the species which comprise the woody vegetation and on the presence or absence of perennial grasses. One type is well illustrated by much of the vegetation of the Tsavo and Amboseli National Parks which also covers large areas of the Central and Lake provinces of Tanzania including much Masai country. The ground cover is made up mainly of woody plants of shrubby habit many of them succulent or thorny and with small decidous leaves but often quite leafless. Prominent among these are species of the genus *Commiphora* (Fig. 2.1) of which forty-six are described for Kenya alone, only two being forest trees and the rest occurring in regions with a rainfall of from 75–500 mm a year, fourteen in the northern province only. The alternate leaves are usually small and have three or more leaflets with serrate edges; but most of the year the trees are leafless. As flowers and fruits are seldom found with the leaves, identification of the species is very difficult. The leaves of some species turn yellow before they fall giving a touch of colour to the landscape. Among other genera many are thorny and of scrambling habit.

Members of the family Capparidaceae are common and include species of *Boscia*, *Cadaba* and *Maerua*. These can be recognised by their showy flowers with many stamens, followed by succulent, berry-like fruits on long stalks. Sometimes they are spiny. Species of *Grewia* belonging to the family Tiliaceae are commonly found and are often in flower. Though the genus includes some forest trees, the bushland species are shrubs or small trees with small, simple leaves with serrate margins and most have yellow or white flowers with many stamens; but one common bushland species (*G. similis*) has bright mauve flowers. Between the woody vegetation the ground is covered with annual grasses and ephemeral herbs. These come up after the rains and are then heavily grazed, so that for much of the year the ground is bare between the shrubs. However, in the rainy season there is a surprisingly beautiful variety of flowering plants, including such genera as *Chlorophytum*, *Anthericum*, *Hypoxis*, *Rhamphicarpa* and *Cycnium*, the last sometimes called the 'paper flower' because its white or pink blossoms look like paper scattered on the ground. *Ipomoea* species of the family Convolvulaceae, both

creeping and erect forms, often provide splashes of white, yellow or purple against the dry banks.

Within this kind of bushland, forming a subtype, there commonly occur scattered thickets of succulent plants. Prominent among these are the candelabra tree (*Euphorbia candelabrum*) (Plate 16) and *E. tirucalli* a tree of quite different habit with long, succulent,

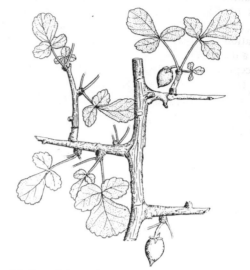

Fig. 2.1 *Commiphora africana*. (Dale and Greenway)

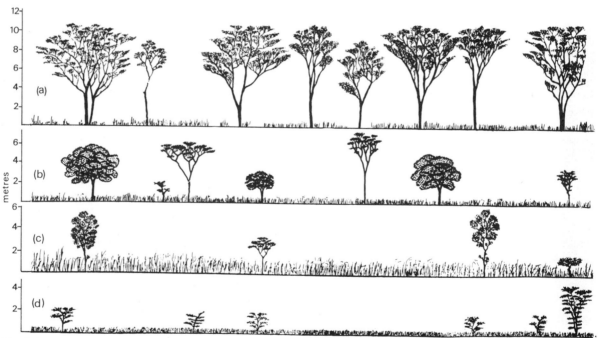

Fig. 2.5 Profiles of bushland types. (Pratt *et al.*, 1966) (a) *Brachystegia* woodland. (b) *Acacia gerrardii—A. seyal* woodland (with *Themeda* understorey). (c) Tall *Hyparrhenia—Combretum* wooded grassland. (d) Medium-height *Pennisetum mezianum—Acacia drepanolobium* seasonally waterlogged dwarf tree grassland.

leafless branches. This plant is often planted as a protective hedge round villages and cattle bomas in many parts of East Africa. Its very milky latex causes blindness for several hours if it enters the eye of a raider.

Another common succulent is *Cissus quadrangularis* a square-stemmed straggling plant belonging to the vine family *Vitaceae*, while the sword-shaped, mottled leaves of *Sanseviera ehrenbergiana*, much liked by rhino, are nearly always to be found among the thicket vegetation. This is the bow-string hemp which provides a useful fibre.

Except in the very driest bushland there are scattered trees, species of *Acacia* being the commonest. *Balanites aegyptica* (Fig. 2.2) also occurs frequently, a tree growing to about 6 m which can be recognised by its rounded crown made up of a tangle of green, thorny twigs bearing pairs of leaflets, the leafless ends of the twigs protruding from the thicket. Its fruit, yellow when ripe, is known as the 'desert date'. The kernel

produces an oil which may have been one of the ingredients of spikenard mentioned in the Bible and an emulsion made from the fruit is known to be toxic to the snails which form the host of the bilharzia parasite. Another common tree is *Lannaea stuhlmannii* with long racemes of cream-coloured strongly scented flowers and leaves with three to five leaflets borne on the ends of the twigs. In damp areas the dichotomously branched doum palm (*Hyphaena coriacea*) may be seen and here and there are baobab trees (*Adansonia digitata*). For a full plant list see Greenway (1969).

Dry Bushland in Karamoja

A somewhat drier type of bushland is described by Thomas (1943) in North East Uganda which has a rainfall of 500–750 mm per annum. Langdale-Brown *et al.* (1964) and Kerfoot (1965) consider it to have been

15   Dry bushland in NW Kenya

61

derived from dry semi-evergreen woodland by degradation through overgrazing and erosion. The intense evaporation together with the erratic nature of the rainfall results in the land being covered with thorny bushes and any annual grass which may appear in the

(Fig. 2.9) and *Cynodon* (Fig. 2.12) as well as annual grasses and ephemeral herbs. Spiny members of the Acanthaceae such as *Barleria* and *Blepharis* are found and many bulbous plants and creeping herbs such as

Tribulus terrestris, well named because its spiny fruits on the ground surface are a problem to barefooted people.

A feature of the Karamoja and the adjacent semi-arid areas is the very tall chimney-like termite mounds, up to 10 m in height. They form a refuge for small animals such as rodents, shrews and pygmy mongoose, elephant shrew and hedgehogs which feed on termites.

Fig. 2.3 *Grewia mollis*. (*Flora Congo Belge*)

Fig. 2.2 *Balanites aegyptica*. (Eggeling)

wet season is so heavily grazed that for nine months of the year the land is bare between the bushes except for a few spiny plants (Plate 17).

In spite of this the studies of Kerfoot (1965) and Wilson (1962) have revealed about 500 species belonging to 30 families. It is therefore difficult to provide a concise guide to this vegetation and the reader is advised to consult the lists given by these two authors. A number of the plants are illustrated in Wilson (1962), but this full, duplicated account of the vegetation of Karamoja is not always easy to obtain. Many of the same genera are represented as in the Tsavo area, including *Commiphora*, *Acacia*, *Euphorbia*, *Grewia* and members of the Capparidaceae. When grazing is strictly controlled, the recovery of the grass after the rains is marked and rapid and it includes perennial grasses such as *Themeda triandra* (Fig. 2.8) and *Hyparrhenia filipendula* (Fig. 2.7) with *Sporobolus*

Thickets

Thickets can be looked upon as extreme types of bushland, woodland or even grassland. They may consist of dense stands of thorny or spiny shrubs or small trees through which taller, usually thorny trees project; or they may be composed of much branched coppicing

16 *Euphorbia candelabrum* in semi-arid country, Tanzania. (Fergus Wilson)

shrubs ranging from 2·5–4·6 m in height with an open appearance when viewed from inside, though the crowns are interlaced to form a thick and often continuous canopy. Sometimes flat-topped trees up to 7·4 m high may project through the canopy (Plate 18).

Perhaps the best known area of thicket of the latter type is the Itigi thicket which lies astride the railway near the eastern edge of the central plateau of Tanzania and occupies an area of not less than 6000 km² of Central Province. Milne (1947) states that the thicket appears to be limited to deep soils of high acidity and light texture which form a 'grey cement'. This is consolidated swamp floor deposit of pre-rift age, termed 'duracrust' by Gillman (1947). Under the same climate and in close juxtaposition a wide range of soils, some of them derived from 'cement' are occupied by various types of *Brachystegia-Julbernardia* and *Combretum-Terminalia* woodland.

A very different type of thicket vegetation is provided by the lowland bamboo (*Oxytenanthera abys-sinica*). This is a giant grass of a different genus to the mountain bamboo (*Arundinaria alpina*) which also forms thickets, but grows at a higher altitude and dies after fruiting. The *Oxytenanthera* thickets are not very close but they occur in more or less dense patches in various types of deciduous woodland. In Tanzania they occur commonly from south of Iringa to the Lindi district where they are found up to an altitude of 615 m; they also occur as rather wide-spaced individual clumps in miombo woodland.

In Uganda large patches of lowland bamboo occur in the Acholi and West Nile districts, frequently in areas of *Combretum* wooded grassland and often on stony hillside soils. In one place, near Padibe in Acholi at 1220 m, *Oxytenanthera* forms a dense thicket with interlaced shoots 4–9 m high and a sparse ground layer.

Flowering takes place over large areas about once in seven years after which the clump dies back to shoot up a year later. *Oxytenanthera* does not occur in Kenya.

Grassland

In East Africa true grassland is rare. Nowhere do we find landscape like the prairies of North America or the Russian steppes. Jackson (1964) considers that, in Africa, climatic climax grassland is confined to desert areas with prolonged drought and low erratic rainfall where a treeless grass sward is dominated by species of *Aristida* and other genera of the Stipeae and Chlorideae.

Most grassland under other conditions has scattered trees and shrubs and falls into two categories—derived or secondary grassland and edaphic grassland. The former has probably been formed from woodland as a result of burning or grazing, both of which reduce the cover of woody trees and shrubs. The latter owes its characteristic composition to features of the soil.

First an example of a dry semi-desert grassland will be described and then a variety of derived and edaphic types. Further examples with special reference to central Africa, including southern Tanzania, will be found in Vesey-Fitzgerald (1963, 70). In our classification we have suggested that where trees and shrubs form more than 2 per cent of the canopy cover the community should be designated wooded grassland.

Semi-desert grassland

The nearest approach to extensive climax grassslands in East Africa are the semi-arid grasslands which cover very large areas of Kenya, extend into Somalia—almost to the Gulf of Aden—and occur to a small extent in north-eastern Uganda. The following description of these grasslands is based on Edwards and Bogdan's (1951) account. Langdale-Brown *et al.* (1964) and Wilson (1962) give some account of the Uganda type.

The climate is characterised by a low rainfall in the range of 250–500 mm per annum and in some places, near Lake Rudolph, perhaps less than 250 mm. The

17   Semi-arid vegetation in Karamoja, Uganda. (Morrison)

Fig. 2.7 *Hyparrhenia filipendula*. (Edwards and Bogdan)

Fig. 2.8 *Themeda triandra*. (Edwards and Bogdan)

Fig. 2.9 *Sporobolus pyramidalis*. (Edwards and Bogdan)

Fig. 2.12 *Cynodon dactylon*. (*Flora of Sudan*)

65

18   Deciduous thicket country between Morogoro and Dodoma, Tanzania. (Fergus Wilson)

rainfall is very erratic and drought of several years duration occurs. There is a single dry season during which dry winds sweep down from the north, trees and shrubs are leafless, and the grasses are dry and brittle. In brief, the entire vegetation persists in a drought-dormant condition. Nothing is yet known in detail of the specialisations of the plants in this xerophytic habitat. We should not, however, think of the plant specialisations only in relation to water, since these areas are characterised also by high temperatures. Griffiths (1967) records the highest air temperature in East Africa as 41°C in shade near Lake Magadi in the Eastern Rift. The high day temperatures and the extremely high temperatures at the soil-atmosphere interface will probably exclude many plants from arid areas.

The ground cover, which ranges from 50–90 per cent is mainly of ephemeral grasses with a few herbs for a short period after the rains. Trees and shrubs occur but they are usually widely spaced. The domin-

ant grass is *Chrysopogon aucheri* var. *quinqueplumis*, but in the coastal area of Kenya to the south, *Chloris roxburghiana* is important, and north-eastwards into Somalia, two species of *Aristida* are dominant.

The scattered trees are not very prominent with the exception of the baobab (*Adansonia digitata*) in the southern part of Kenya. The shrubs, 3–5 m in height, usually branch close to the base and are often species of *Acacia* and *Commiphora*. Edwards and Bogdan (1951) found that in northern Kenya the bushes are spaced approximately 4–10 m apart. It has been thought that competition for moisture by extensive subsurface root systems might produce a regular spacing or pattern of plants in semi-arid vegetation. Greig-Smith and Chadwick (1965) tested this possibility in *Acacia-Capparis* semi-desert scrub in the Sudan, but they found no evidence of such regularity. They suggested that variations in soil factors, resulting in variations in available water, are important in determining distribution.

Table 2.1 Composition of the semi-desert *Chrysopogon aucheri* Vegetation

TREES

Acacia tortilis subsp. *spirocarpa*	Leguminosae subfam. Mimosoideae
Adansonia digitata	Bombacaceae
Delonix elata	Leguminoseae subfam. Caesalpinoideae

SHRUBS

Acacia drepanolobium	Leguminosae subfam. Mimosoideae
A. mellifera subsp. *mellifera*	Leguminosae subfam. Mimosoideae
Balanites spp.	Simaroubaceae
Boscia coriacea	Capparidaceae
Commiphora spp.	Burseraceae
Cordia sp.	Boraginaceae
Dichrostachys cinerea subsp. *cinerea*	Leguminosae subfam. Mimosoideae
Gossypium longicalyx	Malvaceae
Terminalia spp.	Combretaceae

HERBS

Abutilon spp.	Malvaceae
Aerva spp.	Amarantaceae
Aristida kelleri	Gramineae
Barleria spp.	Acanthaceae
Cenchrus ciliaris	Gramineae
Chloris pycnothrix	"
C. roxburghiana	"
C. virgata	"
Chrysopogon aucheri var. *quinqueplumis*	"
Coleus sp.	Labiatae
Commicarpus pedunculosus	Nyctaginaceae
Crossandra stenostachya	Acanthaceae
Dalechampia scandens var. cordofana	Euphorbiaceae
Digitaria macroblephara	Gramineae
Dinebra spp.	"
Disperma spp.	Acanthaceae
Enneapogon cenchroides	Gramineae
Eragrostis cilianensis	"
Eriochloa nubica	"
Hermannia alhiensis	Sterculiaceae
Hibiscus spp.	Malvaceae
Indigofera spp.	Leguminosae subfam. Papilionoideae
Ipomoea spp.	Convolvulaceae

67

Justicia exigia	Acanthaceae
Latipes senegalensis	Gramineae
Melothria spp.	Cucurbitaceae
Monechma debile	Acanthaceae
Pavonia spp.	Malvaceae
Pennisetum mezianum	Gramineae
Polygala petitiana	Polygalaceae
Senecio discifolius	Compositae
Setaria pallidofusca	Gramineae
Solanum zanzibarense	Solanaceae
Sorghum purpureo-sericeum	Gramineae
Sporobolus spp.	"
Stipagrostis uniplumis	"
Tetrapogon cenchriformis	"
T. tenellus	"
Thunbergia annua	Acanthaceae
Vigna sp.	Leguminoseae subfam. Papilionoideae

Despite the dryness of the region it supports nomadic pastoral tribes together with their cattle, goats, sheep, and camels. The carrying capacity of the grassland is low but the total area is so large, covering about two-thirds the area of Kenya, that the total number of animals is very great. Edwards and Bogdan (1951) remark, too, on the healthy appearance of the animals on these sparse pastures, and suggest that it may be due to freedom from certain parasites and to the nutritive richness of the herbage and browse resulting from the rapid drying and small amount of leaching of nutrients. Utilisation is limited to some extent, particularly in the southern baobab area, by tsetse fly. Also the habitat degenerates very easily and overgrazing in some places has led to invasion and increased density of *Acacia mellifera*. In some parts of East Africa from 80–90 per cent of the dry bushland, between 300 and 900 m altitude, is dominated by *Acacia mellifera*.

The life-form of the grasses would seem to be at least in part a reflection of climatic conditions. A feature of nearly all East African grasses is the tufted nature of the growth so that there is bare ground between the tussocks, a habit probably due to the alternation of dry and wet seasons causing competition for moisture when rapid growth follows the rains. In very dry seasons, genera of Chlorideae, Agrosteae, and Stipae tend to occur commonly, and some of the last, such as *Aristida*, may invade overburned or over-grazed grassland previously dominated by Andro-pogoneae. Tufted perennials of the Andropogoneae and Paniceae predominate on low and medium altitude grassland while, at higher altitudes, species of the more temperate subfamilies, such as Agrosteae,

Festuceae and Aveneae, commonly occur. Though tufted, they are of lower stature than the grasses of lower altitude and are more leafy giving a better ground cover. It is in the swamp grasslands that stoloniferous and rhizomatous species occur more frequently.

Derived grassland

Much of the grassland of East Africa is only maintained in that condition by regular burning and by the grazing of wild or domestic animals. Without these it would revert to thicket, woodland or forest, except in places where the soil is seasonally waterlogged. It occurs mainly at low and medium altitudes sometimes reaching up to the edge of the upland forest. Elephant grass thicket may be considered to have developed in the lowland forest zone following cultivation and is here included in the section on forests. On many mountains, above 3000 m low stature woodland has given place to short, tussock grassland and this is described under mountain vegetation.

Good examples of derived grassland are to be found in some of the National Parks and full descriptions are given by Osmaston (1971) for those in Uganda. It should be noted that edaphic factors are also of importance in regard to the distribution of species.

Queen Elizabeth National Park, Uganda

The soils of the park are derived from relatively young lacustrine and volcanic deposits and are fertile but limited in productivity by lack of rainfall which varies from 600 mm on the floor of the Rift at Mweya to 1000 mm at Katwe. There are two rainy seasons, March–May and August–November, and, although there may be weeks of drought, there is a good vegetation cover which supports a high biomass of wild animals among which elephant, buffalo and hippopotamus play a dominant role. Some areas, especially round the crater lakes, are quite heavily wooded, while others have big expanses of grassland with only scattered trees and often interrupted by termite mounds bearing thickets. Much of the grassland is composed of a mosaic of tall, tufted species such as *Hyparrhenia*, spp. *Imperata cylindrica* (mainly in damp valley bottoms) and *Cymbopogon afronardus*, while in other parts there are rolling plains of *Themeda triandra* with scattered trees of *Acacia gerrardii* and only a small admixture of other grasses. *Hyparrhenia* grasslands are, on the whole characteristic of the moist areas and occur on average to good soils. Most species have a low nutritive value (Table 2.2) and the land on which the grass occurs outside

Table 2.2   Chemical analyses of *Hyparrhenia* species (Figures as per cent of dry weight)

Name	Ash	Crude Protein	Crude Fibre	Carbohydrate	Ca	P
Hyparrhenia diplandra	9·0	5·7	36·4	47·2	0·40	0·09
Hyparrhenia dissoluta (*Hyperthelia dissoluta* Syn.)	8·8	12·9	33·7	41·6	0·25	0·19
Hyparrhenia filipendula	5·8	6·6	36·3	49·4	0·30	0·10
Hyparrhenia lintoni	8·6	11·5	31·8	43·2	0·49	0·26
Hyparrhenia rufa	7·2	5·2	38·3	48·0	0·22	0·06

Fig. 2.4 *Capparis tomentosa.* (*Flora Rhodesia*)

the game parks is often used for cultivation of annual crops. Where this is not so, the grass is burnt annually and grazed by cattle. *Themeda* is thought to be characteristic of areas where the rainfall is below 1000 mm and the dry season well marked. It grows on soils of average quality, is burnt frequently and extensively grazed. By the lake shore a much shorter type of grassland prevails where such species as *Heteropogon contortus*, *Sporobolus stapfianus* and *Cenchrus ciliaris* prevail. Where they have been heavily overgrazed, by hippo, *Capparis* thicket is encroaching.

Throughout the park there are scattered *Acacia* trees which sometimes form patches of woodland especially round the crater lakes. *Balanites aegyptica* occurs fairly frequently, especially in the southern parts, and other common trees are species of *Albizia* and *Ficus*. The termite mounds are often crowned by trees of *Euphorbia candelabrum* and among the common shrubs of the mounds are *Capparis tomentosa* and the mauve-flowered *Grewia similis*.

Field (1968) has shown that, in comparison with some other East African ecosystems, much of the vegetation of the Queen Elizabeth National Park is mature and the food habits of the wild ungulates are best described in terms of plant species rather than of their growth stages. In other areas, where there is more marked seasonal fluctuation in the abundance of food, herbivores 'manage' the vegetation by migration to areas where suitable growth occurs. This is the case in some of the national parks in Kenya and Tanzania.

Comparison of the proportions of fragments of grass species taken from different herbivores showed that, by their specialised food habits, they usually avoid competition, but this may occur when food supplies have been depleted by the larger animals. Observations confirm that, by selective grazing, herbivores consume a diet of much higher nutritive value than that of the average grass sward.

Serengeti National Park, Tanzania

The grasslands of the Serengeti National Park are of a different nature. The western part of the park is occupied by broken, hilly country carrying *Acacia Balanites—Commiphora* bushland. In the east are the Crater Highlands, a volcanic block rising to 4250 m, and in between are the Serengeti Plains covering about 2500 km^2 at an altitude 1600–1800 m. These plains probably represent the nearest thing in East Africa, outside the arid areas, to true grassland and, in the wet season, when new grass and surface water are available, this and other grassy areas are visited by huge herds of wild animals. It is estimated that there may be over a million animals in the region including thirty species of ungulates and eight species of larger predators and carnivores. Doubtless with lower grazing pressure these grasslands would return to woodland.

From east to west there is a sequence of grass types with gradually merging boundaries. On the east side where there are recent deposits of volcanic ash from the volcano Lungi, the soil varies in depth from a few inches in some parts to one foot over a limestone pavement built up by the rise and fall of the available soil moisture. Here the grasses are of a shorter type and include species of *Sporobolus* often mixed with the Cyperaceous *Kyllinga*. The soils towards the west are heavy and deep, partly derived from granite hill erosion and an old lake level (Greenway, personal communication). Taller grasses including *Themeda*

19   Thicket clumps of *Capparis tomentosa* and *Euphorbia candelabrum*. Queen Elizabeth National Park, Uganda

triandra and *Pennisetum mezianum* are characteristic of these base-rich soils. Among other common grasses of the plains are *Digitaria macroblephara*, *Cenchrus ciliaris* and many species of *Hyparrhenia*.

Anderson and Talbot (1965) studied the principal soil factors which relate to grass types and their utilisation by wild life. They consider that grazing markedly affected the species composition and basal cover of the vegetation both directly and through burning and that the suggestion that the apparent seasonal preference of wild animals for *Sporobolus-Kyllinga* grassland might be due to shortage of essential elements in soils further west appears to lack real support. When the surface of the long grassland is relatively dry and grasses are at a palatable, early stage of growth they are heavily grazed. In wet periods these grasses rapidly grow coarse, probably as a result of good nitrogen supply and steady moisture availability, and then the animals move south and west where many of the same species are available in smaller and apparently more palatable form. Thus in contrast to the Queen Elizabeth National Park, the animal grazing pattern appears to be determined primarily by the palatibility associated with the stage of growth of the food plants.

In relation to the prevalence of locusts, in the Rukwa Valley, Dean (1967) discusses the distribution there of grassland types and the influence of fire and flooding.

Themeda triandra—Pennisetum clandestinum grassland (Table 2.3)

We must next consider grassland at a rather higher altitude, which is derived from upland forest and reaches right up to its edge. *Themeda triandra* and *Pennisetum clandestinum* are the main constituents, though other species, including *Hyparrhenia* spp., frequently occur. This type of grassland covers large areas of the Kikuyu country of Kenya and of the highlands west of the Rift Valley. In Uganda, highly productive *P. clandestinum* grassland predominates in limited areas between 1600 and 2500 m in South Kigezi and in Bugishu. Above 2000 m on the mountains of Karamoja there are less productive grasslands of such species as *Exotheca abyssinica* and *Setaria sphacelata*. In Uganda, on the whole the *P. clandestinum* grasslands are very heavily cultivated and little used for grazing, while the *Exotheca* and *Setaria* grasslands are uncultivated and only slightly grazed. As trees, especially *Acacia*, are frequently present in these higher altitude grasslands in Kenya and Uganda, more attention will be paid to them under the heading of wooded grasslands.

In Tanzania the grazing lands of the Ngorogoro crater have *Pennisetum schimperi* as the main species of the ground cover on the crater floor, with *P. clandestinum* on the ridges and slopes and large areas of *Eleusine jageri* in the forest glades.

Fig. 2.10 *Hyparrhenia rufa*. (Edwards and Bogdan)

Protea-Dombeya highland grassland

In the northern and central provinces of Tanzania and in the central highlands where upland forest has given place to grassland, there is an interesting and very attractive community named by Burtt (1942) '*Protea-Dombeya* highland grassland'. Its beauty is due to the presence of a number of trees with showy flowers and to the gaily coloured flowers which abound among the grasses. We cannot do better than quote Burtt's description of this vegetation. A full list of the grasses is given by Vesey-Fitzgerald (1963).

The traveller ascends through cedar-like forests of *Brachystegia microphylla* clothing fairly steep hillsides and at about 1700 m leaves this forest abruptly behind him to emerge into fairly open grassland with scattered trees and shrubs hitherto unfamiliar to him. The silvery-green leaves of the sugar bush (*Protea*) with its enormous heads of creamy-white flowers will be seen scattered commonly about the rich pasture. On the Mpwapwa hills, the Usagara Mountains and in the Southern Highlands he will see commonly the tree *Uapaca kirkiana* with its large, leathery, laurel-like foliage tending to be massed in terminal whorls and in season bearing abundant

Table 2.3   Composition of the *Themeda Triandra* grasslands

TUFTED PERENNIALS

More than 150 cm high at flowering	*Beckeropsis uniseta*	Paniceae
	Cymbopogon afronardus	Andropogoneae
	(a) *Hyperthelia dissoluta*	"
From 90–150 cm high at flowering	(a) *Cymbopogon excavatus*	Andropogoneae
	(a) *Hyparrhenia filipendula*	"
	Setaria sphacelata	Paniceae
	Sporobolus pyramidalis	Sporoboleae
	Themeda triandra	Andropogoneae
From 0–90 cm high at flowering	(a) *Aristida adoensis*	Aristideae
	(a) *Bothriochloa insculpta*	Andropogoneae
	Brachiaria eminii	Paniceae
	B. platynota	"
	Cenchrus ciliaris	"
	(a) *Chloris roxburghiana*	Chlorideae
	Digitaria macroblephara	Paniceae
	D. scalarum	"
	Enneapogon elegans	Pappophoreae
	Eragrostis exasperata	Eragrosteae
	Harpachne schimperi	Eragrosteae
	Heteropogon contortus	Andropogoneae
	Hyparrhenia lintoni	Andropogoneae
	Microchloa kunthii	Chlorideae
	Panicum poaeoides	Paniceae
	Pennisetum massaicum	"
	Setaria sphacelata	"
	S. trinervia	"
	Sporobolus angustifolius	Sporoboleae
	S. fimbriatus	"
	S. homblei	"
	S. marginatus	"

RHIZOMATOUS PERENNIALS from 0—90 cm high at flowering

Pennisetum stramineum	Paniceae
Sporobolus confinis	Sporoboleae

STOLONIFEROUS PERENNIALS from 0—90 cm high at flowering

(a) *Bothriochloa radicans*	Andropogoneae
Cenchrus ciliaris	Paniceae
Chloris gayana	Chlorideae
Cynodon dactylon	Chlorideae
	Sporoboleae
Sparobolus spicatus	Sporoboleae

ANNUALS from 0—46 cm high at flowering

(a) *Chloris pycnothrix*	Chlorideae
Eragrostis barelieri	Eragrosteae
E. macilenta	Eragrosteae
Sporobolus festivus	Sporoboleae
Tragus berteronianus	Zoizieae

From 46 cm—90 cm high at flowering

(a)	*Aristida adscensionsis*	Aristideae
(a)	*Chloris virgata*	Chlorideae
	Digitaria ternata	Paniceae
	Eragrostis cilianensis	Eragrosteae
	E. tenuifolia	Eragrosteae
	Eriochloa nubica	Paniceae
	Pennisetum mezianum	Paniceae
	Setaria pallidifusca	Paniceae

fruit resembling and tasting like loquats. Here and there will be seen flat-topped Acacias. Rocky eminences may be clothed with the bay-scented evergreen shrub *Myrica salicifolia* and above Mpwapwa he will see everywhere the cypress-like shrubs of *Aeschynomene burttii*, endemic in this region.

In December he will marvel at the flaming red of *Erythrina abyssinica* a common feature of the highland pasture, and in March and April the whole countryside is decorated with the 'may-blossom' of *Dombeya quinquisete*. Here and there are seen the *Euphorbia*-like trunks of *Cussonia arborea*, large specimens of *Combretum molle* and the willow-like foliage of *Faurea rochetiana*.

In the spring, the hillside becomes a blaze of unusual flowers including orchids, the colours blue, white and red predominating. Perhaps the most beautiful of all are the four-petalled flowers of *Clematopsis* which mostly hang their heads in satin whiteness. Later in the season the wealth of flowers is swallowed up in tall grasses (*Hyparrhenia rufa*) and banks of bracken fern (*Pteridium aquilinum*), while the pungent odour of *Helinus* will be noticed as the traveller forces his way across ravines endeavouring to avoid contact with the dreaded pods of the buffalo bean (*Mucuna*). (These have hairs which cause violent irritation of the skin.)

He will find the freshness of the highland pastures restful after the lowland heat; superb views are everywhere to hand. If he chooses to ascend to higher elevations he will come to sub-tropical evergreen forest with its sombre greens, capping the hilltops in small forest remnants.

The highland pasture is favoured by zebra, roan, eland, hartebeest and the mountain reedbuck and is the favourite resort of buffalo and rhinoceros which lie up in the mountain forest in the daytime.

In much of this grassland which adjoins *Hagenia* forest and includes wide stretches of *Themeda*, the poorer soils are now being used for the establishment of very large plantations of conifers (Parry 1953a and b, 1954), mainly *Pinus patula*, together with wattle, and there are plans for extensive wheat growing in these upland areas (Plate 20).

Edaphic grassland

Prominent among edaphic grasslands are those of permanent and seasonal swamps. These are usually dominated by tall, tufted species such as *Miscanthidium violaceum*, *Loudetia phragmitoides* and *Echinochloa pyramidalis*. They are described more fully in the chapter on aquatic vegetation and, for Tanzania, reference should be made to Vesey-Fitzgerald (1970). The saline grasslands round the shores of salt lakes often contain more Cyperaceae than grasses.

Loudetia kagerensis grassland (Uganda)

As an example of an edaphic grassland we will consider the type found in Uganda round the shores of Lake Victoria and Lake Nabugabo and on the sandy flats of the old lake basin about twenty-five miles from Masaka. The landscape is characterised by extensive meadows of the delicate grass *Loudetia kagerensis* interspersed with large and small patches of forest. Scattered throughout are termite mounds and in the more sandy areas water-lily pools abound in the pits from which sand has been removed. Round the lake shore there are numerous laterite outcrops.

The soil is shallow, sandy, lacking in clay and with a very low base content, and it is surprising that anything will grow there. Yet it has quite a rich flora, though of limited species, and can even carry forest. The extensive Jubiya forest near Bukakata is situated on soil of this nature. Although the sandy, leached and very poor soil is the most important factor controlling the vegetation, the influence of other factors is strong enough to produce within even a modest area a very complex pattern of vegetation and the following description is necessarily somewhat simplified. Similar complex vegetation mosaics characterise most of the grasslands mentioned in this book. The presence of grassland of limited species (Table 2.4) together with swamp, forest, and rock outcrops in a limited area makes this type of country a valuable place for field study by students.

The dominant grass, *Loudetia kagerensis* (Fig.2.11) is short and very different from the tall, tufted *Loudetia* species of swamp and wooded grassland. When the grass is in flower the *Loudetia* meadows are very attractive. The cover is sparse and is associated in the wet season with a number of beautiful, often tiny

20 Kwatere pine planting project near Mbeya, Tanzania. (Tanzania Ministry of Agriculture, Forestry and Wildlife)

herbs. Other common grasses include *Hyperthelia dissoluta*, *Hyparrhenia filipendula*, *Eragrostis* spp. and the small *Microchloa kunthii*. Noticeable among the herbs are the scarlet spikes of *Striga asiatica* which is a hemi-parasite on the roots of grasses. Other species of *Striga* parasitise some of the cereal members of the grass family causing serious disease. Two other members of the Scrophulariaceae are often seen among the grass, *Lindernia nummulariaefolia* with blue flowers and *Ilysanthes* sp. with rather similar blue or white flowers. In some places plants of *Scilla edulis* a small relative of the English bluebell but mauve rather than blue, form a sward and among several members of the sedge family (Cyperaceae) the golden heads of *Kyllinga chrysantha* are conspicuous. *Biophytum petersianum*, with small sensitive compound leaves at the head of an otherwise leafless stalk, may also

be seen in bare places among the grass. Among the larger herbs *Murdannia simplex* with three-petalled pale blue flowers and the showy blue labiate *Solenostemum latifolium* are common especially round the edge of scrub patches where they are accompanied by yellow and white species of *Justicea*, *Crotalaria* spp. with inflated pods and tall plants of *Leonotis nepetifolia* recognised by its groups of orange flowers at intervals along the upright stem. Here, too, may be found *Sesamum angustifolium* and two of the commonest herbs in Uganda, *Aspilia africana*, a rather straggling yellow composite, and *Asystasia gangetica*, belonging to the Acanthaceae with white, two-lipped flowers spotted with purple.

The shallow pockets of soil on the laterite outcrops carry a rather uniform plant cover with *Microchloa* as a common grass and two plants with fleshy leaves often

74

forming large patches, namely, *Cyanotis* sp., a mono-cotyledon with blue flowers enclosed between folded bracts and *Aeolanthus repens*, a showy member of the Labiateae with grey-green foliage which seems to be able to flourish under the unfavourable conditions of the rocky outcrop.

The termite mounds vary from quite bare conical structures to somewhat flattened mounds almost covered by scrub and trees. The stages in the colonisa-

Fig. 2.11 *Loudetia kagerensis*. (Edwards and Bogdan)

tion of the mounds make an interesting study (Plates 21a–e). Usually, few plants grow on them while the termites are active, but when they are deserted they provide a home for a number of herbs, shrubs and trees. The first plants establish themselves round the base where soil washed down by rain accumulates. They include *Sida* spp., *Urena lobata* of the Malvaceae, *Triumfetta rhomboidea*, and the sedge *Kyllinga*. Strag-ling over the mounds are larger plants, such as *Lantana camara* and *Solanum incanum*, with very often tufts of the grasses *Panicum maximum* and *Sporobolus pyramidalis*. Gradually woody plants appear and the herbs are lost among a rich growth of shrubs and small trees common among which are *Grewia trichocarpa* with yellow flowers, *Teclea nobilis*, *Securinega virosa* and *Acacia hockii* with balls of golden yellow flowers. These scrub patches grow and coalesce and other trees appear. Among the colonising trees are *Rhus natalensis*, *Phyllanthus* sp., *Gymnosporia* sp., and *Harungana*, recognised by the brown backs to its leaves and its clusters of white flowers. Climbers such as *Cyphostem-*

ma adenocaule and the leafless, fleshy stems of *Sarcos-temma viminale* clamber over the shrubs and trees.

From time to time fire sweeps across these grass-lands, but as most of the grasses are low-growing the flames do not reach plants which have become established on the termite mounds. Undoubtedly this lakeside area was originally occupied by high forest which has been gradually destroyed to make room for cultivation. A portion of such forest has been preserved in the botanical gardens at Entebbe, though a number of exotic species have now been introduced there. Around the lake shore odd mango trees and coffee bushes bear witness to its former use. Forest regenera-tion can be seen in many places beginning on the termite mounds which later become engrossed in woodland containing large forest trees. An account of similar vegetation on one of the islands in Lake Victoria will be found in Jackson and Gartlan (1965).

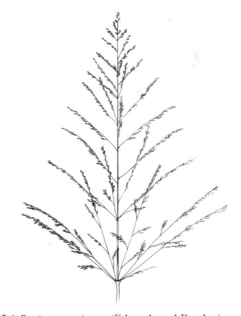

Fig. 2.6 *Panicum maximum*. (Edwards and Bogdan)

Termites and termite mounds in *L. kagerensis* grassland (Table 2.5)

As most termite mounds are found in grassland, this would seem an appropriate place to enlarge further on the subject of these insects which, as we have seen, through their mounds facilitate the regeneration of forest where it has been replaced by grassland. For fuller account of termites reference should be made to Harris (1940b, 1948, 1961) and Lee and Wood (1971).

Table 2.4 Composition of *Loudetia kagerensis* grassland at Lake Nabugabo, near Masaka, Uganda

Andropogon dummeri	Gramineae
Bulbostylis densa	Cyperaceae
Crotalaria glauca	Leguminoseae subfam. Papilionoideae
Ctenium concinnum	Gramineae
Cyanotis longifolia	Commelinaceae
Cyperus tenax	Cyperaceae
Eragrostis mildbraedii	Gramineae
E. blepharoglumis	,,
Fimbristylis hispidula	Cyperaceae
Helichrysum sp.	Compositae
Hyperthelia dissoluta	Gramineae
Hypharrenia filipendula	,,
Indigofera asparagoides	Leguminosae subfam. Papilionoideae
I. congesta	,,
I. drepanocarpa	,,
Kyllinga aurata	Cyperaceae
Lightfootia abyssinica	Campanulaceae
Loudetia kagerensis	Gramineae
Mariscus submacropus	Cyperaceae
Microchloa kunthii	Gramineae
Murdannia simplex	Commelinaceae
Oldenlandia herbacea	Rubiaceae
Polygala luteo-viridis	Polygalaceae
P. spicata	,,
Pycnospharera lutescens	Leguminosae subfam. Papilionoideae
Pycreus macranthus	Cyperaceae
Sopubia mannii var. *tenuifolia*	Scrophulariaceae
Striga asiatica	,,
Tephrosia linearis	Leguminosae subfam. Papilionoideae
Thesium sp.	Santalaceae
Vernonia aemulans	Compositae

Though the mound builders are the ones which attract attention, there are also very many inconspicuous species living in the soil everywhere in East Africa, boring in the branches of dead trees or building their nests on trees in the forest.

The effect of termites on the vegetation is twofold, this results from the fact that their diet is entirely plant material, mainly cellulose, and that they use soil in making the mounds in which their nests are situated. Four types of feeding habit are known:

Wood-eating termites have an intestinal fauna of protozoa which assist in digesting cellulose. They make nests whose walls are of 'carton' of partly digested wood protected by an outer covering of similar carton if it is on a tree, or by earth and carton if it is in a mound. None of the termites with intestinal protozoa are mound builders.

Soil-feeding termites, which ingest soil containing broken-down vegetable matter, roots, etc. Their nests are built of excreta and fine clay and sand.

Fungus-growing termites which feed on wood and vegetable debris some of which has been subject to fungal action. They build with particles of soil cemented together with fine clay and saliva and no excreta. Theirs are the largest mounds.

Harvester termites which feed almost entirely on

(a)

(d)

(b)

(e)

21 Colonisation of termite mounds. (K. Lye)

(a) Active, uncolonised mound.
(b) and (c) Colonisation by bushes and grasses: *Panicum maximum*, *Eragrostis*, *Erlangea tomentosa*, *Sida* sp.
(d) Colonisation by trees: *Maesopsis eminii*, *Sapium ellipticum*, *Harungana madagascariensis*, *Bridelia micrantha*, *Bersama abyssinica*.
(e) Fire in *Loudetia kagerensis* grassland; plants on the mound have escaped burning.

(c)

grass; this is cut into short lengths and carried down into their subterranean nests.

The detail of nest construction depends on the species of termite and the materials available. Harvester termites in dry grassland make nests below ground in a series of chambers with intercommunicating galleries. On the surface only dumps of earth are seen. The same is true of some of the smaller fungus-growers. The soil-feeding termites feed on vegetable residues in the soil and in the process large quantities of earth pass through their bodies. They dispose of this excreta by making small rounded mounds of a type common in the Kenya Highlands.

The nest itself consists of a simple chamber or cluster of chambers excavated in the soil and partitioned into cells with clay or carton. Around this there is a protective covering of dense soil. The largest and best known mounds are built by the group Macrotermitineae which carry sand to the site and cement it with clay and saliva. Their large chambers are grouped round the queen cell and a maze of passages leads to a 'fungus garden'. This consists of a sponge-like mass of chewed wood built up to fit the chambers of the nest. It becomes permeated with fungal mycelium. The fungus garden produces fungi of the genus *Termitomyces*, but only after the termites die or are removed (Batra, 1967). There is much speculation as to their function in the nest, but the fact that as they grow old they are replaced, suggests that they may be an essential part of the food supply. These fungi are commonly eaten by many African tribes.

Below the nest, complex corridors lead into the surrounding country through which the worker insects travel in search of wood, dead grass and other leaves for food, and soil for building. The interior of the nest has a high humidity. Some species move down into the underground galleries in the dry season or to escape from fire.

The large mounds with a tall narrow chimney seen in the dry parts of Kenya and in Karamoja belong to the termites *Macrotermes bellicosus* and *M. goliath*. Some of them reach a diameter of 12 m with steeples up to 9 m. In wetter areas, rainfall limits the growth of the mounds.

It has been found that the mounds consist of subsoil, and it is this that governs their colour. Many are built of the prevailing red earth but in damp places the mounds are often grey. Heavy rain smooths down the surface and leaves a deposit of infertile poor soil round the base which is often uncolonised by vegetation. In some parts, as on the hillsides of Ankole, one can see erosion channels originating at the bare bases of termite mounds.

The question arises as to why the mounds are colonised by shrubs and trees in contrast to the surrounding grassland. This happens when the mound is stabilised either because the termite colony has died or left. First grasses, then common herbs and shrubs, and finally trees appear. It has been suggested that the growth of new vegetation is due to greater fertility of the soil and it is a well-known fact that tobacco and sisal show better growth on soil of old mounds; but any difference is more probably due to better drainage (Hesse, 1955).

In parts of Tanzania, near Tabora, mounds were found to contain nodules of lime which it was worth while to extract for lime burning, and it was suggested that termitaria are richer in bases than the surrounding soil. However, this does not seem to be so in other parts of Africa. An examination of inhabited and abandoned mounds in ten different soils in East Africa showed them to be of no greater fertility than surrounding subsoil. It was suggested that the system of chambers perhaps provided a large evaporating surface and that calcium accumulated there as calcium-charged water evaporated. The fertility of the soil in the mounds depends on its position in the catenary sequence. One would expect the mounds on red earths to be infertile because the subsoil is generally low in bases. On the other hand, mounds over black earths could be rich in calcium because many of them contain this mineral in excess and this is deposited as calcium carbonate in the lower part of the profile.

Milne (1947) suggested that where calcareous mounds occurred in an area of non-calcareous soils, calcium carbonate accumulated in the mounds either by upward capillary movement of ground water containing calcium carbonate or from organic matter brought into the mounds by termites. Watson (1969) investigated these hypotheses of water movement using a radioactive tracer. He found that the tracer within the termite mounds spread upward and downward during the wet season, while in soils adjacent to the mound the tracer moved predominantly downward. He therefore concluded that the termite mound is subject to less leaching than the surrounding soils. This would favour the retention of bases within the mounds, but it does not explain how they enter. The distribution of free carbonates within the mound, the low base content of the ground water, and the Penman evaporation rate indicate that the free carbonates in the mound could not have come from the ground water unless the termite mound is about 5000 years old, whereas recent estimates of the age of termite mounds in central Africa do not exceed about 700 years (Watson, 1967).

Termites are serious pests in forestry and agriculture. In East Africa most of the damage to living trees

Table 2.5   Plants of termite mound thickets in *L. kagerensis* grassland at Lake Nabugabo, Uganda

TREES	*Aningeria* sp.	Sapotaceae
	Antidesma meiocarpum	Euphorbiaceae
	Bersama abyssinica	Melianthaceae
	Blighia unijugata	Sapindaceae
	Canarium schweinfurthii	Burseraceae
	Croton megalacarpus	Euphorbiaceae
	Ekebergia senegalensis	Meliacea
	Erythrina abyssinica	Leguminosae subfam. Papilionoideae
	Ficus spp.	Moraceae
	Maesa lanceolata	Myrsinaceae
	Maesopsis emini	Rhamnaceae
	Phoenix reclinata	Palmiae
	Polyscias fulva	Araliaceae
	Pseudospondias microcarpa	Anacardiaceae
	Pycnanthus angolensis	Myristicaceae
	Sapium ellipticum	Euphorbiaceae
	Teclea nobilis	Rutaceae
SHRUBS		
	Acanthus pubescens	Acanthaceae
	Alchornea cordifolia	Euphorbiaceae
	Allophylus africanus	Sapindaceae
	Bridelia scleroneuroides	Euphorbiaceae
	Canthium vulgare	Rubiaceae
	Capparis sepiaria	Capparidaceae
	Carissa edulis	Apacynaceae
	Clausena anisata	Rutaceae
	Clerodendrum myricoides	Verbenaceae
	Craterispermum laurinum	Rubiaceae
	Dombeya sp.	Stericuliaceae
	Dovyalis macrocalyx	Euphorbiaceae
	Grewia similis	Tiliaceae
	Harungana madagascarensis	Hypericaceae
	Hoslundia opposita	Labiateae
	Lantana spp.	Verbenaceae
	Maytenus sp.	Rubiaceae
	Morinda lucida	Rubiaceae
	Mussaenda arcuata	Rubiaceae
	Pittosporum ripicola sp. *mannii*	Pittosporaceae
	Psychotria fernandopoensis	Rubiaceae
	Rhus vulgaris	Anacardiaceae
	Syzgium sp.	Myrtaceae
	Saba florida	Apocynaceae
	Scutia myrtina	Rhamnaceae
	Securinega virosa	Euphorbiaceae
	Tarrena sp.	Rubiaceae
	Uvaria	Annonaceae
VINES		
	Abrus canescens	Leguminosae subfam. Papilionoideae

Cissus quadrangularis	Vitaceae
Cyphostemma adenecaule	Vitaceae
Cyphostemma sp.	,,
Ipomoea cairica	Convolvulaceae
Jasminum pauciflorum	Oleaceae
Paullinia pinnata	Sapindaceae
Rhoicissus erythrodes	Vitaceae
Rubia cordifolia	Rubiaceae
Smilax kraussiana	Liliaceae
Teclea simplicifolia	Rutaceae
Tetracera potatoria	Dilleniaceae
Thunbergia sp.	Acanthaceae
Toddalia asiatica	Rutaceae

HERBS

Asparagus africanus	Liliaceae
Cymbopogon afronardus	Gramineae
Eragrostis blepharoglumis	Gramineae II
Erlangea sp.	Compositeae
Haemanthus sp.	Lilidceae
Melinis minutiflora	Gramineae
Microglossa pyrifolia	Compositeae
Microsorium	
Pseudarthria hookeri	Leguminosae subfam. Papilonoideae
Sansevieria sp.	Liliaceae
Tephrosia nana	Leguminosae subfam. Papilionoideae
Triumfetta sp.	Tiliaceae
Urena lobata	Malvaceae
Vernonia karaguensis	Compositeae

and plants is caused by the fungus growers. Damage to cotton, sugar and tea plants is occasionally serious locally when the termites' common food is in short supply. The mounds themselves are serious obstacles to mechanised modern agriculture.

The magnitude of the termites' beneficial role in the tropical community remains unmeasured. Those termites with an activity focused on litter removal and breakdown must surely be of great benefit in assisting the rapid turnover of nutrients and the maintenance of an effective nutrient cycle. This is especially important in many tropical soils where the total nutrient content is not high and the stability and luxuriance of the community rests heavily upon rapid breakdown of litter and immediate capture of the released nutrients. But termites of the group Macro-termitineae, which export large volumes of relatively infertile soil to the surface, contribute perhaps to lower levels of fertility, though this trend would be offset by the fact that their large mounds, once deserted, become

the foci for thickets with forest saplings and accelerate the succession to ecologically preferable forest. The role of the termite is clearly complex and until further precise data become available further generalisations on their effect seem undesirable.

Bushed and wooded grassland and woodland (Fig. 2.5)

Woodland refers to stands of trees reaching a height of 15 m, with the crowns just touching to form an open canopy, thus contrasting with the lofty forest and its complex, interlaced, deep canopy. Where the trees or bushes are scattered and the canopy cover is less than 20 per cent, the vegetation is described as wooded or bushed grassland. These terms are satisfactory for general descriptive purposes, but it should be remem-bered that they do not represent natural classes or types, there are no sharp boundaries between them, and they should generally be viewed as overlapping parts of a vegetational continuum, which, within quite a small area, may range from grassland, through bushed or wooded grassland to woodland.

All other factors being equal, it seems that the competition between the trees, bushes and grasses for the available or effective rainfall determines their relative abundance. The occurrence of woodland in Uganda and western Kenya coincides, on the whole, with fairly high, evenly distributed rainfall and further increase in available rainfall leads to forest. In both these areas the potential extent of woodland is probably quite high, but grazing and burning have reduced it so that the trees are usually found in relatively small patches with large stretches of wooded grassland in between. In Tanzania, however, woodland is widespread and it occurs under relatively low annual rainfall coupled with a long and harsh dry season which weakens the grass cover, apparently leaving most of the available rainfall to be taken up by the trees. In Uganda and western Kenya, with higher rainfall than Tanzania, reduction of the grass cover by grazing does not appear to push the development of wooded grassland towards woodland. Trampling of the ground causes increased run-off and reduced percolation so that the succession is headed towards bushland or bushed grassland. Clear examples of this succession have been recorded in heavily grazed areas of north-eastern Uganda with a rainfall of about 800 mm (Thomas, 1943). Grazing, fire, and effective rainfall are interrelated in such subtle and complex ways that often it is impossible to predict the consequences of variations in their relative strength.

The *Brachystegia-Julbernardia* or miombo woodland is one of the most extensive vegetation types in Africa apart from the forests of Zaire. Curiously, its principal trees scarcely extend into Kenya and Uganda. In these two countries it is *Acacia-Themeda* and Combretaceous wooded grasslands which cover large areas and there are smaller patches of the respective woodlands. The nearest relative in Uganda to the miombo woodland of Tanzania is found in the north-western corner of West Nile. It is characterised by such trees as *Isoberlinia doka, Daniellia oliveri* and *Afzelia africana* and is the eastern limit of a West African Guinea type.

It is not always necessary to describe the wooded grasslands and their corresponding woodlands separately. For this reason the following groupings of the main types is a convenience:

Vegetation with predominantly compound-leaved trees: *Branchystegia-Julbernardia* or miombo woodland; *Acacia* woodland; *Acacia-Themeda* wooded grassland.

Vegetation with predominantly simple-leaved trees: Combretaceous woodland and wooded grassland; *Butyrospermum* woodland and wooded grassland; *Borassus* palm grassland.

Vegetation with predominantly compound-leaved trees

Brachystegia—Julbernardia woodlands (Miombo) (Plate 22)

Among the woodlands of this category miombo takes first place. It covers about two-thirds of Tanzania and extends a thousand miles southwards through Zambia and Mozambique into Rhodesia and westwards to Angola. In Tanzania, it extends from sea level up to 1600 m in areas of annual rainfall of 500–1200 mm, where there is a single dry season. It does not occur in similar climatic areas in Uganda or Kenya except locally near Mombasa.

In Tanzania miombo occupies the central plateau to the north and the south-eastern plateau. It is divided into these two great blocks by a miombo-free corridor some 500 km long and widening from 60 km in the south-west to 120 km in Masai land in the north. These plateaux at an altitude of about 1200 m are the remnant of Miocene and Tertiary peneplains, and must have existed in very nearly their present form as a substrate for vegetation for approximately two million years.

On this moderately undulating peneplain where broad, flat, gently sloping ridges alternate with shallow, low-grade, flat-bottomed and seasonally inundated valleys known as 'mbugas', the well-drained ridges bear miombo on their upper and middle slopes, the valley bottoms grassland, and the narrow marginal region in between, bushland or wooded grassland with *Combretum* and other species. Termite mounds and thickets are found in all regions, scattered baobabs occur along the valley edges and on the hills, a mixture of more luxuriant woodland thicket, tufted grass and rock. Stands of *Borassus* palm occur where there is shallow ground water.

The term miombo comes from the Kinyamwezi name 'muyombo' (plural 'miyombo') which refers to the tree *Brachystegia boehmii*. *Brachystegia* is one of the commonest genera in this kind of woodland, but *B. boehmii* is not as widespread as *B. spiciformis* (Fig. 2.13) which has the Kinyamwezi name of 'mtundu'. The other very common genus is *Julbernardia*, species of which make up more than half the trees in any one area. Both genera belong to the

Caesalpinioideae section of the Leguminosae and have compound leaves; there are fourteen species of *Brachystegia* and two of *Julbernardia* in the miombo. Although floral features distinguish the genera, field identifications are usually based on vegetative charac- ters of leaf-form, crown-form, bark and slash. A guide to these will be found in Burtt's field key (1953–57) and fuller taxonomic treatment in White (1962) and, for Caesalpinioideae, Brenan (1967).

The general stature of trees of *Brachystegia* and

22 *Brachystegia boehmii* woodland, West Lake region, Tanzania. (Tanzania Ministry of Agriculture, Forestry and Wildlife)

Julbernardia varies with site and species. Often they have relatively slender trunks and much branched, light crowns with very feathery-looking foliage due to the delicacy and small size of the individual leaflets. Sometimes they are more massive, especially *B. boehmii* and *B. microphylla*, very large trees of the latter being often conspicuous and a memorable feature of hill and rocky outcrops (Philips, 1929).

Miombo trees are typically deciduous, though there are some evergreen species. The length of the leafless period varies from year to year and according to site, but there is a marked flush of new shoot growth just *before* the rains in September and October. At this time the woodlands which are always attractive become especially beautiful. Burtt (1942) gives the following excellent description:

A month before the rains set in, the miombo-covered hills burst all at once into flaming reds, salmons, pinks and coppery tinges of all hues as the *Brachystegia* trees flush into young leaf and, within a week, all this riot of colour has blended into a forest of the freshest green carpeted with legions of flowers.

In full leaf the forest is delightfully shady and the scant grass pleasant to walk through. All the tree leaves are pinnate while, in November, the activity of honey bees among the spikes of green flower heads in the tree tops of *B. spiciformis* is quite phenomenal.

In the dry season, what a change! The whole forest becomes entirely leafless while grass fires burn up all the grass and leaf litter. The sun beats down unmercifully and whichever way one turns one sees the same view, the grey stems of the miombo trees fading into shimmering distance. The buzz of insect life has vanished except for the sharp hiss of tsetse; brown, dry pods in the tree tops split suddenly and scatter their seeds while, overhead, the bark of Batteleur's eagle will momentarily catch the attention.

The quantity and variety of other trees in the miombo differs much from place to place and there are sometimes local concentrations of one species. Among the genera most frequently encountered and most characteristic are *Afzelia, Albizia, Burkea, Dalbergia* (Fig. 2.14), *Erythrophleum, Ostryoderris, Pterocarpus, Swartzia, Combretum, Monotes, Strychnos, Sterculia, Pericopsis angolensis* and *Uapaca*. Of these, the first nine belong to the Leguminosae. *Monotes*, though not common, is of interest as being the only representative here of the family Dipterocarpaceae so characteristic of the Malayan forests.

Two most important features of miombo trees are their resistance to fire (although most *Brachystegia* species are less fire resistant than other miombo components) and the annual dieback of seedlings which will be explained later. It may be these characteristics that have enabled miombo to become one of the most dominant and vigorous of tropical elements under present conditions of climate. Lawton (1963) believes that all forest elements under these conditions have tended to break down and be replaced by it.

The ground flora consists of a large variety of small shrubs and herbs including many grasses nearly all being perennial. This is important since it means that they provide grazing after fire, unlike the annual grasses which tend to predominate in bushland and provide a flush of grazing only after rain. It is noticeable that many of the better-known East African grasses are apparently not true miombo species. Thus, of the genus *Hyparrhenia* only *H. filipendula* is at all common in ecologically mature miombo, and *Cynodon, Chloris gayana* and *Panicum maximum*, common in other open woodland, rarely occur here. Although they are seldom to be seen in the daytime the miombo is the haunt of many wild animals including rhinoceros, buffalo, lion and elephant.

The monotony of the miombo is broken by sinuous networks of green, shallow valleys which can be seen most clearly from a hillside or from the air. These damper depressions and valleys are usually part of an extensive drainage system and are locally known as 'mbugas'. They are floored by greyish-black clay-rich soil which shrinks and cracks noticeably in the dry season, but expands in the wet season and becomes very slippery. Most East African soils are very intensively weathered, are red owing to their richness in iron oxide and the dominant clay fraction is kaolin which swells very little and absorbs very small amounts of other minerals. The mbuga soils are less degraded and contain a much higher proportion of montmorillonite clays which take up water and expand as a consequence; they ordinarily absorb much larger quantities of other minerals. The dark colour is believed to be due to clay-humus complex, whereas neighbouring kaolin soils, even on poorly drained sites, are much less affected by comparable amounts of humus. Mbuga soils often overlie a hard pan of

Fig. 2.13 *Brachystegia spiciformis*. (*Flora of Rhodesia*)

limestone concretion.

Sometimes the mbugas are covered with grasses (Vesey-Fitzgerald, 1963), sedges, and a few herbs; but in other, probably less waterlogged places, there are the usual miombo trees though less densely distributed. There may be some trees additional to the usual assemblage or which occur more frequently than they do in the woodland, for instance, *Combretum* and *Terminalia* species, while the gall *Acacias* are often found in the damp hollows.

Scattered about the mbuga are termite mounds with a very characteristic vegetation of well-grown trees. These mounds are not of the high, conical type so common in Uganda but low and spread out. The effect of the better-drained soil results in good growth over the whole area. The mound vegetation is very often fireproof, and such places therefore provide refuge for small mammals, monkeys, leopards, lions and thicket-loving antelopes, especially in the dry season. With the rains some of these animals repair to higher ground.

Of the miombo trees, very few are of any use; there is only one really important timber tree, namely, the *Pterocarpus angolensis* (muninga) which is a codominant in many parts. It is one of the most fire-resistant and produces high quality timber which is much in demand. The tree forms a single trunk with a spreading crown and the yellow flowers are borne for a very short time at the end of the deciduous period just before the leaf flush. The distinctive bark, which fissures into small rectangles and squares, is sometimes described as 'crocodile bark' because of its resemblance to the scaly surface of crocodile leather. The inner bark contains a blood red juice which darkens on exposure and these features together with the distinctive fruit help to identify it among the very many pinnate-leaved trees of the miombo. The fruit is not the usual pod common among the Leguminoseae, but consists of a winged circular disc about seven centimetres across with a raised, bristly or prickly centre enclosing seeds. The wing aids dispersal at the season when the woodland is leafless and much of the grass has been removed by fire. Under these conditions, muninga fruits have been observed to have been blown as much as a mile from the nearest parent tree. *Pterocarpus* though widely distributed, has a tendency to occur in relatively well-defined patches and research is now in progress to try to discover how this excellent timber tree can be grown on a commercial scale.

One interesting point is the growth habit of the seedling which helps to explain how the young plants withstand successive annual grass fires and eventually become established. Boaler (1966) gives the following details. Germination usually takes place inside the fruit after it has opened along one side. The radical first grows out and into the soil and then the growth of the hypocotyl draws out the cotyledons from the fruit. Development of the seedling is much more rapid below ground than above; the taproot growing rapidly downwards with rather sparse side roots. Root nodules appear in the first few weeks. Several shoots are formed during the growing season and by the end of it the plant has established an extensive root system, with a tap root from 45 cm to 90 cm deep. The possession of a deep root system is essential if the plant is to compete successfully for water for, even during the rainy season, the upper soil layers dry out from time to time. At the end of the rains or early in the dry season the shoots die. The process is repeated in the next and subsequent rainy seasons, the length of the shoot increasing each year but always dying back in the dry season to an inch or so below the soil surface. Figures given by Groome (1955) suggest that an average of 7 to 7·5 years may be spent by *Pterocarpus* plants under normal miombo conditions before they are able to pass into the sapling stage. During this period, food materials are passed down and stored in the root so that, even if the young shoots are destroyed by fire, new fresh shoots can still be thrown up. When the roots and laterals are well established, the shoot system is capable of strong, rapid growth to form a sapling which may escape the fatal effects of further grass burns, at least if these occur in the early dry season while the grass is short (Groome *et al.* 1957).

Boaler (1966) considers that only a small proportion of the young seedlings ever reach sapling stage; he calculated that in an area of 5 ha of miombo where he counted the *Pterocarpus* plants only about 105 per ha would develop into mature trees. It is likely that the vigour of the seedling root system and its ability to throw up successive strong shoots is a character of immense importance to other trees and shrubs of this habitat and most miombo woody plants have a seedling which is woody at least at the base. The reason for the dieback of the seedling is still not known. It should be noted that many trees of annually burnt grassland, including *Erythrina abyssinica*, spend their juvenile years developing a deep root system before producing stem and crown.

Another tree of some economic importance is *Dalbergia melanoxylon* the African blackwood or African ebony, known in the timber trade as 'mpingo' or 'poyi' (Haughton-Sheppard, 1958). It is a much-branched, usually many-stemmed tree or shrub 3–9 m high, spiny and untidy in appearance (Fig. 2.14). Though it occurs in miombo, it is more common in the woodland of the coastal belt from Dar-es-

Salaam southwards. Its fine black wood is in small but steady demand for walking sticks, carved ornaments, wind instruments such as clarinets, and ornamental inlay work. While the volume exported from East Africa is small, it is, volume for volume, the most valuable wood in the territory.

Two other trees *Swartzia madagascarensis* and *Pericopsis angolensis* have been much favoured as charcoal woods and the latter which has very hard wood, has been used for railway sleepers. In Zambia the miombo timber provided a valuable source of fuel for the copper industry.

Present interest now centres on developing techniques for regenerating and increasing timber production from *Pterocarpus*. The tree is not suited to

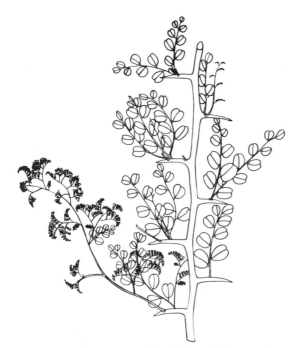

Fig. 2.14 *Dalbergia melanoxylon*. (Dale and Greenway)

plantation forestry because, in order to succeed in a forest nursery, it needs such a large root system that it is awkward to handle and costly to produce. Production of more mining timber must rely on the discovery of methods by which more seedlings can be brought through the sapling stage to maturity.

Experimental plantations of *Pinus* and *Eucalyptus* on miombo soils show some promise, but it is too early yet to say if they will be successful.

The soils of miombo woodlands, except where manuring or fertilising is practised with rotational cropping and resting, are poor and have inadequate

nitrogen for any but subsistence farming. Maize, millet and sorghum are grown in clearings and cashew nuts are a feature of some parts of Southern Tanzania, particularly near the coast. The mango and date palm are important round Tabora and the latter along the coast. Some of the areas chosen for the unfortunate groundnut scheme were in the miombo region. Its failure was largely due to the lack of a thorough study of the ecological relationships of the area before a scheme to cost £37 million was embarked upon. The cost of removing the woody vegetation proved to be very high; some soils were subject to waterlogging and others to excessive compacting and drying out; germination was poor; yields low and there was much loss from disease. Though the larger scheme was abandoned, a pilot scheme continues. There is also hope that tobacco may become a useful crop in this area especially round the edge of the mbugas. The prevalence of tsetse fly over much of the miombo makes it unsuitable for ranching or settlement.

There is one other valuable product of the unexploited miombo woodlands, namely, beeswax. The honey and beeswax are collected by the local miombo dwellers and since the building of a refinery by the Dutch White Fathers, the formation of a cooperative, and the establishment of a Bee Research Station (Smith, 1956, 58, 60, 66) the export of beeswax for incorporation into paints, polishes and cosmetics has brought in a considerable income. Some honey is exported, mainly to Holland, but most of it is used locally. The total value of bee products is probably just over £1 000 000 a year.

The African honey bee is *Apis mellifera unicolor* var. *adansoni*. It is a yellow-banded insect not unlike the Italian bee used by commercial beekeepers in Europe and America. The bees collect pollen, take it to the hive and store it in its unaltered form for use as their main protein food. The nectar which they also collect probably already contains wild yeasts and various fungi. This is mixed with salivary enzymes thus altering the original sugars into a syrupy liquid or honey. This is the insects' main carbohydrate food and is stored in the cells of the comb.

Most of the nectar and pollen comes from the common tree *Julbernardia globiflora* and from the closely related *J. paniculata*. These two trees flower in the later part of the rainy season, blossom appearing in April, reaching a peak in May and fading in June. At the end of the dry season and the beginning of the rains (September to November) most of the other trees of the woodland come into flower, including many species of *Brachystegia*, from which the bees collect abundant nectar. In the dry, central part of Tanzania various species of *Acacia* and *Combretum* are visited, and it is known that in the mountain forest bees collect from *Albizia*, *Cordia*, *Croton* and *Dombeya*.

Beeswax is produced by special glands. The bees gorge themselves with honey and hang quietly in the hive for about twenty-four hours. At the end of this time the wax glands, four on each side of the abdomen, begin to secrete tiny scales of wax which are then chewed by the bee and shaped into the hexagonal storage cells of the comb. Normally each hive produces honey and beeswax in excess of its needs and this is taken by the bee farmer.

The type of hive usually found in the miombo is very simple (Plate 23). It takes the form of an elongated cylinder made from a strip of bark obtained by ringing a tree. This is bound with grasses and fibre and fitted at each end with a circular piece of wood. A hole is bored in one end to give access to the insects. They are suspended in the crowns of trees and can easily be opened for the removal of the surplus combs without damaging or disturbing the rest of the colony. Similar hives can be seen in other parts of East Africa and are common in Kigezi in western Uganda. The hives are lined with wax to attract swarms, but only about one-third are successful. The ringing of the tree bark to make a hive can kill the tree and sometimes *Pterocarpus* meets its end by being used for this purpose. These hives, however, have their disadvantages: sugar ants can destroy a colony by eating the honey and biting the wings off the bees, while honey badgers climb the trees and rob the hives. Moreover, hives on the ground are destroyed by termites and bee houses of mud and wattle in the bush prove too expensive. Thus, at the Bee Research Station near Tabora, efforts are being made to introduce a better type of hive which is more proof against attack, and it is essential that this research should continue, for clearly the bee products are an important source of

23 Beehive in *Brachystegia* tree. (Morrison)

24 *Acacia tortilis* subsp. *spirocarpa*, Serengeti, Tanzania. (Tanzania Ministry of Agriculture, Forestry and Wildlife)

income in an area where forestry and agriculture are not likely to be very productive (Ntenga, 1967).

Acacia woodland (Plate 24)

The other genus of trees with compound leaves, which is common in grassland and sometimes forms stands close enough to be classed as woodland, is *Acacia*. The genus is readily recognised by the flat or mushroom-shaped crowns of the trees, the compound bipinnate leaves on branches invariably armed with spines or prickles, and the elongated spikes or rounded heads of white, yellow or pinkish flowers with many stamens. *Acacia* belongs to the Mimosoideae section of the Leguminoseae and the shape of the pod helps in the identification of the species.

The distribution of the various species is closely related to habitat conditions and the following are among those most frequently forming woodland. *Acacia xanthophloea*, derives its common name, the fevertree, from its habit of growing where its roots can easily find water, the sort of conditions in which fever used to be prevalent. It is a graceful, flat-topped tree with yellow bark and yellow or pinkish globose heads of flowers. It is not listed for Uganda, but is common in many parts of Tanzania and Kenya where it is seen at its best round the shores of Lake Naivasha. *A. tortilis* subsp. *spirocarpa*, sometimes called the 'umbrella acacia' has a more mushroom-shaped crown than usual, cream, rounded flower heads and spirally twisted pods. It favours a rather drier environment and is, in the northern part of its range, often associated with *A. abyssinica* subsp. *calophylla*; and it is the tree

87

which forms the background of so much wildlife photography, and animals eat its pods. A common tree in riverine areas and in land subject to flooding, where it often forms a close canopy of flat-topped trees, is *A. gerrardii*. It has almost black bark and globose, white flower heads. It is often associated with *A. poly-cantha* subsp. *campylacantha* Fig. 2.15 with white flower spikes and sometimes called the 'falcon's claw Acacia' because of its strong, recurved spines. Both these species are common in parts of Ankole and western Uganda where *A. gerrardii* can be seen clothing the banks of the crater lakes in Queen Elizabeth National Park. A particularly attractive tree of the margins of lakes and rivers, especially in Uganda is *A. kirkii* subsp. *mildbraedii*. It is often found in the same habitat as the wild date palm (*Phoenix reclinata*) and has a very delicate leaf tracery and flowers with reddish petals and white stamens. Among the species of *Acacia* which form almost a woodland at higher altitudes in the Rift Valley of Kenya, round Ngorogoro in Tanzania and in parts of eastern Uganda is the flat-topped *A. lahai* with white, axillary flower spikes. *A. clavigera* woodland is characteristic of Tanzania coastal hinterland as far inland as Kilosa. It is also a component of the *Acacia* woodland of Serengeti, the southern highlands, etc.

Fig. 2.15 *Acacia polycantha* subsp. *campylocantha*

Acacia-Themeda wooded grassland

The greatest development of this important vegetation type is the broad belt which encircles the Kenya Highlands 1200–1500 m and contains some of Kenya's best ranching country. The rainfall of 500–750 mm a year is erratic and unreliable and droughts occur. With differing soil types and varying rainfall the species composition varies from place to place, but always *Acacia* is the commonest tree and *Themeda triandra* (red oat grass) the dominant grass.

The commonest species of *Acacia* in dry grassland

are *A. tortilis* subsp. *spirocarpa*, *A. seyal*, *A. nilotica*, *A. gerrardii*, *A. nigrescens*; also *A. senegal* from which gum arabic is obtained. In the semi-arid grasslands of North East Uganda and North West Kenya *A. mellifera* is widespread. In black cotton soil which is frequently waterlogged, the gall acacias are common, *A. drepanolobium* with black galls and *A. seyal* var. *fistulosa* with white. The perennial *Themeda triandra* reaches a height of about 1·4 m and in the fullness of the growing season appears to produce a complete cover easily recognisable by its reddish flowering heads, though in fact, it is a fairly open association with quite a proportion of bare dry ground. *Themeda triandra* is probably the most widely distributed grass in East Africa and occurs in all zones except semi-desert and at very high altitudes on the mountains. In parts of Tanzania it is the main constituent of the high-level grassland below the upland forest. In Uganda *Acacia-Themeda* wooded grassland occurs south of the Katonga Valley and extends southwards as a zone 100–150 km broad, crossing the Kagera Valley and merging finally with the miombo of Tanzania. Again, this is a potential ranching area but it has been out of use for some time because of the rinderpest epidemic of 1891 which was followed by the rapid spread of tsetse in 1910. Prior to these outbreaks immense herds of long-horned cattle were pastured here by the Bahima and the Batutsi. This grassland community occurs also in Karamoja where it is heavily populated by game animals and where tsetse fly occurs in local patches in many places.

In all regions where *Themeda* is dominant, grass

Fig. 2.16 *Acacia sieberiana*. (*Flora of Sudan*)

fires sweep through the herbage in the dry season (Edwards, 1942), and the species probably owes its success at any rate in part to its ability to withstand these conditions. A high proportion of the tufted, mature plants survive grass fires and, in addition, the seed crop is buried in the soil beyond the range of lethal heat from surface fires. This is accomplished as in certain other grasses by means of a movement of the awn under changing moisture conditions and by the hard, sharp point of the fruit which is fringed with stiff bristles and acts as a barb.

Rainfall and grazing modify the grassland. *Themeda* will not withstand heavy grazing and disappears when grazing is intensified to a degree just sufficient to prevent grass fire. It may be replaced by tougher grasses and there may be a spreading and thickening of some *Acacia* species, particularly *A. hockii*. This is a small, yellow-flowered species between 3 and 4·5 m in height. At this height, its seeds are exposed to and killed by the flames of grass fires. However, when the area becomes more heavily grazed, the fire hazard is reduced and regeneration from seed becomes possible. It may be necessary for the seed to pass through the gut of cattle where bacterial action helps to break down the resistant seedcoat making it sufficiently permeable to gases and moisture for germination to take place (Turner, 1967; Harker, 1959). In some districts of southern Ankole in Uganda *Themeda* with its accompanying pasture grasses appears in many places to be losing ground to *Cymbopogon afronardus*, a light-green, tufted grass smelling of lemon when crushed. It is of no grazing value and is often the first step in a succession towards thicket.

Termite mounds frequently occur in the grassland often crowned by a *Euphorbia* and surrounded by a thicket of thorny bushes including some plants not found elsewhere in the area. Sometimes the mounds coalesce to form wooded patches up to a quarter of a hectare in size. Patches of forest occur here and there in this community in Uganda.

25 Combretaceous wooded grassland, Uganda: *Combretum* sp., *Albizia* sp., *Grewia* sp., *Loudetia arundinacea*

Vegetation with predominantly simple-leaved trees

Combretaceous wooded grassland and woodland

In Uganda and western Kenya increase in effective rainfall favours combretaceous wooded grassland rather than the *Acacia-Themeda* type. In both countries combretaceous wooded grassland covers large areas (Plate 25), and in many places, as a result of grazing and burning, it has advanced over countryside which formerly supported forest and woodland. In Tanzania it is not so widespread and is found mainly as a zone in the catena of the miombo woodlands. Combretaceous woodlands are locally common but not extensive in any part of East Africa. Those of Uganda have been described by Langdale-Brown (1959–60) and, by Langdale-Brown *et al.* (1964), Snowden (1953) and by Turner (1966), and those of Tanzania by Burtt (1942). *Combretum*, *Xeroderris*, *Sclerocarya*, etc., woodland occurs in the Kenya and Tanzania coast hinterland between coastal forest and *Acacia clavigera* woodland.

These wooded grasslands and woodlands contain a variety of woody genera (see Table 2.6) most of which are fire tolerant to some extent. They are designated combretaceous because of the common occurrence and sometimes abundance of two genera of the Combretaceae, *Combretum* and *Terminalia*. Most *Combretum* species are stoutly branched trees reaching a height of about 12 m; a few (*C. paniculatum* and *C. racemosum*) are upright shrubs or woody vines in thicket and forest edge. The genus is recognised by its woody, four-winged fruits (Fig. 2.17). The leaves are simple, with entire margins, sometimes sub-opposite but often almost in whorls of three and four. The

racemes of flowers, apart from the red and purple flowered vines, are pale yellowish and greenish white. *Terminalia* is similar in appearance to *Combretum* but its woody fruits are two-winged (Fig. 2.18). None of the species of the combretaceous woodlands yield useful timber. However, large areas of these wooded grasslands have been demarcated as forest reserve and are being converted to plantations of exotic conifers with *Maeopsis* and *Chlorophora*. Some valuable forest

Fig. 2.18 Fruits of *Terminalia mollis*. (*Flora of Rhodesia*)

species of *Terminalia* occur in West Africa and have been planted experimentally in East Africa, but here the bulk of the species of *Combretum* and *Terminalia* occur in the wooded grasslands and woodlands.

In Kenya and Uganda there are about twenty-two species of *Combretum* of which the most common are probably *C. collinum* subsp. *binderianum* and *C. molle*. Details of distribution are given by Eggeling and Dale (1951) and by Dale and Greenway (1961).

In Tanzania, of the many species of *Combretum*, *C. molle*, *C. ghasalense* and *C. zeyheri* are fairly common in the miombo woodlands. In the Lake Province *ghasalense* covers large areas in ironstone hills. Elsewhere in East Africa *C. molle* seems especially common on stony hills even within the high rainfall zone of the elephant grass thickets.

There is very little shrubby undergrowth below the trees but shrubs are common on and around the large

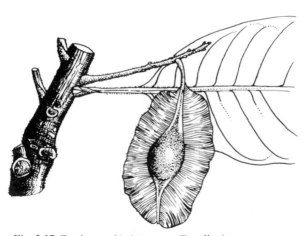

Fig. 2.17 *Combretum binderianum*. (Eggeling)

Table 2.6   Composition of the Combretaceous woodland near Mile 60 Kampala—Masindi Road, Uganda

TREES AND LARGER SHRUBS IN THE STRATUM 9—15 m

Albizia coriaria	Leguminosae subfam. Mimosoideae
A. zygia	Leguminosae subfam. Minosoideae
Allophylus africanus	Sapindaceae
Bridelia micrantha	Euphorbiaceae
Combretum collinum subsp. *binderianum*	Combretaceae
C. ghasalense	,,
Euphorbia candelabrum	Euphorbiaceae
Ficus capensis	Moraceae
Ozoroa reticulata	Anacardiaceae
Hymenocardia acida	Euphorbiaceae
Lannea kerstingii	Anacardiaceae
Stereospermum kunthianum	Bignoniaceae
Terminalia glaucescens	Combretaceae
T. mollis	,,
Vitex doniana	Verbenaceae

SHRUBS IN THE STRATUM 0—9 m

Acacia hockii	Leguminosae subfam. Mimosoideae
Annona senegalensis	Annonaceae
Capparis sp.	Capparidaceae
Clausena anisata	Rutaceae
Commiphora sp.	Burseraceae
Dombeya sp.	Sterculiaceae
Gardenia ternifolia	Rubiaceae
Grewia mollis	Tiliaceae
Grumilea	Rubiaceae
Harrisonia abyssinica	Simaroubaceae
Maytenus sp.	Celestraceae
Pavetta crassipes	Rubiaceae
Piliostigma thonningii	Leguminosae subfam. Caesalpinioideae
Protea madiensis	Proteaceae
Rhus sp.	Anacardiaceae
Securidaca longipedunculata	Polygalaceae
Schefflera barteri	Araliaceae
Steganotaenia araliaceae	Umbelliferae
Ziziphus abyssinica	Rhamnaceae

HERBS

Acalypha villicaulis	Euphorbiaceae
Aframomum sanguineum	Zingiberaceae
Amorphophallus abyssinicus	Araceae
Asparagus pauli-guilelmi	Liliaceae
Aspilia africana	Compositae

Berkheya spekeana	Compositae
Cassia sp.	Leguminosae subfam. Caesalpinioideae
Clerodendion myricoides	Verbenaceae
Costus. sp.	Zinigiberaceae
Crotalaria spp.	Leguminosae subfam. Papilionoideae
Cyphostemma adenocaule	Vitaceae
Dioscorea quartiniana	Dioscoreaceae
Echinops amplexicaluis	Compositae
Ectadiopsis oblongifolia	Asclepiadaceae
Gomphocarpus sp.	,,
Hoslundia opposita	Labiateae
Indigofera dendroides	Leguminosae subfam. Papilionoideae
I. emarginella	Leguminosae subfam. Papilionoideae
I. paniculata	Leguminosae subfam. Papilionoideae
Justicia betonica	Acanthaceae
Kamferia sp.	Zingiberaceae
Pseudarthria hookeri	Leguminosae subfam. Papilionoideae
Rhynchosia sp.	Leguminosae subfam. Papilionoideae
Sida alba	Malvaceae
Tephrosia sp.	Leguminosae subfam. Papilionoideae
Tinnea aethiopica	Labiateae
Vernonia smithiana	Compositae
Vigna sp.	Leguminosae subfam. Papilionoideae
Wissadula amplissima var *rostrata*	Malvaceae

termite mounds constructed by *Macrotermes*. The woody genera of these termite mounds are often different from those in the surrounding wooded grassland and include genera and species more characteristic of drier areas. As in all types of grassland vegetation, these termite thickets shelter small animals which, in turn, help to disperse certain species of plant. Many of the shrubs have brightly coloured and edible fruits which may be eaten or carried away by birds.

The grass layer which reaches a height at flowering of between one and two metres consists of vigorous perennial tussocks of many genera including—*Andropogen*, *Brachiaria*, *Chloris*, *Eragrostis*, *Hyparrhenia*, *Loudetia arundinacea*, *Panicum*, *Setaria*, *Sporobolus* and *Themeda*. In many places low nutritive value species of *Hyparrhenia* predominate (Table 2.2, p. 69). *H. diplandra* and *Hyperthelia dissoluta* are grazed by cattle the latter being fairly nutritious. The tall leafy culms of *Hyparrhenia* give a dense cover which restricts the species richness of the vegetation. The actual basal cover of the tussocks is much lower and after a burn there is considerable bare ground between them. Rapidly growing herbs and bulbous monocotyledons flower at the onset of the rains before the new leafy cover thickens. Some of East Africa's most brilliantly coloured and attractive flowers are found in the wooded grasslands at this time. There are tall yellow and orange *Gladiolus* and many types of lily. Among the latter we find yellow *Hypoxis*, *Crinum* lilies, yellow and white *Chlorophytum*, the brilliant flower heads of *Haemanthus* (sometimes called the 'fire-ball lily') and the large, pink and white inflorescences of *Ammocharis*. Yellow and purple species of *Eulophia* and other ground orchids are perhaps most common in damper places.

Being much grazed, burnt, and cultivated, the combretaceous wooded grasslands contain many seral stages which are most readily identified from the composition of the grass layer. As well as these seral stages there are the soil-vegetation catena sequences some of which, in southern Uganda, have been well described by Thomas (1945–46) and in Chenery (1960).

Where small areas of combretaceous wooded grasslands in Uganda have been enclosed experimentally for a number of years, the reduction in grazing and burning has led to a marked thickening of the trees. It is clear that these two factors retard development towards woodland and forest.

Most combretaceous woodland is of limited area and occurs as patches within larger areas of wooded grassland. Good examples of *Combretum* woodland are known from rocky hills and from the lower, southern slopes of Mt Kilimanjaro. Well-developed *Terminalia* woodland existed formerly in part of Murchison Falls National Park in Uganda, but it has been almost destroyed by elephants. The elephants rip off the bark exposing the cambial layer of the trunk which is then killed by the grass fires which sweep nearly every year through the woodland (see Plates 26 and 27). The changes which have occurred are described by Buechner and Dawkins (1961).

Butyrospermum wooded grassland

Under a rainfall of 1000–1250 mm per year in parts of Teso, Lango, Acholi, and Karamoja in Uganda there is an open grassland vegetation (Plate 28) in which the dominant tree is *Butyrospermum paradoxum* (shea butter nut) often accompanied by *Combretum* and associated woody species. It can be distinguished from members of the Combretaceae by its latex-loaded twigs. *Butyrospermum* has been an important food tree

26 *Terminalia velutina*; beginning of elephant damage. (Osmaston)

27 *Terminalia velutina* wooded grassland devastated by elephant debarking and fire. (Osmaston)

for some of the Nilotic tribes, and shea butter and fats which can be extracted from the fruits are exported from West Africa.

Palm grassland

Though this community is limited in area, it is so noticeable that it must be mentioned. The tree is the African fan palm (*Borassus aethiopum*), and its tall spindle-shaped bole with the thickest portion usually above the middle may reach a height of 24 m (Plate 29). It tends to occur in Uganda on sands and sandy loams with mobile ground water, with a ground cover of *Hyparrhenia* spp. and there are notable stands to the north-west of Lira, north-east of Arua and quite a large group of these trees west of the Kampala-Hoima road. They are also found on the Lake Albert flats. In Kenya *Borassus* is scattered through the coastal belt and is noticeable on the Shimba Hills. In Tanzania extensive stands are described for the flood plains of the Igombe and Ugala rivers near Tabora and in other riverine communities where it is sometimes accompanied by species of *Hyphaene*, the doum palm. The fruits are the size of a football and are sought after by elephant which, after chewing them, spit out the seed and may be instrumental in spreading the tree. The leaves are used for making mats, the fruits may be eaten raw, and the fresh sap on fermentation yields the best kind of African palm wine.

Hyphaene coriacea, superficially similar to *Borassus* is the dominant palm on the Warmi and Mkata flood plain and on the Ruaha–Usangu plain complex. It is the 'mikumi' of Mikumi National Park. *Borassus* stands are commoner than *Hyphaene* in parts of the coastal plain, but small trees of *Hyphaene* (up to 3 m) form extensive stands in grassland on poorly drained sands sometimes mixed with other woody species.

Utilisation of the rangelands

Rangelands comprise bushland, grassland, bushed and wooded grassland and woodlands and could be one of East Africa's most valuable assets. According to Davies (1961) the tropical rangelands, as a whole, sustain about half the domestic animals of the world and produce about one-fifth of the milk products consumed. With no substantial reduction of the world birth rate in sight, it is clear that there will be growing pressure to upgrade the rangelands to give high, sustained yields of milk and beef, at economic outlay. Rangeland development in East Africa has been held in check by livestock diseases, by tsetse, and by the small local demand. However, local demands for milk and meat are increasing, livestock diseases can be controlled, and, in some places, the battle against tsetse is showing favourable results. Once the tsetse is controlled, the exploitation of the rangelands will depend on demand and on various factors such as the accumulation of ecological information on rangeland potentials and behaviour, the correct choice of livestock, whether wild or domestic, the careful control of grazing and burning, the improvement of pastures chiefly by grass-legume mixtures, the provision of feed

28   Wooded grassland in Teso, Uganda. (Fergus Wilson)

29 *Borassus* palm, Sukumuland, Tanzania. (Fergus Wilson)

to maintain cattle healthily through the dry seasons, and on a careful programme of grazing and burning.

Let us first consider the control of tsetse. Three methods are in use. First, as wild animals form the reservoir of the trypanosomes which are transmitted to men and animals, it is obvious that their extermination would wipe out the fly-borne disease. This method has been successful in Rhodesia where *Glossina morsitans* is near its southerly limit. Hunting has also been used in parts of Uganda where it was necessary to take urgent action against the spread of the fly. But as it results in the destruction of both animals used for food and of game, which is the main tourist attraction, it is not looked upon with favour. Moreover, where game is removed in quantity it is necessary to plan resettlement quickly, otherwise the natural woody vegetation will reassert itself and the country become unsuitable and difficult for human habitation. Shooting of game to eliminate tsetse has not been widely practised in Tanzania, but has been used experimentally on a small scale in Uganda.

The second method is to attack the vegetation which provides the home and breeding ground of the fly. Most species of *Glossina* require at least two types of vegetation for breeding and feeding; moreover, grassland areas free of trees are not invaded by them. Research has sometimes seemed to indicate that in a particular area the elimination of certain elements of the woody vegetation is sufficient and accordingly discriminative clearance has been practised. This, however, has not always proved satisfactory and so total clearance has been tried. Tractors with heavy chains traverse the woodland to knock over bushes and pull over trees, leaving a scene of devastation. The intention is to eliminate the fly with the tree cover leaving only the grass, and to introduce men and cattle quickly to encourage grass and to keep the area free of fly. Unfortunately, even this drastic treatment of the landscape is not always successful especially in regions of adequate rainfall where the reestablishment of woody vegetation is relatively rapid. In some parts of Uganda young *Acacia* returns before settlement can be accomplished. In drier areas, such as the Central Province of Tanzania, it meets with better success, for regeneration of the woodland is much slower. Another difficulty about this method is that it deprives the population of wood which is needed for fire and building and removes a source of browse for the livestock. Furthermore, ruthless clearing is very expensive. It is clear that there is bound to be conflict between forester and entomologist when it comes to bush clearance, either by late burning or by cutting. This may be a great problem in Tanzania where very large areas of miombo woodland are not forest reserves

and are regarded as necessary to supply the valuable timber of *Pterocarpus* and various softwoods.

The third method is aimed at the destruction of the fly, but not the vegetation, by the use of insecticide spray and this has been used recently quite extensively, for example, in Ankole, Uganda. Tsetse have well-defined resting site preferences and if an insecticide is sprayed on these sites the residue is usually effective for at least four months in killing tsetse that land on it. But such treatment must be followed at once by settlement and the removal of woody vegetation. Carefully planned settlement eliminates bush and discourages game. But it is difficult to persuade people to settle in areas known to have been infested and to use the land in such a way that the fly does not return.

The problem of tsetse elimination illustrates clearly the complex interplay between the vegetation which is its home, the wild animals which provide its food and men and cattle to whom it transmits disease. This subject is very fully explained by Ford (1971).

Given freedom from tsetse the method of exploiting the rangelands depends on whether they are to be intensively managed on a small scale for milk cattle, extensively managed for beef animals, or confined to wild ungulates which will be cropped.

Where the production of milk cattle is the object, progress so far made is very encouraging (Hennings, 1961; Haarer, 1951). It has been shown clearly that the natural pastures can be improved to double and triple their original productivity by irrigation, fertilisation (Poultney, 1959), the introduction of grass-legume mixtures, the use of better grass varieties and the provision of dry-season feeding, since irrigation and fertilisation are generally too expensive, even on small-scale work. A vigorous stand of legumes in a grass mixture can fix between 45 and 50 kg of nitrogen per half hectare and, taking into account the cost of legume seed, this works out at about one-quarter the cost of half a kilogramme of artificial nitrogen. Many legumes have been tried and species of *Desmodium* and *Stylosanthes*, mainly those indigenous to tropical America, have shown much promise in many conditions. They grow vigorously, withstand grazing and drought, and are able to establish themselves and compete successfully with the tall tufted grasses (Morrison, 1966). Legumes are even used with elephant grass ley, but it is believed that their beneficial action there is mainly in reducing competition from other weeds. Legumes also increase the protein content of the graze.

Dry-season feeding can be provided with fenced fields of reserved hay or *Sorghum* or with silage. It is claimed (McIlroy, 1964) that silage is the most suitable form of fodder conservation under tropical conditions and that, if well made and the crop harvested at the right time, the product approaches closely to the nutritive value of the crop from which it was cut. Data on the nutritive value and chemical composition of East African grasses are given by Bredon & Horrell (1961 and 62) Bredon and Wilson (1963) and Marshall and Bredon (1963 and 67). Trees and shrubs, the leaves of which are rich in minerals often provide a valuable dry-season food, though their use may be incompatible with tsetse control. The introduction of exotic breeds of cattle has much improved milk production, but they are more susceptible than native cattle to prevalent cattle diseases.

Where the object is the raising of beef cattle, recent ranching schemes have demonstrated our lack of basic ecological knowledge of such matters as the optimum carrying capacity of the different range lands, their response to different stocking intensities and frequency of burning. Naveh (1966), Naveh and Anderson (1966 a, b), Walter (1963), Vos and Jones (1968), Walker and Scott (1968).

Turner (1967) describes a study of the ecological problems of ranch management in Combretaceous woodland of Uganda. The area was formerly infested by tsetse and when this problem seemed to have been overcome, largely by drastic reduction of game, a cattle ranch of 250 km² was established and the less intensive trampling and grazing of game animals was replaced by the more intensive and uniform effect of domestic stock. It was noticed that some areas were no longer grazed, some plants were left untouched, and the spread of plants by cattle and herdsmen was changing the composition of the vegetation. Well-manured kraal sites were colonised by nutritive, stoloniferous grasses such as *Cynodon dactylon* and *Chloris gayana* while disturbed and compacted ground was liable to infestation by grasses of poor fodder value (Fig. 2.21).

The use of bushland for ranching schemes may often require the removal of much of the woody vegetation and sometimes irrigation as well. Clearance of the bush is a big undertaking which involves an understanding of fire tolerance (see Figs. 2.22 and 2.23), depth and type of rooting and the effect on the soil of uprooting and chemical poisoning (Adams, 1967; Thomas and Pratt, 1967). It is generally considered that burning is a necessary factor in the maintenance of good rangeland, but the composition of the vegetation is the result of a very delicate balance between various biotic factors of which grazing and burning are of the greatest importance.

Jackson (1964) describes the effect of burning as follows:

Grass in the savannah areas reacts to burning by putting out new shoots after the fire, drawing on reserves of moisture and food in the rhizomes and root systems and naturally depleting them prior to the onset of the rains. If the grass is then defoliated severely by grazing there is a tendency for the tufts to be weakened and to deteriorate or even be killed and replaced by annual or unpalatable grasses. . . . Used in a sensible way, fire is useful especially if employed once in several years. It will clear up all the old, accumulated dead stems of the grasses which are unpalatable and very resistant to breakdown, will prevent scrub formation and bush encroachment, set back the coarse tufted grass species and give the seedlings of other species a chance to grow early in the season if they are vigorous enough to become well established.

Overgrazing occurs where there are too many cattle and the carrying capacity is exceeded. Grasses are heavily defoliated and since they are then unable to produce a healthy root system, the result is deterioration and death and replacement in the sward by grasses which are either less demanding or less palatable to grazing animals. Undergrazing, on the other hand, leads eventually to poor, uneven grassland and to bush encroachment.

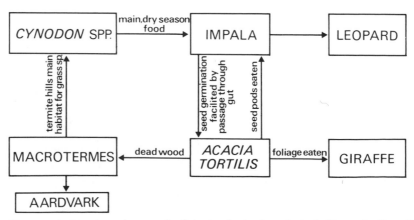

Fig. 2.19 Diagram to illustrate the interdependence cycle of plant and animal species in the Tarangire Game Reserve (Lamprey)

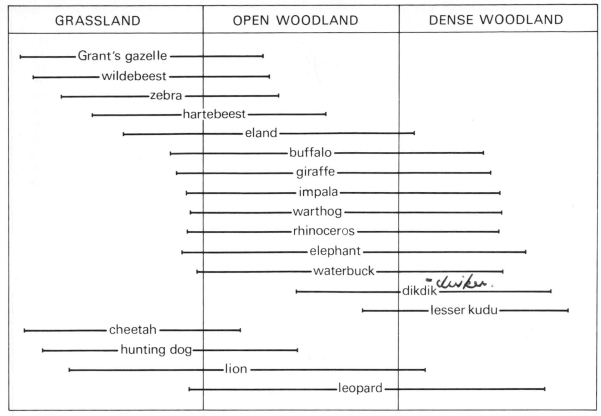

GRASSLAND	OPEN WOODLAND	DENSE WOODLAND

Grant's gazelle
wildebeest
zebra
hartebeest
eland
buffalo
giraffe
impala
warthog
rhinoceros
elephant
waterbuck
dikdik ~duiker.
lesser kudu
cheetah
hunting dog
lion
leopard

Fig. 2.20 Food preferences of animals in the Tarangire Game Reserve (Lamprey)

Greenway and Vesey-Fitzgerald (1969), in an account of the vegetation of Manyara National Park in Tanzania, point out how a close analysis of the dynamic ecology of the natural vegetation of an area, as distinguished from the changes induced by man, renders it possible to make recommendations for improved land use, provides a sound basis for management practice and makes possible the assessment of the vegetation as a habitat for wild animals. McClean (1971) summarises some of the ecological problems which must be studied if the most efficient use is to be made of the land either for forestry, livestock or game animals.

When national parks were first established in East Africa very little was known about the interplay and relationships of animals, plants and soils. It was clear that this had to be put right if the parks were to conserve their animals as well as their soil and scenery. In 1960, through a generous grant from the Nuffield Foundation, a research project in tropical animal ecology was set up in the Queen Elizabeth National Park in Uganda. One of its chief concerns has been the utilisation of rangeland by large animal communities especially hippopotamus and elephant. This has

necessitated a close study of the ecology of the grassland types (Lock, 1967, 1970). Similar but smaller units are now at work in other national parks in Kenya and Tanzania, including Tsavo and the Serengeti.

Wild ungulates have evolved in the rangeland habitat over many millions of years and their physiology and graze feeding habits are highly adapted to the rigours and uncertainties of that environment. Figures 2.19 and 2.20 illustrate the grazing habits of animals in Tarangire game reserve in Tanzania and the interdependence there of plant and animal communities. This work of Lamprey (1963) shows clearly how wild herbivores spread their grazing over a wide variety of vegetable material in contrast to domestic cattle which are confined to grass sward and not physiologically adapted to the dry seasons.

It has been suggested that more might be done to use the wild ungulates as a source of food as has been the practice in other parts of Africa. But so far, cropping in East Africa has been confined to game parks and reserves where hippo and elephant have been cropped at intervals to check grassland erosion and woodland destruction.

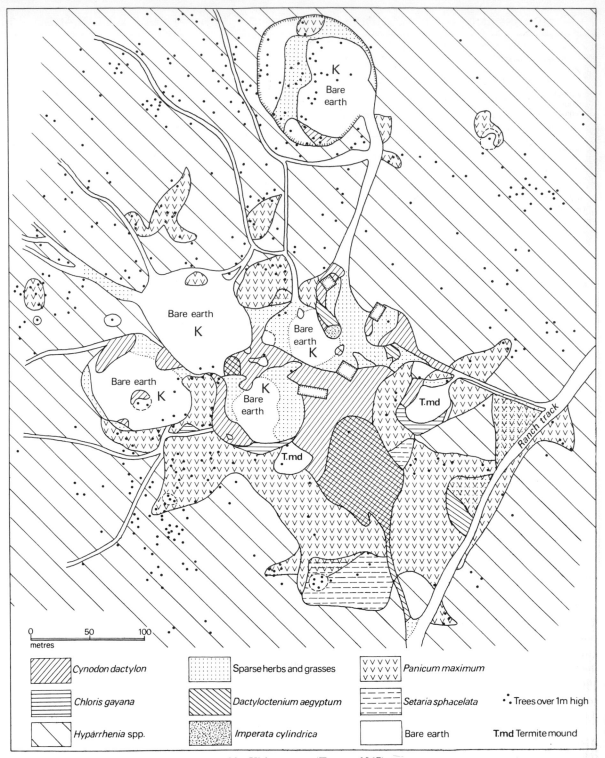

Fig. 2.21 Colonisation of kraal sites (designated by K) by grasses (Turner, 1967)

Legend:

- *Cynodon dactylon*
- *Chloris gayana*
- *Hyparrhenia* spp.
- Sparse herbs and grasses
- *Dactyloctenium aegyptum*
- *Imperata cylindrica*
- *Panicum maximum*
- *Setaria sphacelata*
- Bare earth
- Trees over 1m high
- **T.md** Termite mound

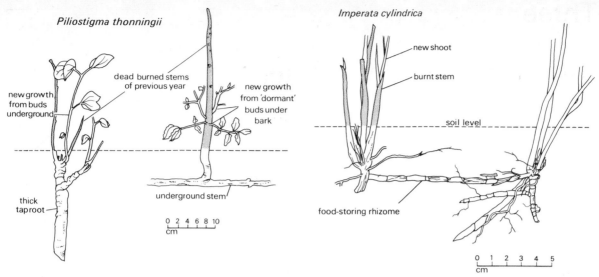

Piliostigma thonningii

new growth
from buds
underground

dead burned stems
of previous year

new growth
from 'dormant'
buds under
bark

thick
taproot

underground stem

0 2 4 6 8 10
cm

Imperata cylindrica

new shoot

burnt stem

soil level

food-storing rhizome

0 1 2 3 4 5
cm

Figs. 2.22 and 2.23 Regrowth after burning of *Piliostigma thonningii* and *Imperata cylindrica*

Three

Inland aquatic vegetation

The lakes of East Africa are numerous, varied and beautiful. Whether it be the broad, shallow waters of Lake Victoria, lakes of the Western Rift, like Lake Edward and Lake Albert, the saline waters of the Eastern Rift Valley in Kenya, or the deep waters of Lake Tanganyika, each has its own beauty and its own biological interest. The lakes differ considerably in size, depth (Table 3.1) and richness of plant and animal life, and many of them support a fish population which provides a valuable source of food in a country where traditional diet is short of protein.

It was largely the lakes and rivers which interested the early European explorers and the accounts of their travels (Langlands, 1962) make fascinating reading. The search for the source of the Nile brought Burton and Speke to Lake Victoria, and it was Baker who first drew the attention of the outside world to the existence of Lake Albert. Although the lakes have received considerable attention from biologists much less is known about the rivers.

History and origin of the lakes

To understand the variety of the lakes we must know something of their geological history and how they were formed. During the early Tertiary the direction of the rivers was probably determined by a north-south watershed lying to the west of what is now Lake Victoria, and, from here, rivers flowed east by a relatively short route to the Indian Ocean while others, the ancestors of the present Kafu/Nkusi/Katonga and Kagera rivers, drained westward towards Zaire. At this time there were no rift valleys. There is evidence (Beadle, 1962) that these ancient rivers, especially those flowing west and north, supported genera of fish very similar to those still found in the Nile and the larger rivers west of the rift. Lakes were present at this period and warping may have created a basin in the neighbourhood of the present Lake Victoria.

It seems, however, that some of the major lake basins had their origin in powerful and extensive disturbances which fractured the end-Tertiary land surfaces. The major movements which clearly defined the Western Rift have usually been dated to Pleiocene-Pleistocene times, but Bishop (1966) considers that the first major downthrow, at least on the Albertine section, may have been earlier than this. If, as is claimed, the basin of Lake Albert has been in existence since the Lower Miocene, it must be one of the oldest lakes in the world. It is widely claimed that the remainder of the lakes of the East and West Rift Valleys are of Pleistocene age.

It is believed that at this period further major faulting along the boundary faults of the West Rift was associated with a rise of the shoulders of the Rift Valley resulting in an uplift of the land between the Western Rift and Lake Victoria basin. This created a new watershed approximately 64 km east of the Western Rift Valley and dislocated the upper reaches of the ancient rivers, in some cases reversing their flow, so that today the rivers Kafu, Katonga and Kagera have a short section flowing westward fairly rapidly into Lakes Albert, George and Edward respectively, and longer sluggish sections flowing eastwards. This new uplift continued into later Pleistocene times and forced the waters of Lake Victoria to retreat eastwards leaving behind fossil strandlines ranging in height from 66 m to less than 30 m above the present lake surface. There are other strandlines below the 30 m level, but these are not tilted and maintain their height, and would seem to be related to downcutting of the Jinja outlet or possibly to climatic changes. In some places the eastward retreat of the lake waters was very extensive as, for instance, in the Kagera Valley. In mid Pleistocene times a long arm of Lake Victoria extended 112 km westward reaching close to Nsongezi, but eventually this long

Table 3.1   East African lakes

Lake	Area km²	Altitude m	Max. depth m	Conductivity mhos cm⁻¹, at 20 °C
Albert	5350	619	46	700 Fish, 1954
Edward	2250	916	117	878 Fish, 1952
Elmenteita	15	1777	1·9	22500 Fish, unpublished
George	270	916	2·0	165 Fish, 1953
Nakuru	40	1726	2·8	162500 Talling, 1965
Naivasha	180	1992	18·0	250 Lind, 1968
Rudolph	9300	380	73·0	3300 Talling, 1965
Tanganyika	34000	773	1470·0	620 Ricardo, 1939
Victoria	69500	1135	97·0	98 Fish, 1953

From Ross, 1955. Conductivity from source indicated

embayment was emptied of lake water and the Kagera, now having reversed its direction of flow, made its way across, and cut down deeply into, former Lake Victoria sediments.

From the point of view of their origin, therefore, the larger East African lakes fall into three main groups— those of the two Rift Valleys and those of the shallower basins in between. Lakes of a fourth group owe their origin to volcanic activity in more recent times. We must now look at these groups of lakes in more detail, examine the different types of waters and the vegetation both macroscopic and microscopic which lives in and around them.

Lakes of the Western Rift

These form a chain from Lake Albert in the north to Lake Tanganyika in the south. Lake Albert, at whose northern end the Victoria Nile flows in and the Albert Nile emerges to form the White Nile, lies at about 610 m a.s.l. surrounded by some of the lowest land in Uganda but with well-defined boundary scarps. It is about 46 m deep. Lake Albert is separated by the Ruwenzori massif from Lakes Edward and George which are at about 915 m. Further south, in Zaire, lies Lake Kivu at 1460 m which has been formed by volcanic damming, and from it the river Ruizi carries the overflow into the largest and deepest of the rift valley lakes, Lake Tanganyika, some 1189 m in depth with steep rift walls coming close to its shores or descending direct into its waters.

Lakes of the Eastern Rift

Here also we have a chain of lakes reaching from northern Kenya to Tanzania. Very few travellers are likely to reach Lake Rudolph in the very dry northern part of Kenya, though air transport has recently opened up a route to a safari camp on its shore. It is the largest of the Kenya rift lakes. Between Lake Rudolph and Lake Nakuru lie Lake Baringo and Lake Hannington and south of Lake Nakuru are Lake Elmenteita and Lake Naivasha. Geologists believe that in the early Pleistocene, there was probably one lake extending from Lake Naivasha north to Baringo, but that the Naivasha basin has become separated from the Nakuru-Elmenteita basin by subsequent earth movement. All these lakes, except Baringo and Naivasha, are very saline, and it is thought that these two must owe their freshness to subterranean outlets as no surface outlet is visible.

The lake levels have fluctuated considerably and raised beaches well above the present water level have been taken to indicate the much higher water of late pluvial times (Richardson, 1966). In 1931 Dr Leakey could note that the level of Lake Nakuru was then 1·6 m lower than in 1906 and Jenkin (1936), reporting in 1929, could postulate that it was then in danger of disappearance. Yet in 1964 the water was so high that no shore remained and surrounding trees had been killed by flooding; the same was true of Lake Naivasha.

As we follow the Rift Valley south into Tanzania we pass Lake Magadi in Kenya, and Lakes Natron and Manyara, Eyasi and Rukwa, all very saline. These highly alkaline lakes which have no outlet are served by rivers flowing through volcanic deposits rich in alkali salts, especially sodium. Under the strong evaporating power of the sun the concentration of salts gradually increases and at times of low water Lakes Nakuru, Natron and Manyara, and Lake Katwe in Uganda, show salt deposits round their shores, while Lakes Magadi and Katwe in Uganda are actually the site of an

103

important salt industry which will be described later. The dry shores carry only a sparse, semi-arid vegetation.

Lake Victoria

Lake Victoria shares its water between Kenya, Uganda and Tanzania, and is the source of the Nile. This lake is of comparatively recent origin and its waters have accumulated between the shoulders of the Eastern and Western Rifts probably within the last 500 000 years. It is not all of the same age and some of the northern may not be more than 20–30 000 years old. It is nowhere more than 97 m deep and in most parts is much shallower than this. During its history its outline has changed considerably and the water level has fluctuated greatly. Tilting of the land surface in upper Pleistocene times caused some of the border swamps and alluvial land to dry out leaving old strand lines at heights ranging from 30–60 m above the present water surface. These may be seen at various points around the lake including a place in the neighbourhood of mile 26 on the Kampala-Masaka road, where, below the ridge on which the Roman Catholic Mission stands, there is a high beach terrace and the road runs across extensive sand flats. Between this terrace and the lake there are other terraces and cliffs and also the basin of Lake Nabugabo. A comprehensive account of the ecological history of the Lake Victoria basin based on a study of mud cores is provided by Kendal (1969).

Lake Nabugabo deserves special mention as it has been cut off from Lake Victoria within the last 4000 years. Sometime during this period a longshore bar formation of sand and gravel began across the mouth of the Juma river where it flowed into Lake Victoria. This, together with the growth of another sandy spit from the south-west to meet it, impounded the waters of the Juma river to form a shallow lake which is remarkable for the low base status of its water. The conductivity of the stream feeding the lake is only 7·4 mhos cm^{-1}. Lake Nabugabo is now separated from the main lake by a sand bar with swamp along its inner edge, the whole about five kilometres wide, and extensive swamps have developed along the other shores. The cliffs to the west of the lake represent the old shore line of Lake Victoria and remains found in caves in some of them indicate that these ancient beach terraces were occupied by early man.

Lakes of volcanic origin

Some of the beautiful smaller lakes of East Africa owe their origin to comparatively recent volcanic activity.

On the Uganda-Zaire Border stand the Bufumbira volcanoes which were probably thrown up by great earth movements some 30 000 years ago. Lava flows from these craters dammed some of the surrounding valleys to form Lake Kivu and Lake Mutanda. The depth of Lake Bunyonyi may have been increased by the raising of the overflow bars by lava flows. Two of the Zaire volcanoes are still active.

In several parts of western Uganda, including the Queen Elizabeth National Park and in parts bordering on the Eastern Rift in Kenya and Tanzania, crater lakes are to be found. These small, usually deep lakes lie in depressions caused either by the collapse of a small volcanic cone or by the explosive eruption of abortive volcanoes. They usually have very steep sides clothed with forest. Sometimes a lake occupies the actual crater of an extinct volcano.

The plant life of the lakes

From what has been said by way of introduction about the different ages and origins of the lakes, it is not surprising that they show a wide variety of plant and animal life. In considering the plant life we shall have to look first at the macrophytic vegetation of the lake shore and the inshore waters and then at the microscopic organisms of the open water which owe their importance to the fact they form the basis of the food chain of larger forms of aquatic animal life.

The lake-edge plant communities

Most of the shallower lakes, except those which are strongly saline, have a deep belt of reed swamp through which the fishermen have to cut a passage for their canoes. One of the main constituents of the reed swamp, especially in Uganda, is papyrus. This plant, *Cyperus papyrus*, belongs to the sedge family (Cyperaceae) and can grow either on waterlogged mud, or float. Its stout rhizomes push out into the water of sheltered bays giving rise to characteristic erect, leafless shoots bearing feathery heads of bracts and flowers (Plate 30).

Papyrus, so common in parts of East Africa, is very familiar to students of Egyptian history although it no longer occurs in Egypt, except under cultivation (Tackholm and Drar, 1950). At one time there must have been extensive growths of this plant in the Nile for there are records of its use not only for papyri but for boat-making, cordage, matting, food and medicine, may have been left to dry out, or marshes may have been cut off and become too saline for the plant. Papyrus was cultivated as a crop in Graeco-Roman times but probably the growing industry of paper making from other materials during the twelfth century brought its cultivation to an end.

Although very widespread in Uganda, papyrus is less common in Kenya, probably because many of the lakes lie above 2321 m, which is about its altitudinal limit. It is abundant in Lake Naivasha. It is a very fast-growing plant and, after a fire, new stems arise from the rhizome and attain their maximum size in ten weeks. It may seem surprising that it has not become the basis of a paper industry in East Africa. Attempts have been made to use it but it seems only suitable for a coarse brown paper or for fibre board made from the fibrous outer layers, and the fact that it is often floating makes harvesting the plant very difficult.

30   Lake edge with papyrus. (K. Lye)

as well as for formal bouquets and funeral garlands. It is almost certain that it was among the papyrus that the baby Moses was found. Its main use was for papyri. The stems were softened in water, the green rind removed and the soft pith split into strips. These were then arranged in a layer and covered by another layer lying in the opposite direction and the two pressed together. The resulting sheets were dried in the sun and later polished with a piece of ivory. The oldest papyrus paper known was found in the tomb of Hemaka and is about 5200 years old.

Papyrus still grows prolifically in parts of the upper Nile and is one of the main constituents of the Sudd which made passage so difficult for the early explorers. The reason for its disappearance from the lower Nile is unknown, but it may be due to extensive harvesting for centuries followed by the use of the moist flats for raising crops. Also, the Nile has changed its course from time to time and, in this way, areas of papyrus

31   Channel in *Miscanthidium* swamp. (K. Lye)

Papyrus is probably the best known constituent of the lake-side swamps but there are many other plants associated with it. Among these the tall grasses *Miscanthidium violaceum* (Plate 31) with purple plumes and *Loudetia phragmitoides* with brown inflorescences are conspicuous while *Phragmites australis* occurs most commonly in silted areas and round lakes of volcanic origin. The 'bulrush' *Typha domingensis* is particularly common round some of the higher Kenya lakes and together with members of the Cyperaceae is one of the main agents in the fillings in of smaller depressions such as brick and sand pits. A number of other herbaceous plants establish themselves among the reeds including the showy purple heads of *Dissotis* spp. orchids of the genus *Eulophia*, climbing plants such as *Ipomoea*. In some parts there are extensive communities of *Limnophyton obtusifolium*, a monocotyledon with large, arrow-shaped leaves and branching heads of white flowers. These reed swamps are important as the nurseries of young fish. A list of the commoner plants in the Lake Victoria swamps is given in Table 3.2. The flora of the rather higher swamps surrounding the Kenya lakes may be rather different but contains a number of the same genera.

An aerial photo of part of the shore of Lake Victoria (Plate 32) shows vast areas of grass and sedge swamp, usually in sheltered bays and with a fringe of papyrus

Table 3.2   Commoner plants of a swamp-filled bay on Lake Victoria

SUBMERGED ZONE (6)		
	Ceratophyllum demersum	Ceratophyllaceae
	Hydrilla verticillata	Hydrocharitaceae
	Lagarosiphon sp.	,,
	Vallisneria spiralis	,,
	Utricularia foliosa	Lentibulariaceae
	Nitella sp.	Characeae
FLOATING LEAF COMMUNITY (6)		
	Nymphaea caerulea	Nymphaeaceae
	N. lotus	,,
	Nymphoides indica	Menyanthaceae
	Potamogeton thunbergii	Potamogetonaceae
	Trapa natans var. *africana*	Trapaceae
PLANTS ON THE WATER EDGE (6)		
	Cyperus papyrus	Cyperaceae
	Eleocharis acutangula	,,
	Paspalidium geminatum	Graminiieae
	Scirpus inclinatus	Cyperaceae
	Vossia cuspidata	Gramineae
PLANTS OF THE PAPYRUS ZONE (5)		
	Bridelia micrantha (tree)	Euphorbiaceae
	Cissampelos mucronata	Menispermaceae
	Cyphostemma adenocaule	Vitaceae
	Dissotis rotundifolia	Melastomataceae
	Ficus sp. (tree)	Moraceae
	Hibiscus diversifolius	Malvaceae
	Hyparrhenia diplandra	Gramineae
	Hyptis lanceolata	Labiateae
	Impatiens irvingii	Balsaminaceae
	Ipomoea spp.	Convolvulaceae
	Ludwigia leptocarpa	Onagraceae
	Limnophyton obtusifolium	Alismataceae

Melanthera scandens	Compositeae
Melastomastrum segregatum	Melastomataceae
Melochia bracteosa	Sterculiaceae
Pentodon pentander	Rubiaceae
Oldenlandia goreensis	,,
Neohyptis paniculata	Labiateae
Polygonum pulchrum	Polygonaceae
P. strigosum	,,
P. salicifolium	,,
Thelypteris striata	Pteridophyta
Tristemma incompletum	Melastomataceae
Triumfetta macrophylla	Tiliaceae
Vigna luteola	Papilionaceae

MISCANTHIDIUM ZONE (2,3,4)

Andropogon canaliculatus	Gramineae
Biophytum petersianum	Oxalidaceae
Cassia kirkii	Caesalpiniaceae
Cyperus haspan	Cyperaceae
Digitaria scalarum	Gramineae
Eragrostis mildbraedii	Gramineae
Eriosema glomeratum	Papilionaceae
Fimbristylis sp.	Cyperaceae
Fuerina umbellata	Cyperaceae
Hypericum lalandii	Hypericaceae
Loudetia phragmitoides	Gramineae
Mimosa pigra	Leguminosae subfam. Mimosoideae
Oldenlandia goreensis	Rubiaceae
Panicum subalbidum	Gramineae
Paspalum commersonii	,,
Polygonum pulchrum	Polygonaceae
P. strigosum	,,
Rhyncospora subquadrata	Cyperaceae
Sauvagesia erecta	Ochnaceae
Scleria nyasensis	Cyperaceae
S. nutans	,,
Smithia elliotii	Leguminosae subfam. Papilionoideae
Thelypteris confluens	Pteridophyta
Tristemma incompletum	Melastomataceae
Utricularia gibba subsp. *exoleta*	Lentibulariaceae

PLANTS COMMON THROUGHOUT:

Acriulus greigiffolia	Cyperaceae
Cyperus denudatus	,,
Leersia hexandra	Gramineae
Mikania cordata	Compositeae
Panicum parvifolium	Gramineae
Pycreus nitidus	Cyperaceae
P. polystachyos	,,
Torenia thouarsii	Scrophulariaceae

The figures refer to the zones in Fig. 3.3

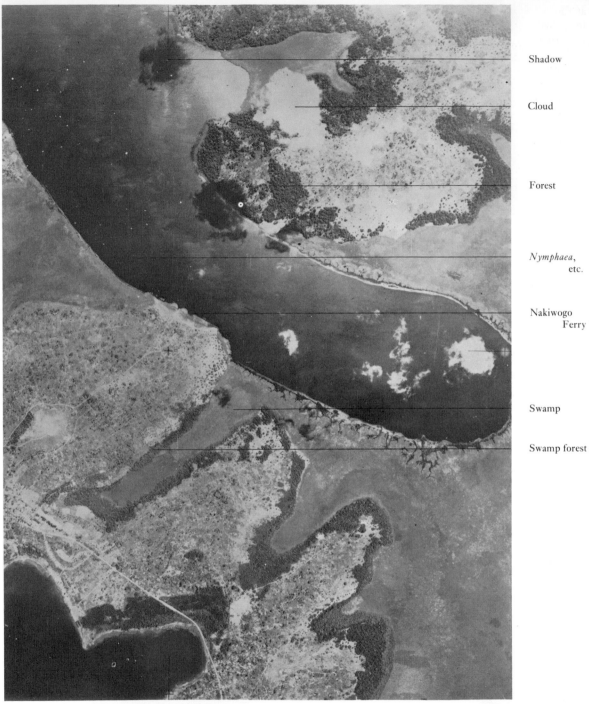

Shadow

Cloud

Forest

Nymphaea,
etc.

Nakiwogo
Ferry

Swamp

Swamp forest

32 Aerial photograph of part of the Entebbe peninsula, showing lake bay and swamps. (Uganda Lands and Surveys)

at the water edge. Where rocky laterite stretches out into shallow water its place is taken by plants more of the 'rush' type such as *Eleocharis* spp. and *Scirpus inclinatus*. The composition of the reed swamp will depend on various factors such as altitude, acidity, degree of silting, depth of water and amount of fluctuation of the water level. The mud round the saline lakes carries little vegetation except certain Cyperaceae which are able to stand the high concentration of alkali.

The lakeside succession: swamp formation

In many of the shallower East African lakes it is clear that the lake-edge vegetation is gradually advancing into the open water and in some places, as, for instance, in the smaller lakes near Lake Nabugabo in Uganda, there is little or no water left and the basin has become a swamp. To see how this process is brought about we must return to the water's edge and study what is known as the hydrosere, the building up by successive phases of vegetation of a terrestrial community in what was formerly open water. In the open water we

find floating plants such as the water lettuce *Pistia stratiotes* and the water chestnut *Trapa natans* with its attractive leaf mosaics and prolonged fruits, and also a variety of submerged plants such as *Ceratophyllum demersum*, *Utricularia* (Bladderwort), *Potamogeton* (Pondweeds), *Najas*, *Hydrilla*, *Nitella* etc. (Fig. 3.2 a–d). As these plants die, their accumulated remains together with soil particles raise the level of the substrate and stabilize it, reducing the effect of wave action so that rooted plants like the water lily *Nymphaea caerulea* are able to colonise it. These, in their turn, are displaced by the pioneering plants of the reed-swamp edge among which the grass *Vossia cuspidata* is important, and papyrus whose rhizomes grow out over the water often preceded by a zone of floating grass. Behind the pioneering fringe, which is floating, there gradually builds up the rich swamp community dominated by *Miscanthidium*.

It is clear that the floating and submerged vegetation play an essential part in building up the mud surface. Some species, such as *Potomageton schweinfurthii*, appear to tolerate shallow water whereas others, like *Ceratophyllum*, can grow in much deeper water. Figure 3.1 shows a transect on the edge of Lake Bunyonyi in South West Uganda at a place where the

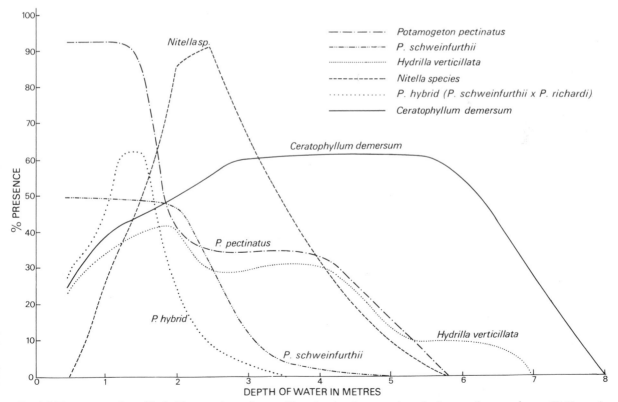

Fig. 3.1 Transect at edge of Lake Bunyonyi, south-west Uganda, showing zonation of submerged macrophytes. (P. Denny)

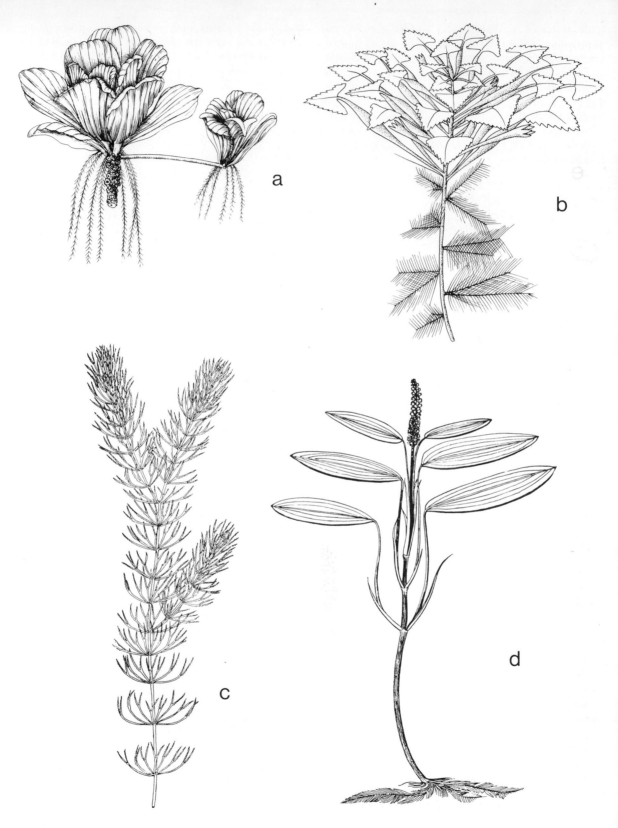

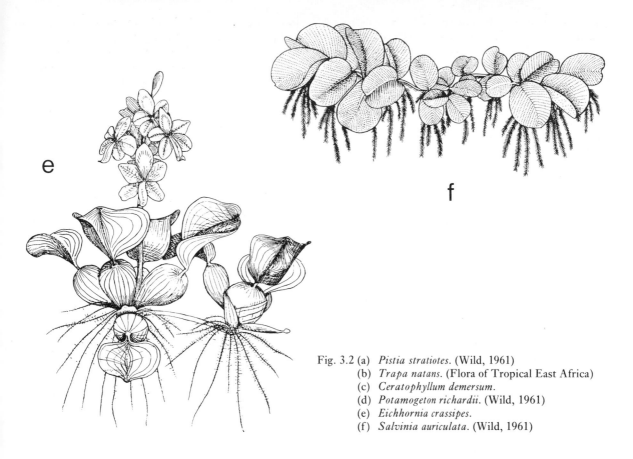

e

f

Fig. 3.2 (a) *Pistia stratiotes.* (Wild, 1961)
     (b) *Trapa natans.* (Flora of Tropical East Africa)
     (c) *Ceratophyllum demersum.*
     (d) *Potamogeton richardii.* (Wild, 1961)
     (e) *Eichhornia crassipes.*
     (f) *Salvinia auriculata.* (Wild, 1961)

shore line was free of the large emergent aquatics, mainly *Phragmites* and some *Cyperus papyrus*, which occur in other parts. Below the surface, in the photosynthetic zone, twenty-six floating or submerged species of aquatic macrophytes were recorded and the figure shows diagramatically the zonation and depth tolerance of the six most abundant. Several factors are concerned in the control of the zonation, including the morphology and flowering habits of the plants, physical and chemical factors of the water and the substrate, and the quantity and wavelength of light at different depths. Not only do the submerged macrophytes help to stabilize the substrate, but they also provide a valuable habitat for the protection of fish fry.

Figure 3.3 shows a diagram of a transect across a corner of one of the large bay swamps near Entebbe Airport on Lake Victoria (Plate 32). Here, papyrus is the pioneer plant at the water's edge, pushing out into the zone of submerged and floating vegetation, and it will be noted that the whole zone where it is dominant forms a floating mat. *Miscanthidium* and its associates occupy the drier area behind the papyrus and form raised tussocks with wet hollows in between. A number

of plants including *Thelypteris confluens* colonise the sides of the tussocks. Nearer the land, *Miscanthidium* gives way to *Loudetia phragmitoides* and finally to encroaching bushes.

Figure 3.4 shows a transect across the lake edge at Lake Nabugabo where there is very little papyrus and the pioneer plant is *Vossia cuspidata*. Its great culms, up to 6 m long, lie on the water and other rhizomatous plants, such as *Ludwigia* and the fern *Thelypteris striata*, weave their way through the *Vossia* stems to form a strong, floating mat which will bear the weight of a man. On this mat, mud, silt and decaying vegetation collect, and grasses and other herbaceous plants seed themselves so that again a mixed swamp vegetation dominated by *Miscanthidium* is built up. At the landward edge of the *Miscanthidium-Loudetia* swamp, especially where it abuts on *Loudetia kagerensis* grassland, there is often a community characterised by a number of plants which do not occur in the swamp nearer the open water. It would seem that this interesting community is favoured by acid drainage from the grassland. Foremost among the common plants is *Mesanthemum radicans* with large clumps of basal leaves between which grow stiff stalks bearing button-like heads of white flowers. Species of *Xyris* are usually found in this situation, another monocotyledon, with terminal heads of yellow flowers; and on the sandy ground where the grassland meets the swamp *Lycopodium affine* will often be seen and the delicate *Polygala filicaulis*. In the stagnant pools, between the *Miscanthidium* tussocks, the bog moss *Sphagnum* gains a footing and gradually spreads throughout the swamp to form a carpet with such associated plants as *Drosera madascariensis*, an insectivorous plant, and the small rosettes of *Syngonanthus*.

The filling-up process will continue till the open water is replaced completely by the advancing swamp vegetation. In temperate regions where a similar filling-in process occurs, when the level of the mud has been raised above the water level of the lake so that the source of water is largely rainfall, a more acid-loving type of vegetation invades the edge of the reed swamp and the bog moss *Sphagnum* makes its appearance building up the surface to form a bog. With further drying out trees establish themselves and what was originally a lake becomes a woodland.

These last stages in the succession are usually lacking in the East African lake basins. *Sphagnum* is never found with papyrus. Where it appears at the landward edge of some *Miscanthidium* swamps filling

in the stagnant hollows between the tussocks of grass and finally forming a carpet through which the grass stems project (Lind, 1953) the moss does not appear to raise the level of the surface sufficiently to allow the entry of other more acid-loving species.

The final woodland stage seems rarely to be reached in the tropical lake succession, although tree seedlings are quite common among the reed-swamp plants. This absence of woodland may be due to the effects of burning, for fires rage through the grass and papyrus even burning the papyrus which is floating at the water's edge. Another reason may be the fluctuating water levels with which the vegetation of the lake edge has to contend. Some of the valley swamps in South Western Uganda are an exception in that they support a dense forest of *Syzigium cordatum* and *Myrica kandtiana*. The reason for this is still obscure.

Although woodland does not usually spread across the swamp, many of the lakes are surrounded by a forest fringe. In Kenya the dominant trees are species of *Acacia* among which the flat-topped, graceful fever tree (*Acacia xanthophloea*) with yellow bark is often noticeable; banks of the deeper crater lakes in Uganda are also often clothed with *Acacia* woodland. Round Lake Victoria, however, where the rainfall is 1250–1500 mm the lake-edge forest is of quite a different nature. The beginning of the forest growth

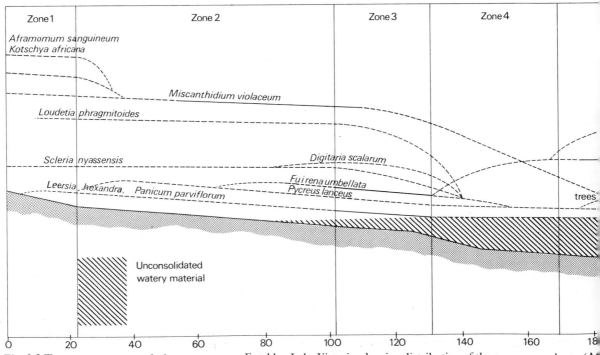

Fig. 3.3 Transect across part of a bay swamp near Entebbe, Lake Victoria, showing distribution of the commoner plants. (A

shows itself in the appearance at the landward edge of the swamp of plants, such as *Aframomum sanguineum*, whose flowers are followed by red edible fruits springing from the rhizome and the wild date palm (*Phoenix reclinata*). Among the commoner trees of the well-developed swamp forest are the raphia palm (*Raphia monbuttorum*) whose large leaves are much sought after for mat and basket making, *Mitragyna stipulosa* remarkable for its breathing roots in the form of 'knees' coming above the surface of the waterlogged mud, the large leaved *Macaranga schweinfurthii* and the much smaller *Alchornea*. Right at the water edge, in places where there is little swamp, and round islands, a common tree is the ambatch (*Aeschynomene elaphroxylon*) with compound leaves and large golden yellow pealike flowers. Its very light wood is used to make floats for fishing nets.

Valley swamps

So far we have attempted to describe the macrophytic vegetation of the lakes and their surrounding swamps. Attention must now be given to the aquatic communities of river valleys. Valley swamps are particularly a feature of Uganda especially in the region north and west of Lake Victoria and in the mountainous country of the South West (Beadle and Lind, 1960). When the direction of flow of the main river system was reversed in the Pleistocene, mud was deposited along the banks of the now sluggish, eastflowing sections. On this mud, papyrus established itself together with *Miscanthidium* and other plants of wet mud. These have advanced across the valley floor so that it is sometimes difficult to distinguish a flow of water through the dense vegetation. Permanent swamps of this nature in valleys and by lakes occupy 6500 km^2 in Uganda alone, and there are extensive swamps in other territories. Where the valley swamps drain into the lakes they merge into the swamps fringing the coastline. In some parts drying out has been encouraged by artificial drainage to enable the planting of *Eucalyptus* trees on the newly created land (Eggeling, 1935).

In the mountains of Kigezi, a volcanic area, valley swamps (Plate 33) have formed where local earth movements have levelled the valley floor so that the stream sometimes discharges over a waterfall into the lower surrounding country. Where lava has blocked flow of the water, lakes such as Bunyonyi and Mutanda have formed, and here *Phragmites australis* is a common associate of papyrus round the swampy edges.

A good example of the influence of drainage on the vegetation is provided by a study of the Kashambya valley in south-western Uganda (Lind, 1956a). There

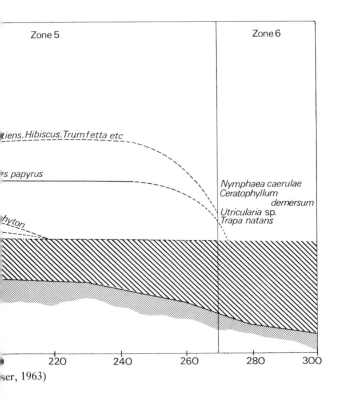

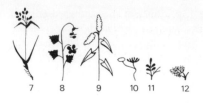

1 *Miscanthidium violaceum*
2 *Limnophyton obtusifolium*
3 *Cyperus denudatus*
4 *Dissotis brazzei*
5 *Dryopteris striata*
6 *Vossia cuspidata*

7 *Pycreus*
8 *Impatiens irvingii*
9 *Polygonum strigosum*
10 *Nymphaea caerulea*
11 *Leersia hexandra*
12 *Panicum*

Fig. 3.4 Transect across edge of Lake Nabugabo, Uganda. (After Milburn and Tallantire, unpublished)

is evidence that this upland valley was once occupied by a chain of lakes all of which have now been converted into swamps. Although there is common drainage, the nature of the dominant vegetation changes abruptly on passing from south to north (Fig. 3.5). At the southern end is *Typha* and papyrus, further north is a zone of almost pure *Miscanthidium* with *Sphagnum* and, beyond a place where the valley narrows, the dominant plant is *Cladium mariscus* without *Sphagnum* and with some open water. Whatever the vegetation in the middle, the margin is always occupied by papyrus. These sudden changes in plant cover can best be explained on the theory that the former small lakes were separated from each other by land bridges or by silt deposited by tributary streams, and that each lake received a rather different form of drainage which favoured a particular plant community. The figures for *p*H and conductivity indicate that *Cladium* favours a rather more base-rich environment than *Miscanthidium* which occurs in the region of greatest acidity. Further studies of valley swamps have been made by Deuse (1959) in Ruanda. Both in Kigezi and in Ruanda, parts of these high-level valley swamps have been drained and cultivated, but care has to be taken that they do not dry too severely as resultant chemical changes in the mud may make it infertile (Chenery, 1952).

High level swamps

At still higher altitudes of 2000 m another type of valley swamp is found. Here there is no papyrus, but there are deep deposits of peat covered with a monotonous mat of grasses and sedges relieved by the bright red heads of *Kniphophia* (red-hot poker) and occasional tufts of *Osmunda regalis* (royal fern). The banks are occupied by mountain bamboo and ericaceous scrub. It is valleys of this kind which have provided the locality for research on vegetational and climatic history described in the last chapter.

Swamps at a much higher altitude, 3800–4400 m

33   Valley swamp with papyrus, Kigezi, Uganda. (Norman)

are described by Loveridge (1968) for the Nyamagen-sani Valley of Ruwenzori where they often surround mountain lakes. On sloping banks, dense ericaceous forest may reach right to the water edge, associated with *Helichrysum stuhlmannii*, *Hypericum bequaertii* and *Senecio adnivalis*, but flatter banks are occupied by the coarse, tufted sedge, *Carex runssoroensis*, sometimes with tussocks a metre in height and small bushes of *Helichrysum stuhlmannii*, *Alchemilla john-stonii*, *A. argyrophylla*, *Lobelia wollastonii* and *Lyco-podium saururus*. On the water itself plants such as *Alchemilla johnstonii*, *Subularia monticola* and *Calli-triche* sp. build up floating islands upon which seed-lings of *Carex* become established. Gradually the *Carex* bog encroaches on the open water as silt and organic matter accumulate. Near the lake edge the *Carex* forms isolated tussocks with pools in the de-pressions which, further from the water, are occupied by *Sphagnum*. The highest of the lakes studied was just above the permanent snow line. Here the rocks were covered with a black lichen (*Umbilicaria*) with *Usnea*

and some mosses. There were few grasses and only small plants of the higher altitude ecotype of *Heli-chrysum stuhlmannii* and *Senecio adnivalis* var. *erion-euron*.

Seasonal swamps

In many parts of East Africa there occur large tracts of country which, in the rainy season, are often flooded quite deeply, but which dry out completely in the dry season and provide quite good pasturage for cattle.

In Uganda, this type of swamp is found mainly on the more alkaline, clayey soils of the north, especially in the valleys draining into Lake Kyoga. There is no papyrus except at the lake edge and its place is largely taken by grasses and members of the Cyperaceae. *Echinochloa pyramidalis* is the dominant grass with the smaller *Leersia hexandra* often spreading out over the surface of open water. In some parts *Oryza barthii* (wild rice) is co-dominant with *Echinochloa*. During the dry season of one to four months, when these swamps dry out and the clayey surface bakes hard, they are used for pasturage for cattle. But in the wet season there are large sheets of open water. Somewhat similar areas of impeded drainage occur in many river valleys. A list of the commoner plants in the seasonal swamps of Uganda is given in Table 3.3.

The mbugas of Tanzania fall into a somewhat similar category. These are seasonally waterlogged depressions in black cotton soil in the miombo woodland. In the rainy season they support a rich pasture mainly com-posed of the tussocky grass *Setaria holstii* with *Themeda triandra* and various flowering herbs; the gall *Acacia* is a feature of these mbugas; *A. pseudofistulosa* occurs on the margins especially in Masai land, and in the depressions *A. drepanolobium* is widespread par-ticularly in Lake Province extending up the east side of Lake Victoria; it flowers in the rains. There is a rather similar species which flowers in the dry season, namely, *A. malacocephala*, the main source of the gum arabic of commerce. *A. seyal* var. *fistulosa* is another wide-spread species. Other trees of the mbuga are men-tioned under woodlands.

Seasonal swamp merges with the flood plain of the valley bottoms and along the course of seasonal rivers in Tanzania. One of the largest areas of swamp covers about 1820 km^2 adjoining the Malagarasi River which flows into Lake Tanganyika. Lowe (1956) gives the following description of the area.

The water level of the swamp varies with the annual rains (Dec.–Mar.). The observations here described were made on the edge of the perennial swamp at Katare, about 160 km upstream from Lake Tanganyika in August 1952. At that

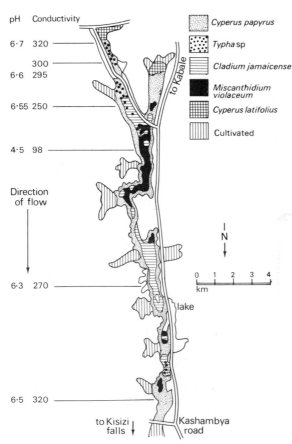

Fig. 3.5 Vegetation of Kashambya swamp, Kigezi, Uganda. (Lind, 1956a)

time, the swamp at Katare was about 7 km wide and rather less than 4 m deep over an area of many square kilometres. Blue water lilies covered the surface and below the surface were dense stands of *Ceratophyllum*, *Chara* and *Utricularia*. Except near the shore the bottom of the swamp consisted of a thick flocculent suspension of algae and vegetable debris. Above this the water was very clear.

In the same paper (part II) observations are described on Lake Salisbury, an arm of Lake Kyoga in Uganda which is said to superficially resemble the Malagarasi swamp but without the thick, flocculent deposit on the bottom. An almost complete absence of phytoplankton and zooplankton is reported from Malagarasi swamps and this poverty of phytoplankton is also noticeable in papyrus swamps in Uganda.

There are occasional large, permanent swamps and countless minor ones on the alluvial depressions of the Kilombero and Wami adjoining Lake Rukwa. They merge outward into wooded grassland and finally into woodland on the higher ground.

Table 3.3 Composition of seasonal swamps S.E. Lango, Lowland West Nile

TREES AND SHRUBS

Aeschynomene elaphroxylon
Mimosa pigra

PRINCIPAL GRASSES AND CYPERACEAE

Oryza longistaminata
Echinochloa pyramidalis
Leersia hexandra

OTHER PLANTS

Cymbopogon excavatus
Cyperus articulatus
C. denudatus
C. dives
C. haspan
C. papyrus
Echinochloa colonum
E. crus-pavonis
E. haploclada
Fimbristylis sp.
Fuirena sp.
Hyparrhenia lintonii
H. rufa
Imperata cylindrica
Ludwigia sp.
Justicia sp.
Lipocarpa pulcherrima
Nymphaea caerulea
Panicum repens
Polygonum sp.
Rhamphicarpa sp.
Saccolepis africana
S. auriculata
Setaria sphacelata
Sorghastrum rigidifolium
Themeda triandra

Ponds and dams

Besides the larger bodies of water classed as lakes there are many smaller ones with interesting vegetation. Some of these are artificial dams made to provide water in drier parts of the country, or, as in the case of Sasumua and Ruiru dams in Kenya and Igombe Dam in Tanzania, to serve as a water supply for towns. Others are pools formed in old diggings for sand or for clay (Plate 34) for bricks, while another group is to be found in the mountains often surrounded by a moorland of grasses and sedges. Lastly, there are the fish ponds which are dug near the village and stocked with young fry; these are described later in the chapter.

In the case of artificial dams, some attempt is made to keep the water clear of weeds by providing steep banks and clearing the vegetation at intervals. This is important because shaded edges form a good habitat for mosquitoes to breed and also for the snails which are hosts for the bilharzia parasite. Natural pools or those forming in brick or sand pits very quickly become filled with plants and offer excellent habitats for continuous study by biologists.

The open water usually becomes colonised by water lilies (*Nymphaea caerulea*) and pond weeds (*Potomageton* spp.) with submerged plants such as *Utricularia* spp. and, especially in sheltered corners, there is often a covering of floating plants including *Pistia*, *Lemna* spp. *Wolffia* spp. and *Azolla*. Encroaching on the open water from the edge there will often be *Typha* spp., various species of *Cyperus* and *Pycreus* and the grass *Leersia hexandra*. This sometimes covers the water so closely as to give a misleading appearance of solidity. Growing out from the bank *Ludwigia stolonifera* will also be seen, with its floating stems and curious white, spongy roots growing up to the surface, and *Hydrocotyle ranunculoides*, with lobed, cordate leaves floating on the water.

The algae of these ponds usually include *Spirogyra*, *Mougeotia*, *Oedogonium*, *Bulbochaete* and many unicellular forms especially diatoms and desmids which are associated with the water weeds. The process of filling in proceeds rapidly in the tropical climate as the mud builds up from the decaying vegetation. After a few years it may be difficult to realise that a given area of marshy ground was once a series of brick or sand pits.

Special features of water plants

We have seen that the macrophytic vegetation of the lake shore and of inland waters includes plants which are totally submerged, others which float, others again which are rooted in the mud but have floating leaves, and many, especially at the water edge, which carry their leaves up into the air though they are rooted below the water in damp mud. These plants show a number of specialised characters associated with the aquatic environment some of which will now be considered. For a fuller account reference may be made to Arber (1920) and Sculthorpe (1967).

Leaves

A feature of many aquatic plants is the possession of two kinds of leaf (heterophylly) the submerged ones differing from those which lie on, or above the water surface. Floating leaves are often leathery, peltate or orbicular in shape, as in *Nymphaea* and *Nymphoides*, with entire margins and stomata only on the upper surface. They sometimes have very long petioles when the orbicular or peltate shape of the leaf ensures that the point of attachment of the leaf stalk is near the centre of gravity of the leaf. These leaves arrange themselves in a mosaic pattern so that a maximum surface is exposed to the sun. This is particularly well illustrated in the case of the water chestnut (*Trapa natans*, Fig. 3.2b). The submerged leaves are sometimes quite different, as, for example, in *Nymphaea* where they are translucent, sinuate, have short petioles and are usually devoid of stomata. The pondweeds (*Potomageton* spp.) show a gradation from long, narrow submerged leaves to much broader floating ones.

Leaves of submerged plants may be entire or dissected. Entire leaves are frequently thin, translucent and in many cases filiform (*Potamogeton pectinatus*) or ribbon-like (*Vallisneria*). However, the stems of a number of submerged plants are clothed with small linear-lanceolate leaves (*Hydrilla*, *Lagarosiphon*, *Najas*). Dissected leaves are well represented by *Ceratophyllum demersum* and by some species of *Utricularia*. In the last case, they are provided with small bladders which trap insects. The seeds of *Ceratophyllum* germinate in the mud and the small plant rises to the surface when it is about 7 cm long. As well as the characteristic dissected leaves, the plant produces, on a special type of branch, delicate leaves which penetrate the mud and act as anchoring organs. In *Utricularia* stem and leaf are indistinguishable and the plant has no roots.

The free floating habit

A number of examples will now be considered where the whole plant floats on the surface. These usually have a rosette form with pendulous submerged roots, as, for example, *Pistia*, *Trapa*, *Eichhornia* (Figs 3.2 a, b and e), and *Lemna* among flowering plants and *Salvinia* (Fig. 3.2f) and *Azolla* among the ferns. Their buoyancy is due to the high development of air space in the blade and petiole and it is noticeable that, when the plants are stranded on the mud, this aerenchyma fails to develop. The leaves of some (e.g., *Pistia*) are clothed on both surfaces with hairs which trap air and help to dispel rain. *Azolla* has only feebly developed roots and in *Salvinia* (Fig. 3.2f) from each node hang finely dissected, root-like, submerged leaves which absorb water and nutrients. Any part of a stem bearing one or two buds can form a new plant and this enables a very rapid spread such as has occurred on the Kariba Dam.

In members of the Lemnaceae (the duckweeds) modification of the vegetative body has been carried so far that it is difficult to distinguish between stem and leaf. The plant body is reduced to a very small thallus a few cells thick which is flattened in *Lemna* with a single root, globose in *Wolffia*, and rectangular or falcate in *Wolffiella*, both these genera being devoid of roots.

Special mention must be made of two floating plants which, though they have fortunately not spread far in East Africa as yet, can become a real danger. These are *Eichhornia crassipes*, the water hyacinth, and *Salvinia auriculata*, a small water fern. (Wild, 1961). *Eichhornia* is widespread in tropical America and in various parts of tropical Africa especially the Congo River. It is an attractive plant with a rosette of leaves and pale violet flowers and may have been introduced for use in tropical aquaria. It is found in the Pangani River in Tanzania and has made its appearance in the Nile north of Juba, where it has increased rapidly till it seriously interferes with navigation. A strict watch is kept on its spread in case it should invade the waters of Uganda. Any broken bit can reproduce

34 Pond filling with *Nymphaea* sp. and *Phragmites*. (Morrison)

vegetatively and seedlings grow on the bottom mud and float to the top.

Salvinia is a much smaller plant with pairs of un-wettable leaves. It is indigenous over Central and S. America and as an aquarium plant is a prohibited immigrant to East Africa. Nevertheless it appeared in 1965 on Lake Naivasha forming a scum at the lake edge among the rotting flooded vegetation. It has been kept in check and has not yet spread sufficiently to be dangerous, but it causes some concern because it is this little plant which multiplied so rapidly in the early days of the Kariba Dam. It was first reported there in 1959, six months after impounding started. By April 1960 it covered 194 km^2, in 1962 it occupied 1040 km^2 of water surface and formed a mat 25 cm thick, which provided a substratum for *Vossia* and forty other species of flowering plant. By 1965 it had declined, but still covered 520 km^2.

Sexual reproduction

As many water plants are in part or wholly submerged their sexual reproduction presents some problems. In some cases the flowers remain submerged and the pollen is carried by the water; but most are adapted to aerial life in regard to pollination which is accomplished by wind or insects. The raising of the flowers so that they are above the surface is carried out by a variety of devices. In the water lily (*Nymphaea*) the sturdy flower stalk provides sufficient support for the large flower and its anatomy is well suited to this function. In *Nymphoides* the long inflorescence axis raises the flower well above the water and the associated leaf helps to keep it afloat. Some species of *Utricularia* have a series of floats on the inflorescence axis.

Other aquatic plants are more highly specialised in their sexual reproduction. In *Vallisneria spiralis* a common plant of shallower waters, the solitary female flowers within a spathe are carried up to the surface by the elongation of the flower stalk. The male spathes remain submerged, each containing over 2000 small flowers, each with two stamens sealed within a perianth which encloses a bubble of air. These flowers become detached and float to the surface where they open and are conveyed by currents to the female flowers; the tips of the stamens bearing sticky pollen touch the stigmatic surfaces and after pollination, the spiral stalk contracts and draws the young fruit down into the water. *Hydrilla verticillata* also produces female flowers which open at the surface revealing dry, fringed stigmas. The male flowers float to the surface where the perianth parts curve back and the three anthers spring erect, explosively discharging the pollen grains. Fertilisation depends on the discharge by the male flowers of the pollen in the neighbourhood of an open female flower. Pollen landing on the water is lost.

A good example of hydrophily is afforded by *Ceratophyllum*. Here the female flowers never reach the surface and the pollen is carried by water. The male flower consists of a group of stamens enclosed in a perianth. It becomes detached and rises to the surface where it opens, releasing the pollen. But the pollen sinks again to reach the stigmas of the female flowers which are still submerged. The fruit ripens under the water. It is not only the hydrophilous plants which ripen their fruits below the surface. This is true also of a number of aquatic plants which carry out pollination at the surface by wind or water. In fact, retention of aerial fruit development is rare among aquatic angiosperms, whether submerged, floating or with floating leaves. The contracting flower stalk of *Vallisneria* has already been mentioned; in *Ottelia* the flower stalk bends over and submerges the young fruit, while in many cases the fruit simply sinks to the bottom. *Utricularia* is an exception in that it retains an aerial method of fruit ripening.

Finally, a word must be said about the very reduced floral structure of members of the Lemnaceae. In *Lemna* the minute inflorescence, made up of one female and two male flowers in a membranous spathe, is borne in a marginal pocket near the base of the thallus. Each male flower has one stamen and each female one carpel. In *Wolffia* and *Wolffiella*, which are among the smallest known flowering plants, the inflorescence arises from a furrow on the upper surface of the thallus and consists of one male flower with a single stamen and one female with a single carpel. All three genera sometimes occur in great quantity, carpeting the water of quiet, stagnant pools. *Wolffiella* is uncommon in East Africa but when it is found it is hard to recognise that its tiny strap-shaped thallus about 1·5 mm long, bearing a minute pouch and two dots which represent the flowers, does indeed belong to the family of flowering plants.

Vegetative reproduction

Dr Patrick Denny has kindly supplied the following information on the vegetative reproduction of some water plants in Uganda.

This method is common among emergent and free-floating and submerged species. Round Lake Victoria, for example, the fringe of papyrus swamp is composed of mats of interweaving rhizomes together with associated plants. These rooted mats extend from the lake shore out over the water surface and form a floating raft of living vegetation. The extremities of the mat are exposed to severe wind and wave action and become easily detached from the more stable inshore swamp to form floating islands (Plate 35). These float across the water and accumulate on the windward shore producing sudd-type conglomerates. Though these never reach the proportions of the Sudd of the Sudanese Nile, they can be troublesome and hazardous to shipping.

Other species of emergent fringe vegetation have also adopted this successful means of distribution and, in Lake Bunyonyi, for example, islands dominated by *Cladium mariscus* with *Typha latifolia* are common. As they were initially attached to the shore, species typical of the swamp and water edge, together with floating-leaved and submerged macrophytes, are included and their distribution thus implemented. The rapid establishment of clones of *Cladium* is further assisted by a second vegetative mechanism in which the vertical inflorescence becomes top-heavy and, where support is lacking, falls to a horizontal position. New propagules then sprout from the nodes of the shoots and ultimately become rooted and detached. The decumbent attitude of the flower head also makes it easy for the fruit to be washed away.

Phragmites australis is a familiar plant of the higher altitude lakes of Uganda but, unlike its fellow emergents (or its counterpart in Europe) the rhizomes advance firmly rooted to the substrate and it is thus rarely, if ever, seen as a component of floating islands. It presumably relies upon its seeds for wide dispersal followed by strong rhizomatous growth in a suitable habitat.

The free-floating species tend to be at the mercy of the prevailing winds, water currents, and some animals including man, for their distribution but, on reaching a suitable environment they have an enormous capacity for vegetative growth. Reference has already been made to *Eichhornia crassipes* and *Salvinia auriculata*. Another plant which is abundant and can reach explosive proportions is *Pistia stratiotes* (the Nile cabbage). The plants have shortened vertical stems from which one or more stolons emerge producing young plants and a dense population quickly develops. In restricted conditions, an interwoven mat of vegetation builds up with dead and dying material under the surface of the water and new plants above. Free plants or the entire mat can be transported when the local climate changes or when they get caught up in water currents or are conveyed by larger animals, like the hippopotamus, from one drinking place to another.

Populations of the smaller, free-floating macrophytes, such as *Lemna* and *Wolffiella* spp., quickly expand by the budding off of new plants from a parent thallus, but colonies of more than three or four thalli are easily broken up thus facilitating their distribution. *Azolla* is a water fern found in sheltered pools and in favourable conditions it can take possession forming a green, or reddish velvety carpet over the surface. The small, fragile plant consists of horizontal many-branched stems which fragment and can easily be carried by birds or animals to other pools.

The way in which many water plants successfully

35 Papyrus islands in the Nile, near Rhino. (A. C. Brooks)

combine sexual and vegetative methods of reproduction is well illustrated by *Potamogeton thunbergii*, a floating-leaved species which fruits abundantly in Uganda. The ripe fruits become detached from the spikes and, as they may float for up to six months before sinking and germinating, they are conveyed by wind and waves to widely diverse habitats. The fruits are also a favourite food of water fowl and can be distributed to geographically different regions glued with mud to the feet and beaks of the birds. However, although fruit formation of *Potamogeton* is an effective means of reproduction and protects the species in times of unfavourable environmental conditions, in favourable conditions vegetative reproduction is responsible for its profusion. The mechanism is simple. During growth and development the stems become brittle and even slight disturbances produced by wind, waves and water currents and by the abundant animal and bird populations easily break them. These fragmented, vegetative parts are buoyant with oxygen-enriched air trapped in the lacunae and aerenchyma and, like the ripe fruits, are then dispersed in the water by the elements. Ultimately, some fragments are washed into shallow water where they root in the substrate and form new plants. As they are already at a mature stage of development they can quickly form flowering spikes as well as extending stolons along the surface of the mud. The stoloniferous habit encourages rapid development of a colony from a single plant.

The microscopic plant life of lakes (Phytoplankton)

Any unfiltered natural water contains minute free-floating plant and animal organisms which constitute the plankton. Where the lake or reservoir serves as a source of drinking water, these micro-organisms are removed by filtration before the water reaches the consumer, but a trawl of a fine silk net through the surface water reveals a great variety of minute plant and animal life. Sometimes the plant forms (the phytoplankton) are few in number, but in other cases, especially in alkaline lakes the growth is so dense that a white plate becomes invisible a few centimetres below the surface. The quality and quantity of the microscopic life varies throughout the year and in different types of lake. High alkalinity and organic impurity both favour an abundance of phytoplankton though the species of which it is composed in each case are quite different. Lakes receiving their drainage from agricultural land are likely to have quite a different

population from those in rock basins or deriving their water from volcanic rocks. These differences depend on the interaction of many factors, including the chemical composition of the water (Lind, 1968; Talling and Talling, 1966), and climatic variation and will be discussed under productivity.

The organisms of the phytoplankton belong to a wide variety of algal classes and form a rather specialist study. For the purpose of this general survey of vegetation it will be sufficient to list the major groups which include those types occurring most commonly in East African freshwater.

CHLOROPHYTA (green algae)

Minute green cells existing either singly or grouped together into colonies. The colonies may be immotile or free-swimming by means of *flagellae* (Fig. 3.6, 12 and 9). Among the immotile single-celled forms the desmids are important. In these each cell consists of two semicells joined by an isthmus and the walls are ornamented in various ways, often bearing spines (Fig. 3.6, 8 and 11).

CHRYSOPHYTA (Yellow-green or yellow-brown algae)

The diatoms are included in this group. Although they contain chlorophyll they are yellow-brown in colour. Each diatom consists of a single cell, often boat-shaped or needle-like with a wall composed partly of silica and bearing very delicate markings. In one common plankton diatom (*Melosira* Fig. 3.6,6) the cells are the shape of a pill box and joined by their flat surfaces to form chains. Also in the Chrysophyta and common in the smaller lakes and reservoirs is *Dinobryon*. Here the tiny, flagellated yellow-brown cells are contained within flask-shaped capsules open at one end and pointed at the other and the capsules are arranged in a branched, tree-like structure. (Fig. 3.6,5).

PYRROPHYTA (Golden-brown algae)

Two very common planktonic forms *Ceratium* and *Peridinium* belong to this group. They are unicellular organisms characterised by the presence of a deep, transverse furrow and by a wall ornamented by a number of plates. Both are actively motile by means of two flagellae (Fig. 3.6, 10 and 7).

CYANOPHYTA or MYXOPHYCEAE (Blue-green algae)

The algae of the Myxophyceae show a great

variety of form and may be unicellular, colonial or filamentous. They are often enclosed in a mucilage sheath. Some are so minute that they pass through the finest meshes of a plankton net, but they are present in such huge quantities that they colour the water deep blue-green. They are especially common in alkaline lakes. (Fig. 3.6,2 and 3).

EUGLENOPHYTA

Euglena and *Phacus* are unicellular motile organisms bright green in colour and especially characteristic of waters of high organic content. The cells are naked, but, while *Phacus* has a rigid periplast, *Euglena* is able to change its shape (Fig. 3.6,13).

At certain times in some lakes these algae form a thick scum or 'bloom' on the surface consisting mainly of Myxophyceae, and under these conditions fish sometimes die. Though some Myxophyceae are known to produce toxic substances it is more likely that the death of the fish is due to competition for the oxygen dissolved in the water.

The Myxophyceae are also of interest because of their ability to grow at much higher temperatures than other algae. They have been found at temperatures of 69°C and can be seen forming a dark-green scum on the mud round some of the hot springs in East Africa. It is known, too, that they have the ability to fix nitrogen and this attribute is used in the rice-growing areas of the world.

Saline lakes are characterised by an abundance of blue-green algae and diatoms. The fresh waters have a much more mixed phytoplankton dominated at various seasons by different species. Sometimes the water will be coloured brown by an abundance of Pyrrophyta and at other times bright green by Chlorophyta. Among well-known plankton genera of the green algae are *Volvox* and *Eudorina* which often become very abundant. Desmids are favoured by more acid conditions and are often found both in open water and in stagnant pools at the swamp edge or among the submerged plants of the lake edge.

It is in the alkaline lakes that the flamingoes are found (Brown, 1959). It is estimated that there may be as many as 3 million of the lesser flamingo and 50000 of the greater on the chain of lakes in the eastern Rift Valley. Many of these lakes are difficult to reach and the birds seem to choose the more inaccessible, such as Lake Natron, for their breeding places. These pink birds, the size of a domestic duck and moving on bright pink legs, fly in flocks from one lake to another. The flamingoes are associated with the alkaline lakes

because the food of the lesser flamingo consists of Myxophyceae (blue-green algae). The bird feeds on the top few centimetres of the alga-rich water, swinging its head from side to side and extracting the microscopic organisms by a remarkably efficient filtering device. They drink near springs of relatively fresh water. Many of the lakes are renowned for their magnificent bird population, and the bird sanctuary on Lake Nakuru is an especially popular tourist attraction. The flamingo is unusual in its ability to feed directly on the minute plant organisms. Most water birds live on fish or on minute animal life in the water or mud.

The Euglenophyta are usually found in the more highly organic waters and in pools where animals have come down to drink. Because of their ability to grow in these conditions they are useful in the purification of sewage. The sewage is run into shallow lagoons where bacterial decomposition of the sewage organic matter progresses swiftly, oxygen is used up and carbon dioxide and ammonia are produced and escape into the atmosphere. After some days, when light and temperature are satisfactory, algal populations appear in large numbers. They absorb carbon dioxide and release oxygen which is then used by the bacteria in the breakdown of sewage. The use of these oxidation ponds has been developed in America, but they are now being increasingly used in East Africa where the process is favoured by the good light and high temperatures (Hunt and Westenberg, 1964). In the Nairobi lagoons and at Entebbe *Euglena* and *Phacus* are predominant in the algal population. In some parts of the tropics the purified effluent from the sewage ponds is led into fish ponds where it stimulates the growth of algae beneficial to the fish.

From what has been said it will be recognised that the minute algae of the phytoplankton play an important part in the economy of lakes and reservoirs, but so far, apart from the work of Rich (1932) on the

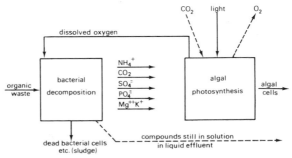

Fig. 3.7 Algal-bacterial oxidation of organic waste. (Hunt and Westenberg, 1964)

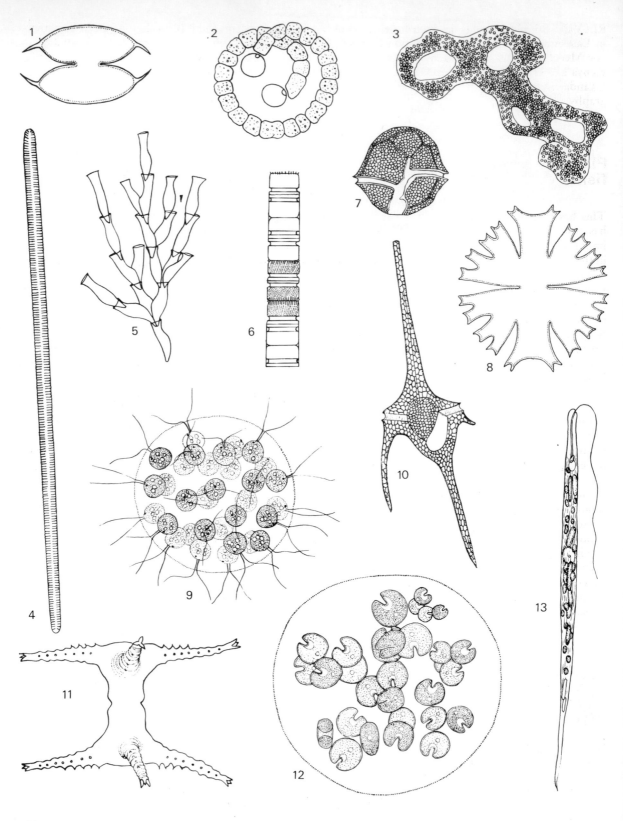

Rift Valley lakes, their study has been largely neglected in East Africa. Further reference should be made to von Meel (1954) on Lake Tanganyika, Lind (1968) on Kenya and Grönblad, Scott and Croasdale (1964) on Uganda, each of these papers providing a full bibliography.

Plankton productivity and fisheries

This has been left to the end because much of what has been said about aquatic vegetation leads to the important consideration of lakes as a source of food in the form of fish.

The minute algae of the phytoplankton play a very valuable part in the economy of the lakes. They absorb salts and gases from the water and by means of the green chlorophyll in their cells and the energy derived from sunlight they build them up into the carbohydrates and proteins of which their bodies are composed. Either they are eaten directly by fish or birds, or they provide the food of small forms of aquatic animal life which are then consumed by fish. The amount and composition of the phytoplankton of any body of water is therefore of great importance when considering its potentiality as a source of fish. It is related to the chemical composition of the water which depends in part on the nature of the drainage area, and on the time of year. In Naivasha, for instance, gatherings made throughout the year showed the following variations: Chlorophyta 4–24 per cent, Diatoms 4–63 per cent, Myxophyceae 23–88 per cent (Lind, 1968).

This periodicity of the plankton throughout the year is in part due to the thermal stratification of the water. As the lake warms up, the colder water falls to the bottom (the hypolimnion) while the warmer water remains on top (the epilimnion). The decaying zoo- and phytoplankton organisms fall to the bottom and release nutrients there on the bottom mud. In temperate waters, which are subject to seasonal variation in temperature, there is an overturn and mixing of the water at the onset of the colder season and by this mixing the accumulated nutrients on the mud are brought again into circulation. This favours the growth of plankton and there is a more or less regular periodicity of development. In East African lakes where there is little variation in temperature throughout the year, two types of thermal stratification may occur. The first, which is temporary, is caused by daily heating by the sun and is destroyed by nocturnal cooling bringing about complete mixing. It occurs in shallower lakes and is probably true for Lake Naivasha (Jenkin, 1936; Beadle, 1932) and for a number of smaller upland lakes investigated by Baxter et al. (1965).

Deep-seated stratification in which complete mixing is infrequent or absent is exemplified by Lake Bunyonyi in South West Uganda and by Lake Nkuguti a small crater lake near Kichwamba in western Uganda both of which lie in steep-sided depressions (Beadle, 1963, 1966). Even here, slight cooling, perhaps following rain or cloudy days, can set up convection currents. It should be remembered that a small change in temperature in warmer water means a large change of density, a temperature range of 24°–26°C in Lake Victoria being equal in this respect to a range of 6·0–12·6°C in temperate lakes. These convection currents may lead to a turnover, but these will be infrequent or at irregular intervals. The water of Lake Victoria is stratified in the open lake and the stratification is broken down seasonally. In the shallow bays there is diurnal stratification and a richer plankton. Very deep lakes like Tanganyika and Malawi and some crater lakes show persistent stratification and the large volume of water is unproductive. The breakdown of stratification whether seasonal or intermittent brings nutrients to the surface and affects the distribution and quantity of the phytoplankton. For a fuller discussion of seasonal succession of phytoplankton reference should be made to Hutchinson (1967) and for detail of stratification in East Africa to

1 *Staurodesmus convergens*	7 *Peridinium cinctum*
2 *Anabaenopsis circularis*	8 *Micrasterias crux-melitensis*
3 *Microcystis aeruginosa*	9 *Eudorina elegans*
4 *Synedra ulna*	10 *Ceratium hirundinella*
5 *Dinobryon sertularia*	11 *Staurastrum pingue*
6 *Melosira ambigua*	12 *Kirchneriella obesa*
	13 *Euglena acus*

Talling (1957a and b); 1963, 64, 65).

Not many fish are direct phytoplankton feeders, and some seem to behave differently in different waters. Most of the fish offered for sale in hotels in the inland parts of E. Africa belong to the genus *Tilapia*. One species *T. nilotica* is considered to be a plankton eater in lakes, but becomes a consumer of detritus in ponds. In Lake Victoria, its main diet seems to be the diatom *Melosira*, blue-green algae being undigested. In other lakes (Lake Edward, George, Rudolph) it eats the Myxophyceae (Fish, 1955b), and in Lake George they are known to be digested. *T. zillii*, on the other hand thrives on submerged water plants.

There is still a great deal to be discovered about the productivity of lakes in the tropics where a good supply of fish would be so valuable in supplying the protein required for a balanced diet.

Under the International Biological Programme a team has been at work on Lake George which produces a very large crop of fish. Here, the Tufmac fisheries utilise the *Tilapia* brought in by fishermen and send it in refrigerated vans to many parts of Uganda and Kenya. On the shores of all the freshwater lakes are fishing villages from which canoes ply, many of them now with outboard motors, bringing in fish of many kinds for local consumption. Many edible species are found in the larger swamp pools. Dried fish is much enjoyed, and in some areas it is an important industry especially where it can be exported in quantity to an industrial population such as that of the copper belt in Zambia.

Recently there has been increasing interest in the construction of fish ponds in East Africa (Bailey, 1966; Wurtz, 1961). Experiments in Zaire had reported annual production of 1000–2000 kg of fish per hectare per annum by this means. The people have been encouraged to make fish ponds near their villages and by 1965 there were estimated to be about 12 000 fishponds in Uganda alone ranging in size from 1 acre to half a hectare. These are stocked by the Fisheries Department with young fry of several species of *Tilapia*. Young *T. zillei* can live on plankton but, from the length of about 4 cm upwards, they require a plant diet which is fed in the form of chopped grass and leaves of cabbage and sweet potato. The other species of *Tilapia* do not require to be fed but can utilize the natural food of the pond. It has not always been easy to control the fish population of these ponds, but well managed ponds frequently produce 286–460 kg per hectare per annum and it is believed that, with expert management, this could be raised to 2300 kg per hectare per annum. This compares with about 100 kg per hectare per annum for highly productive natural freshwater lakes in Uganda.

Although the provision of this protein in the form of fish may help to combat malnutrition, the construction of fish ponds near villages is not without its dangers as they offer suitable habitats for the snail, intermediate hosts of both types of bilharzia (Berrie, 1966) and breeding places for several malaria-carrying mosquitoes. Fish ponds will prove a satisfactory means of increasing fish production only if developed under strictly controlled conditions which will minimise the risk of increasing human disease. There may therefore be an argument in favour of large scale, expertly managed fish farms rather than numerous small village ponds.

Some larger lakes have also been stocked with *Tilapia* and other fish. In Naivasha, for example, *Tilapia nigra* was introduced in 1925, and it multiplied so fast that a predator fish, the black bass a native of America, was brought in to curb its increase. Now, both fish are plentiful and provide food as well as sport for the angler. It was not so easy to introduce trout as they need water below 15°C for breeding and have therefore only been really successful in rivers and dams above 2300 m (Copley, 1948).

The lakes and industry

East Africa is fortunate in having a ready supply of salt provided by its salt lakes. To the north of Lake Edward in western Uganda is a small lake, Lake Katwe, which is the basis of a flourishing salt industry. It is not more than 1 m deep and lies in an old crater with no outlet and is fed by brine springs. Under the tropical sun the shallow water evaporates leaving a precipitate of salt. The salt workers wade out into the lake and cut off slabs of what looks like slightly purple ice. These are stocked on the shore, graded, crushed and sold in the markets throughout the country. Since the sources of supply are few and the demand for salt very great, the salt trade forms an important source of revenue for those parts of Uganda fortunate enough to have these natural deposits.

Special mention must also be made of Lake Magadi in Kenya (Plate 36) which, in times of drought, is a solid mass of alkali (Hill, 1964). The lake lies at about 660 m on the floor of the Rift Valley some 129 km from Nairobi and is fed by a group of hot saline springs. The water of the springs contains a 2 per cent solution of alkali as sodium carbonate and bicarbonate and 1 per cent of sodium chloride. As the water evaporates, crystals of alkali (trona) separate out of the saturated solution and, in dry years, form a crust on the surface which will bear the weight of a man. It is dredged out and forms the basis of the important soda-ash industry operated by the Magadi Soda Company, a subsidiary of Imperial Chemical Industries.

Except in times of drought, the trona is covered by a liquor which contains from 9–12 per cent of sodium chloride and 18 per cent of sodium carbonate. This is run into shallow ponds where it is evaporated till the liquid is nearly saturated with salt. It is then run into the final pans to a depth of 5–7·5 cm where further evaporation results in the alkali being deposited on the bottom while the salt separates out as crystals on top. The thin layer of salt is brushed up and taken to the finishing plant to be refined.

One would not expect any form of life in these strong brine solutions, yet the pink colour so characteristic of these brine ponds is due to bacteria including *Halobacterium cutirubrum* which can live and reproduce in concentrations of up to 30 per cent sodium chloride.

There is also a fish *Tilapia grahami* which lives in the lagoons near the marginal hot springs in water of 2 per cent salinity and at a temperature of 42°C close to the limits possible for living animal organisms.

The last industry to be mentioned is the one most closely connected with plant life. Diatoms are minute unicellular plants with thin silica walls, very delicately marked. In some lakes, Naivasha and Victoria, for example, they occur in great quantities at certain times of year. When the algae die, their silica walls fall into the mud where they accumulate. The early lakes of the Pleistocene Rift must have had a huge population of these minute organisms whose remains now form deposits of diatom earth or kieselguhr. This silica material is used for insulating and other purposes and extensive deposits are now being worked near Gilgil in Kenya. The main diatom is *Melosira* which still occurs in quantity in some of the lakes.

36   Lake Magadi, Kenya. Supplied by Magadi Soda Co.

Four

The vegetation of the sea coast

The vegetation of the coast may be divided roughly into three zones:
The coastal strip of land.
The shore, including the intertidal zone.
The mangrove swamps.

The last two are quite clearly defined, but the first is more difficult to delimit.

The coastal strip

Physiographically this consists of:

1 The coastal plain varying in width from 3–8 km and consisting of deposits of corals and sands with occasional true dune formations of limited extent. It lies generally below 30 m.

2 The foot plateau lying between 60–130 m based on marine shales, mudstones and limestone.

3 The coastal range of intermittent hills, best represented by the Shimba Hills in Kenya, and Pugu Hills in Tanzania.

4 Dry bushland, which begins about 32 km from the sea in Kenya.

Here we shall be concerned mainly with the coastal plain. Reference is made to the coastal range under forests and to bushland in the appropriate chapter.

The weather of the region is largely controlled by the monsoonal currents of the Indian Ocean. The South-east Monsoon brings the 'long rains' in April, May and June; the 'short rains' begin in October or November. As the long rains, bringing more than half the annual precipitation, come from the South, the coastal hills and the Uluguru, Nguru, Usumbara and Pare Mountains of Tanzania are effective in preventing the rain from reaching the back country so that cultivation is not possible 48 km from the coast except along the rivers. The annual average rainfall decreases from south to north, being about 1425 mm in the south

and only 1075 mm at Lamu. Further south as we pass into Tanzania the single-season rainfall of the 'trade winds' climate prevails. The temperature is high all the year, averaging 25°C from December to March, and 24–26°C between June and September.

The vegetation of the coast is closely bound up with its history. The coast was known to the ancient civilisations of Egypt, Phoenicia and Greece, and from AD 900 it was occupied by a succession of Arabian sultans who built a series of stone towns, including Gedi, and were interested in trading by sea especially in slaves, ivory and gold. From 1650 the slave trade was the most important economic factor, approximately 100 000 Africans being shipped each year to the plantations of America. During all this time there must have been considerable cultivation by Bantu and Nilotic tribes and, later, by the settlers to supply food to the large transient population. It is therefore unlikely that much of the natural vegetation is left.

There is now a highly variegated pattern of coastal hinterland between 6° and 11° South best described by Swynnerton's term 'Coastal debris'. Some attempt will now be made to describe aspects of this vegetation especially as it occurs on the seaward side on coral; but it will be appreciated that there is considerable variation in different parts of the coast. The information on the coastal strip is derived mainly from Moomaw (1960) and Birch (1963). Philips (1931) gives a brief description of the coastal region of Tanzania but, without personal knowledge, it is difficult to correlate his zones with those of Kenya. Many species appear to be different.

The littoral

This is the beach immediately above spring tide level and can be up to six metres wide. The seaward edge is briefly inundated twice a year at the highest spring tides, the landward side gets much spray, and both are subject to wind erosion. To seaward it carries *Ipomoea pes-caprae*, a plant of the Convolvulaceae, with stolons 4 m long. To landward is a dense cover of halophytic grasses 1·0–1·5 m high, succulent plants and occasional small, thick-leaved bushes. Inland, and beyond the reach of the highest tides, are species of *Grewia*, *Cadaba*, etc., forming a dense cover 1·0–1·5 m high, with *Cenchrus perinvolucratus* as the dominant grass.

Sand over coral

This is a much disturbed area where dicotyledonous trees of many species are mixed with houses, coconut plantations and doum palms.

Coral cliffs

These may be 7·6 m high and are subject to sea spray. Here there is a *Capparis*—evergreen shrub association, mostly small and stunted, with larger trees at the top including *Allophylus*, *Fagara*, and the baobab, and in some places *Mimusops fruticosa* within a few feet of the cliff edge.

Dunes over coral

In the area under consideration there is one example, at Nyali, but Moomaw (1960) records them north of Malindi and there is some in the Lamu district. Many of the strand species from the southern beaches occur along the foredunes including *Ipomoea pes-caprae* and on the second dune from high tide marks and in the interdunes there is a more stable vegetation of small trees and shrubs. In places the line of the beach is clearly marked by a line of *Casuarina* trees. There are many similar examples in Tanzania, but none has been studied in detail.

Coral and *breccia* vegetation

This occurs on the fossil coral reef where the soils are red humic ones and highly fertile, and Birch (1963) distinguishes three subtypes:

Deeply pot-holed coral is seen only in Shimoni forest, in the south. The forest is of tall clean-boled trees unbranched until two-thirds way up with a sparse shrub and floor cover. Among the main canopy trees are *Antiaris toxicaria*, *Chlorophora excelsa*, *Cussonia zimmermannii*, *Lecaniodiscus fraxinifolius*, and *Terminalia kilimandscharica*, with an under storey of smaller

trees and shrubs. Birch considers that there is little evidence that the high forest ever grew throughout the coral strip, but it may have been restricted to uneroded coral with high rainfall too rough for agricultural use.

A more eroded coral as at Jadini bears forest, but more commonly carries tall bushland of small trees up to 4·5 m.

The breccia proper is gently undulating and has fairly large pockets of shallow soil most of which have at some time been cultivated. It carries an open association which has been so much disturbed that it varies from *Lantana camara* scrub to an association of fire-resistant small trees. Common to the whole coral strip on the breccia are *Adansonia digitata*, *Ehretia petiolaris*, *Fagara chalybea*, *Grewia plagiophylla*, *Heeria mucronata*, but there are marked differences in species composition from south to north. The dominant grasses are *Heteropogon contortus*, *Hyperthelia dissoluta* and *Pennisetum polystachyon* with a variety of associated herbs and small shrubs. Probably the true coral reef is not more than 1 km wide and the breccia 1·5–3 km. As a whole it extends from Shimoni in the south to Mambrui in the north, a distance of some 190 km, it reaches a height of 26 m above datum at Kilifi, but is usually about 15 m above datum.

Moomaw (1960) describes a further type of vegetation growing on coral rag where there is considerable outcropping of coral and coral gravel covered with a thin mantle of sand which he called 'lowland dry forest'; it occurs in a small area north of the Mida Creek. *Combretum schumannii* and *Cassipourea euryoides* are the dominant trees with *Fagara* species. In this area burning and cultivation have not been carried on in recent history and Moomaw considers that this may have been the original type of forest on the coral. When it is destroyed coastal bush quickly dominates.

Once we leave the coral, there occurs a variety of forest types which have been described by Dale (1939). Birch points out that the main canopy trees of the coral forest (except *Lecaniodiscus*) are present in Dale's 'lowland evergreen rain forest' and in Moomaw's 'lowland rain forest', but most of the true rain forest species of trees and shrubs are lacking on the coral.

The intertidal zone

Until recently very little had been written about the plants which grow on the coral reef and between high and low tide level in the sublittoral zone. But owing to the work of Professor and Mrs W. E. Isaac we are now beginning to get a clearer picture of the marine vegetation of East Africa (Isaac, W. E., 1967, 1968, 1971).

Fringing coral reef is almost continuous except where it is replaced by mangroves. The coral platform slopes gently from shore to reef edge, dissected by deeper water channels and containing pools rarely more than a foot in depth. At low tide it is possible to walk through shallow water to the reef edge. Much of the rocky shore is overlaid with sand.

The algal flora of the coast varies greatly from place to place. There are three main groups: green algae (Chlorophyta), brown algae (Phaeophyta), and red algae (Rhodophyta). In general, the green algae are found in the upper tidal zone, where fresh water seeps in from the land and the rocks are exposed at times to rainfall. Here are to be found, among others, several species of the foliaceous genus *Ulva* which also extends well down the shore, the filamentous *Chaetomorpha*, and *Struvea anastomosans* (Fig. 4.1a) and *Boodlea composita*, the latter two forming densely tangled clumps. At the pool edges are species of *Caulerpa* especially *C. sertularoides* (Fig. 4.1g). These coenocytic algae have a stoloniferous portion from which arise delicate upright branchlets. The flattened, pointed thallus of *Halimeda* (Fig. 4.1e) is also common in the pools. The massive green alga *Codium* often dichotomously branched, is found washed up on the shore and, though widespread, is only locally common or abundant.

Among the brown algae of the coral reef Fucales are prominent in some parts although *Fucus* itself is not present being a genus of temperate waters. Probably the commonest larger species belong to the genera *Sargassum* (Fig. 4.2e) and *Turbinaria* (Fig. 4.2f). They are present in the pools and again are to be found among the seaweeds washed up on the shore. Bushy small plants of *Cystoseira myrica*, often iridescent, with a branched body arising from a basal disc and bearing small inflated vesicles, is common in pools at all levels.

Of the smaller brown algae, *Padina gymnospora* is widespread and *P. commersonii* (Fig. 4.1d) in places dominant especially in the larger pools. *Dictyota* spp. (Fig. 4.1f) are sometimes common while the inflated bladders of *Colpomenia sinuosa* (Fig. 4.1c) are more often seen towards low tide level.

The red algae tend to occur at rather deeper levels where most of the light which penetrates is red, and are not usually dominant in the intertidal area. Among the more common ones round the edges of shallow rock pools are species of *Gracilaria* and *Laurencia* (Fig. 4.1h) with various members of the Corallinaceae, including the beautiful, plumed *Corallina mauritiana*, and *Jania capillaceae* (Fig. 4.1b) both usually epi-

phytic on larger brown algae or on *Cymadocea*.

Other common Rhodophyta are species of *Galaxaura*, *Amphiroa* and *Hypnea* while among the algae growing below tide level and often washed up on the shore are *Halymenia* spp. and *Chondrococcus hornemannii*.

A feature of the intertidal zone is the presence of marine angiosperms often referred to as 'sea grasses' though only *Zostera* and *Halodula* are grass-like. They occur mainly where rock or coral is covered with sand and extend from low water of neap tides out beyond the reef into deep water. After stormy weather they are often cast up in heaps on the shore.

Seven genera are represented on the Kenya coast and, with the exception of *Zostera* which occurs only in a few isolated places, they are all confined to the tropics. The commonest are *Cymadocea ciliata* (Fig. 4.2a) and *Thalassia hemprichii* (Fig. 4.2b), plants rooted in the mud with erect stems and ribbon-shaped leaves in two rows. They are dioecious and have very inconspicuous flowers. For a full description of the Kenya genera reference should be made to Isaac, W. E. (1968) and Isaac, F. M. (1968).

SOME COASTAL LICHENS
(Dr. T. D. V. Swinscow)

ON TREES AND ROCKS	*Dirinaria* spp.
	Pyxine spp.
	Parmelia reticulata
	Roccella linearis
	Roccella spp.
MANGROVE SWAMPS	*Anaptychia boryi*
	Candelaria concolor
	Parmelia austrosiniensis
	P. reticulata
	Ramalina farinacea
	Ramalina spp.

Mangrove forest (Plate 37)

Mangrove forest is found on tropical and subtropical coasts and is widely distributed, but it is most luxuriantly developed in parts of the Malayan peninsula. Species of mangrove are found from 27°N. in Egypt to almost 30°S. near Durban; but mangrove forest is intermittent throughout this range and much of it is

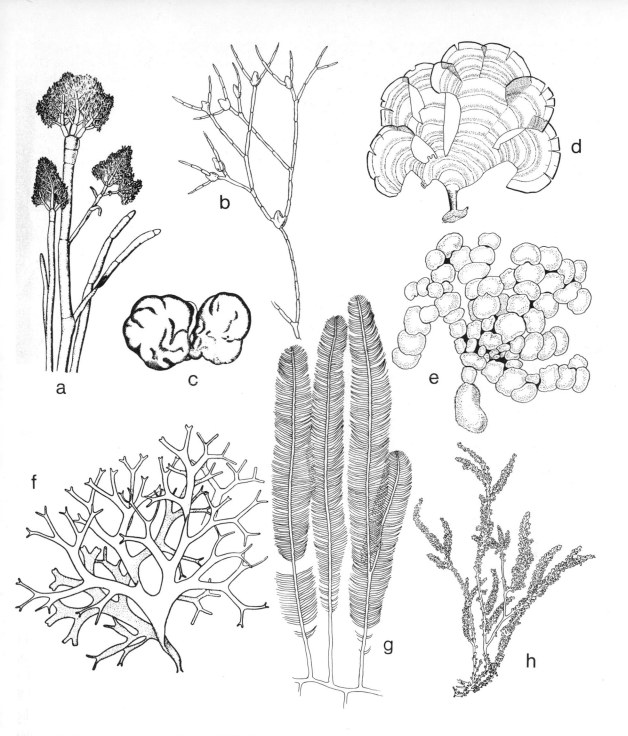

Fig. 4.1 (a) *Struvea anastomosans* (Taylor, 1960). (b) *Jania capillacea* (Taylor, 1960). (c) *Colpomenia sinuosa*. (Taylor, 1960). (d) *Padina commersonii*. (After P. Hall, in W. E. and F. M. Isaac). (e) *Halimeda macroloba*. (After P. Hall, in W. E. and F. M. Isaac). (f) *Dictyota divaricata*. (Taylor, 1960). (g) *Caulerpa sertularoides* f. *longiseta*. (Taylor). (h) *Laurencia papillosa*. (Taylor, 1960).

37 Mangrove swamp with *Rhizophora mucronata*. (Fergus Wilson)

poorly developed. The best East African mangrove forests occur round the mouth of the Rufiji River in Tanzania and the mouth of the Zambesi near Queli-more in Mozambique, and well-grown forest rivals in stature and general appearance the high forests of the land. The ecology of mangroves in the southern section is described by MacNae (1963) and MacNae and Kalk (1962). Some of the off-shore islands, such as Lamu, also have well-developed forest.

Mangroves grow in sheltered estuaries and on the exposed coast but, in the latter case, only if the force of the waves is broken by coral reefs or islands. At high tides only the crowns of the trees emerge, while at low water their unusual roots are visible. Floristically, the species-rich forests of the Indian and Pacific Oceans must be distinguished from those, much poorer in species, of the Atlantic Ocean: the genera are the same, but the species differ. Of the twenty species listed by Walter (1964) for the eastern mangrove swamps, only

Java has all while East Africa has only twelve. Of these the commonest are:

Rhizophora mucronata	Rhizophoraceae
Ceriops tagal	,,
Brugiera gymnorrhiza	,,
Avicennia marina	Avicenniaceae (formerly in Verbenaceae)
Sonneratia alba	Sonneratiaceae
Heritiera littoralis	Sterculiaceae

A few other species are commonly included in a list of plants from the mangrove community, although they are not true swamp dwellers but are plants of a coastal situation and not found in inland sites. Of these, *Terminalia boivinii* and *Casuarina equisetifolia* occur on sand and could be classed as strand plants; *Lumnitzera racemosa* occurs on the brackish fringes of mangrove and on dunes above the level of normal spring tides; *Pemphis acidula* is a white-flowered shrub growing usually on otherwise barren coral outcrops covered only at spring tides.

Each tree species of the mangrove forest grows best under slightly different conditions which depend on factors such as the amount of water in the mud, the salinity and the ability of the plant to tolerate shade. This means that the various species are not mingled together in a haphazard way, but occur in fairly distinct zonation. The pioneer species at the seaward edge help to stabilise the site and make it more suitable for other species which will invade and replace them, and this

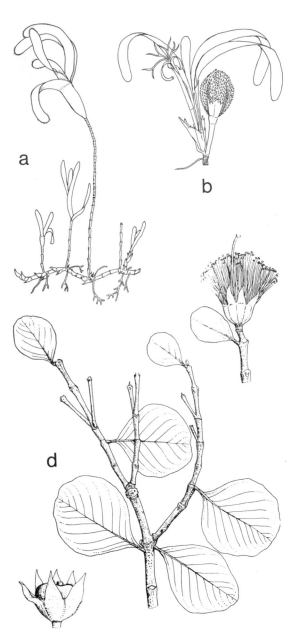
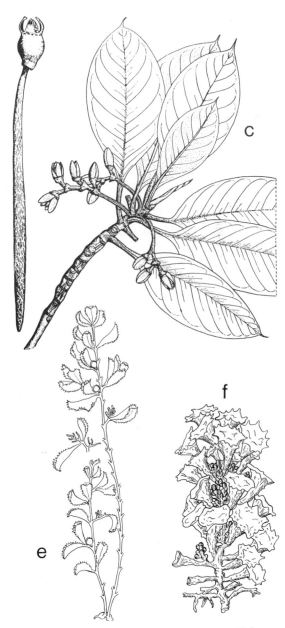

Fig. 4.2 (a) *Cymadocea ciliata*. (After F. M. Isaac). (b) *Thallasia hemprichii*, with female flower and fruit. (After F. M. Isaac). (c) *Rhizophora mucronata*, flowering branch and germinating seed. (M. E. Church). (d) *Sonneratia alba*. (Dale and Greenway). (e) *Sargassum duplicatum*. (After F. M. Isaac). (f) *Turbinaria ornata*. (F. Anderson in W. E. and F. M. Isaac)

process of invasion and replacement continues until a climax community of forest is formed. We have here an excellent example of an ecological succession which requires fifty or more years to reach its climax. Doubtless the details of the succession vary a little from place to place along the coast, depending on the type of mud and the rate of its deposition.

Near Tanga (Walter, 1964) (Fig. 4.3), the pioneer plant on the open coast is Sonneratia (Fig. 4.2d), but on mudbanks, creeks and delta tributaries *Avicennia* is more often the pioneer. *Sonneratia* does not appear to require fresh water at any stage, and it is sometimes found growing in pockets of silt in the coral reef. Once the pioneer plant has appeared, the deposition of silt is greatly accelerated by the retarding action of the vegetation and, as new soil and shade conditions develop, other genera come in and *Sonneratia* is eventually replaced by Rhizophora (Fig. 4.2c). Assuming new deposition of mud occurs on the seaward side, there will always be a zone of *Sonneratia* between the *Rhizophora* and the open water. The *Rhizophora* grows up to a pure stand and, when mature, can produce straight, cylindrical, clean stems up to 25 m in height.

A mixture of *Ceriops* and *Brugiera* next invades the *Rhizophora* and occupies the landward side of the swamp. By this time, the soil level has built up above the reach of all but the highest tides and the *Rhizophora* dies out. At the landward side of the climax forest, mangrove species drop out one by one as the salinity becomes unsuitable for them and/or competition from non-mangrove plants becomes too great. Where freshwater is lacking, salt-marsh tends to develop behind the mangroves and here *Avicennia* is most prominent as it is tolerant of drought and high salinity. To the landward are palms and coastal forest and between this and the salt-marsh is often a sandy plain devoid of vegetation and only flooded twice a year at the time of the equinoctial spring tides. Evaporation, following this, causes a rise in salt concentration, but for the rest of the year the salt is washed out by rain. Under these conditions, it appears that neither halophytes nor non-halophytes can survive.

Walter and Steiner (1936) give data on the concentrations of sodium chloride in the mangrove swamp soil solution and in the sap of various parts of the mangroves. They attempt to relate these data to the zonation of the mangroves and to the movement of water into the root system. The concentration of sodium chloride in the soil solution increases landwards, and, without a corresponding increase in the solute concentration inside the plant, this would reduce the rate of water movement into the plant. It appears from their results, that there is an increase in the concentration of sap solutes, mainly perhaps as sugars and organic acids, so that the energy gradient, which causes water movement into the plant, is comparable with that recorded in rain forest trees. They claim that the mangroves, in spite of the high sodium chloride content of the soil solution, are no worse off in respect of water intake than trees of non-saline soils. The intake and accumulation of salt, however, poses a problem which some of the mangroves, such as *Avicennia*, apparently solve by secreting the salt through special glands so that, on dry, windy days, the undersides of the leaves become covered with salt crystals. The exuded salts are 90 per cent NaCl and 4 per cent KCl, which more or less corresponds with that of seawater. In a day of nine hours, Walter and Steiner found that $0 \cdot 2–3 \cdot 5$ mg/ 10 cm^2 were excreted. This is little compared with the amount of water transpired, so that it appears that the root takes up a very dilute salt solution. There is much experimental evidence in favour of discrimination by roots against some solutes. Those mangroves which do not have salt-secreting glands can, presumably, get rid of salt only through old fallen leaves.

The mangrove roots are specialised to secure adequate oxygen and achorage in the soft, waterlogged muds (Graham, 1929). In *Rhizophora*, these are in the form of prop roots which arise from the stem and support the trees and help, perhaps, to aerate the root

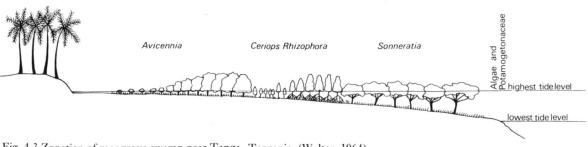

Fig. 4.3 Zonation of mangrove swamp near Tanga, Tanzania. (Walter, 1964)

system. In all the other species, including the two remaining Rhizophoraceae, the roots send up erect, aerating roots or pneumatophores. In *Sonneratia* and *Avicennia* they arise vertically from a horizontal root 20 or 30 cm below the ground. In places where the plants are not too numerous, a row of pneumatophores arising from the same horizontal root can often be seen radiating out from the base of the tree.

Ceriops and *Brugiera* have knee roots which form when a horizontal root grows gradually upwards until it breaks through the soil or mud surface. Soon, it turns sharply downwards again, forming a bent knee which later becomes woody. Pneumatophores are very spongy just below the ground level and often have lenticels on their aerial parts. These specially constructed roots, which occur also in the freshwater inland swamp trees, *Mitragyna* and *Uapaca*, provide an outstanding example of convergent evolution, that is, similarities which have evolved between taxonomically distantly related forms. Evolution usually leads to diversification but sometimes certain habitats have elements in common which produce morphological similarities between distantly related organisms.

The land of the mangrove forests is shifting and laced with intricate channels. Soil and silt are carried away from one place and built up elsewhere. In this unstable environment it is not surprising to find curious specialisations which assist the establishment of seedlings. Thus, *Rhizophora*, *Brugiera* and *Ceriops* are viviparous, that is, the seed germinates on the parent tree (Fig. 4.2c). In the Rhizophoraceae, one seed develops from each fertilized flower and its root and hypocotyl project out of the fruit like a green finger or dart which sometimes is slightly thickened at the end. In *Brugiera* this projection may be only a few centimetres in length, while in some species of *Rhizophora* it occasionally achieves a length of 60 cm, the ultimate size being typical of the species. These seeds, with their projections, hang on the trees and sway in the wind. Later the seedling drops off the tree and falls like a dart to lodge upright in the soft mud into which it penetrates to some depth. Roots grow from the end of the dart and the seedling becomes established quite quickly. Upward growth, carrying the seedling leaves above the level of the flood tide, is very rapid and as much as 60 cm growth in 24 hours has been observed in *Rhizophora*. Later, the growth rate falls off. The seedlings of these trees will also float and, on coming to rest on soft mud, take root and curve upwards becoming vertical in a very short time. The other genera also have fruits which are highly suited to dispersal by floating, but are otherwise less specialised than the Rhizophoraceae.

The mangrove forests are of considerable commercial value and they have been exploited almost constantly since the Arabs secured a footing on the East African coast. Although we are usually told that they came to East Africa for slaves and ivory, it may well have been a more pressing necessity to find wood for the ceilings and roofs of their houses which sent them in this direction, for their own land is virtually treeless. These building poles which are cut from the mangrove forests are known in Swahili as 'boriti'. Arab and Persian dhows loaded with tiles, carpets and spices, sailed south with the north-eastern trade winds in December and January and many of these dhows put into the Rufiji delta on the Tanzanian coast to load mangrove poles. A large dhow—some were from 200 to 275 tons burden—was able to carry as many as 4,000 long poles. The loaded dhows then made their way slowly to Zanzibar and there waited for the south-eastern monsoon which carried them back to ports on the Red Sea or in the Persian Gulf, and in this way they. had an extremely profitable trade, for the boriti fetched good prices.

For some years the Tanzanian Forest Department, and more recently the Kenyan, endeavoured to increase the yield from the mangrove swamps by planting and by a management which would hasten the natural succession, and it has been claimed that it was, in some places at least, possible to get the succession through to the *Rhizophora* stage in twenty instead of fifty years. However, the policy of attempted management of the swamp forests was not long pursued, although the boriti trade was for many years important in the coastal economy and in 1956 some £80 000 worth of boriti were cut from the mangrove swamps of the Kenya coast alone. But from 1956 onwards the trade in boriti declined because oil companies of the Persian Gulf—which had recently been big importers of poles—became sufficiently wealthy to afford the more convenient metal tubing for construction and scaffolding. Fortunately for parts of the coast, especially those areas such as Lamu where the economy depended on the export of boriti, it has been possible to develop the export of charcoal made from the mangrove woods, Hunt (1962), but the Rufiji trade has virtually ceased. The Persian Gulf peoples formerly imported vast quantities of Somalia charcoal, where there was a large industry producing about 40 000 tons per year, but the Somalia production depended on fuel from land forests which have been depleted, and importers in the Persian Gulf are now turning more to East Africa.

The bark of mangroves can be used for tanning leather, but much of the bark produces a red-coloured tan which makes marketing a problem because most leather tanners prefer their tans colourless. Nevertheless, mangrove bark was at one time in great demand for tanning and for many years the Rufiji and other mangrove forests were exploited as much for bark as for poles. The species most sought was *Rhizophora*, and it is worth noting that this is one of the few examples in East Africa of two distinct products being obtained from the felling of a single tree. Various attempts are now underway to obtain fresh markets for the large amount of bark available. Egypt is probably the best customer, but some bark reaches also India, Europe and the U.S.A.

Five

High mountain vegetation

Distribution and origin of the flora

The equatorial African mountains may be envisaged as an archipelago of small islands of cool climate separated by extensive warm lowlands. The lowlands range from 1000–2000 m and many of the mountains reach into altitudes between 3000 and 6000 m. The East African equatorial group includes three of the highest mountains of Africa, namely, Mt Kilimanjaro (5895 m), Mt Kenya (5200 m) and the Ruwenzori range, whose highest peak is Margherita (5108 m). The cool climate of these mountains has a diurnal range which carries the night temperatures below freezing. It might be expected that under these harsh thermal conditions the vegetation would be stunted, but some of the most characteristic plants of these high mountains—such as Dendrosenecio and Lobelia—are tall and luxuriant (Plate 38). It is only close to the thermal limit of plant growth, just above 5000 m, that the vegetation is reduced to a few vascular plants with bryophytes and lichens.

Nevertheless the high mountain vegetation is distinct from the lowland vegetation. It has some remarkable growth-forms, a very high percentage of endemic species, and it includes only 280 taxa, very few of which extend into the surrounding lowlands. On account of its distinctiveness Hauman (1933, 1955) considered that these high mountains in tropical eastern Africa should constitute a phytogeographical region distinct from the surrounding Guinean and Sudano-Zambesian regions. He termed this region, which is interpreted as including eastern Zaire, Uganda, Kenya, Tanzania, Rwanda and Burundi, the Afroalpine region. However, some taxa of the eastern African Afroalpine region occur also at high altitudes in Ethiopia, west tropical Africa and Malawi. Hedberg (1965) considers that the Afroalpine region could be extended to include the high mountains of Ethiopia because of the rich representation of eastern African

taxa there. But, because of the absence on the Ethiopian mountains of all *Dendrosenecio* and almost all shrubby *Alchemilla*, and the presence of such genera as *Rosa* and *Primula*, he suggests that the Ethiopian high mountains might be treated better as a separate subregion.

Of the 278 taxa of vascular plants in the Afroalpine flora some 81 per cent are, according to Hedberg (1961a) endemic to the high mountains of eastern Africa; that is to say they occur nowhere else. This is a remarkably high percentage of endemism, similar to that recorded in the very isolated floras of the Hawaiian Islands and Juan Fernandes Island. Endemic species may either have arisen after the isolation of the habitat and have had no opportunity to become more widely distributed or they may be relics of forms originally more widely distributed in the region but elsewhere extinct. At the generic level, the endemism of the Afroalpine flora drops to about 19 per cent and the majority of the Afroalpine genera have most of their closest living relatives elsewhere. Thus, according to the region where these closest living relatives are most abundant we may distinguish the so-called genetic elements shown in Table 5.1. The variety of these elements indicates clearly the diverse origin of the flora and would seem to imply that there have been relatively easy land routes giving these plants access to eastern Africa. It is customary to admit the possibility that some of the genera may have dispersed in Africa before the complete break up of the Gondwanaland continent. However, it has been claimed that, by the early Cretaceous, when the Angiosperms were spreading substantially, fragmentation of the continent had already outlined Africa in much its present form. As well as the problem of dispersal of the flora there is the problem of its survival. Where in eastern Africa could it have survived prior to the origin of some of the present mountains in and around Tertiary times?

Some of the Afroalpine flora may have come into eastern Africa since Tertiary times by long-distance dispersal by birds and by wind. Birds are able to carry seeds and other parts of plants capable of vegetative reproduction externally attached, for example, to feathers, feet and beaks. Wind is undoubtedly an important long-distance disperser, but it is likely to be effective only for very small diaspores, namely, with no diameter, in any direction, over 0·2 mm, a density of not more than about 2·6 g/cm^3 and the largest diameter under 0·6 mm. Most plant species have diaspores outside these limits, the principal exceptions in the Afroalpine flora being species of the Orchidaceae and the Campanulaceae; the Afroalpine Lobelias belong to this family. Lobelias, similar to the columnar giants of the Afroalpine flora are, it seems, found in parts of Asia and elsewhere (Croizat, 1952; Good, 1953). Most spores of Afroalpine Pteridophytes are suitable for long-distance wind dispersal and their arrival on the eastern African mountains by this means seems to be borne out by the wide distribution of the same species outside eastern Africa, namely, species of the clubmoss *Lycopodium*, and of the ferns *Aleuritopteris, Asplenium, Cystopteris, Dryopteris, Elaphoglossum, Pleopeltis, Polypodium* and *Polystichum*.

Dispersal of species between the main mountains in eastern Africa may have been facilitated during several phases of the Pleistocene period when low temperatures caused a depression of the vegetation belts by, perhaps, as much as 100 m (Moreau, 1963a). He claims that a temperature reduction of 5°C would have brought the lower limit of upland vegetation down from 1500 m to 700 m thereby causing an enormous extension of the upland biome, which he suggests 'would have occupied a continuous block from Ethiopia to the Cape with an extension to the Cameroons, comprising by far the greater part of tropical Africa (except West Africa) as well as southern Africa. At the

38   *Dendrosenecio adnivalis, Lobelia wollastonii*, Ruwenzori, 4000 m. (Morrison)

Table 5.1   Floral elements on the paramos or Afroalpine belt (from Hedberg 1965)

Group		No. of taxa	%
1	Endemic Afroalpine element	52	19
2	Endemic Afromontane element	35	13
3	South African element	17	6
4	Cape element	10	4
5	South—hemispheric temperate element	11	4
6	North—hemispheric temperate element	43	15
7	Mediterranean element	18	6
8	Himalayan element	5	2
9	Pantemperate element	87	31
	Total	278	100

same time the strictly lowland biomes, comprised of species that today do not enter areas above 1500 m would, outside West Africa, have been confined to a coastal rim and to one or two isolated inland areas, namely the middle of the Congo basin and most of the Sudan'. This hypothesis implies that the upland vegetation, which exists today on more or less island refugia on the high mountains, would in some part of the Pleistocene past have been more typical of tropical Africa than the present lowland vegetation. Moreau (1963a) believes this idea is supported by biogeographical evidence, namely, that the upland plants as well as the upland birds of the Cameroun highlands are largely those of the Eastern African highlands though separated by some 1930 km of unbroken lowland. Thus, he claims free interchange between the two highland areas must have been possible at some stage in the past and perhaps at more than one stage. He believes that a reduction of temperature amounting to less than 5°C would not have produced the necessary connection. Keay (1955) points out that any theory attempting to explain the similarities of the floras of the eastern African mountains and the mountains and highlands of the Camerouns must also explain why a rather similar flora is also found on some of the islands, notably Fernando Po, of the Gulf of Guinea which lies about 81 km off the Cameroun coast. If at the height of the Pleistocene glaciations, the eustatic fall in sea level had amounted to more than 100 m, which is sometimes claimed, this would have reduced the extent of Fernando Po's isolation.

Climate

It is impossible in this short account to describe adequately the macro- and microclimates of the high altitudes. The subject is extensive and complex for, despite fundamental similarities, there are substantial differences between the East African mountains. We do no more than give some known salient points and readers requiring more information must consult the memoirs of Hedberg (1964a) and Geiger (1965).

Four features are outstanding. First, the low temperatures at altitudes above 3400 m; second, the large diurnal temperature range swinging back and forth across 0°C; third, the immense annual variation in total rainfall on some of the mountains; fourth, the approximately twelve-hour day. Integrated into these fundamental factors are many other local variations of rainfall, frequency of clouds and mist, degree of slope, soil types, and biotic influences.

The fall in temperature is approximately 1·6°C for each 300 m increase in altitude, and this gives a mean annual temperature of −1·0°C around 5000 m which is the approximate limit of plant growth. We should not expect a uniform decrease of temperature with altitude on the actual mountains. These are often topographically varied and provide habitats with contrasting exposures. Differences in the temperature regimes of valley slopes and valley floors may often exceed the differences between neighbouring belts of vegetation. The mountain habitat usually constitutes a climatic and vegetational mosaic with highly contrasting temperature zones in close juxtaposition. This is why it is difficult to generalise about the vegetation in terms of belts and zones, and why schemes in such terms are so artificial. More important perhaps than the mean annual, monthly, and daily temperatures, is the relatively large diurnal range. This is caused by large quantities of incoming shortwave radiation in daytime and large quantities of outgoing longwave radiation at night. Close to 4000 m on Mt Kenya, the diurnal range is just over 16°C; the maximum temperature of about 12°C is reached usually in the later part of the morning and falls to a minimum, of about −5°C,

39 *Dendrosenecio adnivalis*, Ruwenzori, 4000 m. (Morrison)

close to 05.00 hours in the early morning. After sunrise, especially on eastern facing slopes, the atmospheric saturation deficit will—except in mist—be great enough to establish a substantial vapour pressure gradient within the saturated air spaces in the leaf. The plant could have great difficulty in meeting a water deficit since, at this time of the day, the upper soil water will still be frozen.

Growth forms

In these high-altitude climates some plants have growth forms which are believed to assist their survival. Some of these shield the plant from very low temperatures; others are xeromorphic adaptations preventing excessive water vapour loss; and others give protection from intense solar radiation.

The giant rosettes or night buds of the Dendrosenecios and some of the Lobelias are leaf arrangements which protect the shoot apices from temperatures at and below 0°C, while the cork and sheaths of persistent dead leaves on the trunk and branches insulate the internal water-storage and water-conducting tissues (Plate 39). These internal aqueous tissues of the pith and cortex are probably important in supplying water in the early morning when the upper soil water is frozen, although the ambient air temperature will have risen high enough to create considerable evaporative demand. The temperature inside the trunk probably never falls to freezing even though the ambient atmosphere may, for part of the night, be as low as —5°C. Not all Dendrosenecios have a layer of persistent leaves; it is missing in *Senecio barbatipes*, although this grows at very high altitudes 3650–4300 m. Its trunk is, however, protected by a thick layer of cork. In the daytime the outermost leaves of the Dendrosenecios rosette spread outwards and flatten, but at night most of the leaves bend inwards towards the centre and cover the delicate primordia. Hedberg (1964a) compares the whole rosette, which is sometimes a metre in diameter, to a huge night bud giving the shoot apex excellent thermal insulation. On a frosty morning, after a clear night, he recorded that the temperature inside such a rosette (at the surface of the dense central cone) was $+1 \cdot 5°C$ while the temperature among the outer leaves was $-4°C$. Inside the central cone, the temperature surges of the environment will certainly be more effectively buffered and the shoot apex and young leaf primordia will be well-protected against frost.

The giant Lobelias which, like the Dendrosenecios, occur in the paramos belt and in the ericaceous belt, have similar night buds. Some of these, of the *L. deckenii* group (Plate 40) have further specialisation. Liquid is secreted at the base of the fleshy leaves and trapped between these to a depth of about 10 cm when the rosette is closed. Each night the surface of this liquid freezes to a depth of between $0 \cdot 5 – 1 \cdot 0$ cm, but below this it remains unfrozen, and Hedberg (1964a) points out that, since water has its greatest density at $+4°C$, the interior of the rosette and the shoot apex will therefore remain a couple of degrees centigrade above freezing.

Very similar stem morphology and giant rosette night buds are known in the species of *Espeletia* (Compositae) which grow in similar climates in the South American paramos belt.

The grass tussock has a high adaptive value due in part to the good temperature and moisture insulation given by the leaf sheaths, and also by the dead and decaying leaf and culm bases to apices in the central part of the tussock. Hedberg (1964a) found that, before sunrise on a frosty morning, the outermost leaves of a tussock of *Festuca pilgeri* spp. *pilgeri* had a temperature of $-5°C$. Among the innovation shoots, in the central part of the tussock, the temperature was, at the same time, $+2 \cdot 5°C$ and in the upper part of the rhisosphere it was $3 \cdot 0°C$. Also, he observed that, after several days of dry weather, the central interior of such a tussock had free moisture. Thus, it is probable that the tussock protects and buffers some of the growing apices from harmful temperature and moisture levels and variations. Xeromorphic adaptations include woolly pubescence, as on the leaves of *Helichrysum newii* and *Senecio brassica*, the filiform leaves of the grasses, and the ericoid leaves of *Stoebe* and the Ericaceae. The ericoid leaf is small, more or less cylindrical, with revolute leaf margins and the stomata lie in a central groove so that escape of water vapour is retarded. A leaf of this shape and size probably also gives rapid heat exchange with the surrounding

atmosphere and this would avert damaging heating of the mesophyll during daytime. Thus, the detailed anatomy of the leaf and its overall morphology combine to lower water vapour loss.

Strong solar radiation coupled with low night temperatures retards elongation of the internodes so

40 *Lobelia deckenii*. (Morrison)

that cushion and rosette plants result. Some ordinary plants, such as the ragwort (*Senecio jacobea*), will develop rosettes if they are exposed to powerful illumination during the day and put into ice-chests at night. Many of the examples of acaulescent rosette plants in the high African mountain flora have the habit genetically fixed and do not lose the habit if grown at low altitudes. Examples are *Carduus chamoecephalus*, *Haplosciadium abyssinicum*, *Dianthoseris schimperi*, *Oreophyton falcatum*, *Ranunculus cryptanthus*, and *Wahlenbergia pusilla*. It is not clear if the acaulescent rosette habit confers physiological or ecological advantage. It is true that such plants will probably enjoy higher temperature close to the soil surface during the day, but at night the soil surface is frozen and they would then seem to be very unfavourably placed.

Finally, we ought to mention the curious life-form known as 'the solifluction floater' which occurs among a few cryptogamic plants. Solifluction soils are unstable soils where alternate thawing and freezing causes the soil surface to heave and move in various ways. This phenomenon is commonest on Mts Elgon, Kenya and Kilimanjaro, where the diurnal temperature range is very great, Zeuner (1950). It is very difficult for any plant, especially seedlings, to survive on solifluction soils because before they can become anchored they are pulled out. One way round the problem is for the plant to ride on top of the soil. For example, Hedberg (1964a) describes finding moss balls of species of *Grimmia*. These had diameters of between 1 and 7 cm and the central part of each ball was found on dissection to contain a core of soil from which the moss stems radiated. Hedberg suggested that the balls had arisen by erosion, the margin of a moss carpet being broken up by needle-ice formation. Afterwards the pieces had been lying on top of the restless soil surface, being lifted by needle ice at night, to subside again in daytime to a different position. By these perpetual movements they were kept shifting around, so that the moss could grow equally well on all sides. In this way some of them had become almost perfectly globular.

Agnew and Hedberg (1969) draw attention to the number of Afroalpine plants whch have developed geocarpy and bury their fruits while still attached. Among strongly geocarpic plants are *Haplocarpha rueppellii*, *Haplosciadium abyssinicum*, *Limosella africana*, *L. micrantha*, *Ranunculus cryptanthus*, *R. oreophytus*, *R. stagnalis*. Twelve other species are 'depositers' where the fruit is brought into contact with the soil but not buried.

As well as obvious growth-form adaptations, some plants at high altitudes probably obtain immunity from temperatures at and below freezing by converting their insoluble food reserves, such as starch, to soluble sugars. Cells with a high sugar concentration, will be more difficult to freeze. Pearsall (1950) reports that Scott Russell had verified the existence of high sugar concentrations in arctic plants collected on Jan Meyen Island and in the Karakorum mountains. No research has yet been done to show if temperatures approaching the freezing point induce conversion of food reserves to sugars in the plants of the high East African mountains.

Vegetation belts and zones

There are seven main types of vegetation of the eastern African mountains, namely, ericaceous woodland and wooded grassland, *Dendrosenecio* woodland and wooded grassland, tussock grassland, *Helichrysum* scrub, and swamp or mire vegetation. It is difficult to generalise about the distribution of these types because there are substantial climatic and biotic differences between the several mountains. Further, the climate of the cardinal faces of the mountains are different and there are considerable microclimatic variations. Hedberg (1951) has attempted to define the distribution of mountain vegetation in terms of belts and zones (Fig. 5.1). His scheme is founded mainly on the physiognomy of the vegetation, using the dominant plant communities (Hedberg, 1955). A belt is defined as an 'altitudinal region' which can be traced on all (or most) mountains of sufficient height in a definite part of the world as, for instance, the upland forest belt on the East African mountains, though the zones representing it on each mountain may differ as to number and appearance. Hedberg describes three vegetational limits which recur on all the mountains investigated. They are the lower limit of the upland against the savanna or steppe; the limit between the broad-leaved or coniferous upland forest and the ericaceous belt; the upper limit of the continuous ericaceous vegetation. He accordingly distinguishes three vegetation belts occurring on all the mountains: upland forest belt; ericaceous belt; alpine belt. The Afroalpine region of the mountains includes principally the ericaceous and the alpine belt, although a few of the Afroalpine taxa listed by Hedberg (1957) are found as low as 1000 m.

On the least disturbed and wettest mountains the vegetation *is* in broad belts. For example, on the Virunga volcanoes and the Ruwenzori range there is well-grown ericaceous woodland in an irregular girdle

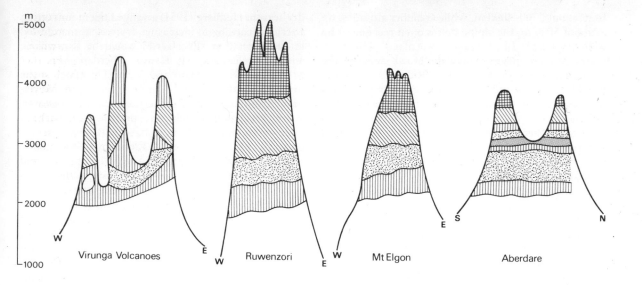

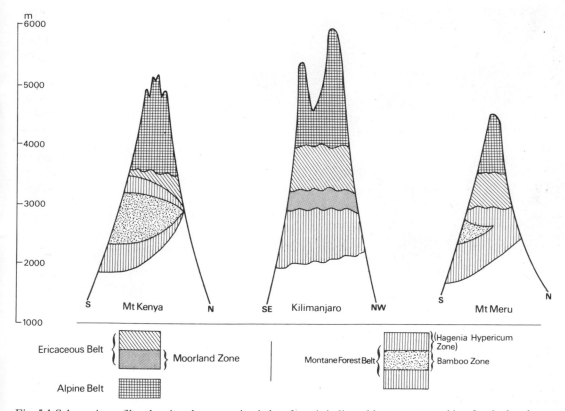

Fig. 5.1 Schematic profiles showing the vegetation belts of
the mountains investigated. The wettest side of each
mountain is turned to the left; the letters at their lower
parts denote the quarters (W.–E., S.–N., etc.). Where the
upland forest belt is differentiated into distinct zones this
is indicated by separate marking for the bamboo zone. The
zone next below is then an upland rain-forest zone, the one
next above a *Hagenia—Hypericum* zone. Only the vertical
distances are drawn to scale. (Hedberg, 1951)

143

from about 2700–4000 m, while at higher altitudes, up to about 5000 m, the alpine belt is often recognised by well-developed *Dendrosenecio* woodland. However, there are many places where the broad sweep of the vegetation belts is interrupted. Ridges and valleys provide edaphic and microclimatic contrasts which frequently cause vegetation boundaries to be complex, but the factors determining them are often obscure. Sometimes there are marked departures from the ideal zonation such as the impressive exception which has been recorded on a steep east-west ridge connecting several peaks of the Virunga range. *Senecio adnivalis* woodland, which is considered part of the alpine belt vegetation, extends down the south face of this ridge, while ericaceous woodland of *Philippia johnstonii* covers the northern face; the top of the ridge is about 6 m in width. The topography of the two sides is similar and the difference in vegetation can be attributed probably to climate, the northern face being more exposed to desiccation during the June–August dry season. Hedberg (1951) arranged the mountains in a series according to increasing dryness, as manifested in the vegetation. That series would be Ruwenzori, Virunga volcanoes, Mt Kenya, Aberdare Mts, and Mts Elgon, Kilimanjaro and Meru. On many of the drier mountains, drought and/or grazing and/or fire have profoundly modified ericaceous and *Dendrosenecio* woodlands which may previously have been more continuous and these have been replaced extensively, for example, on Mt Elgon, by tussock grasslands or moorland (Plate 41). Even on the wet and relatively undisturbed Ruwenzori range, the broad zonation of Hedberg's scheme is often interrupted by large areas of *Helichrysum* scrub which is widespread there in the alpine belt because of the abundance of rocky slopes. Indeed, the alpine belt is *so* variable on the different mountains that the unsatisfactory term 'alpine' was used by Hedberg who found no floristic name which would suit all the mountains. Alpine suggests, *too* strongly, affinities with the Central

41   Tussock grassland with *Dendrosenecio*'s, Mt Elgon. (Morrison)

144

European Alps, and, for this reason—in company with Mildbraed (1933), Richards (1963) and Troll (1959)—we believe that the term 'paramos' would have been better. This term is used for the upper vegetation of the high mountains of South America which support growth-forms very similar to those seen in eastern Africa. Indeed, the broad similarities of climate and vegetation between the high mountains of South America and eastern Africa were noted as far back as 1909 by Warming (1925) after he had read Volkens (1897) account of Mt Kilimanjaro. There is much to be said for emphasising this similarity rather than the alpine one which is much feebler.

The following descriptions of the main types of vegetation in the high mountains is based largely on Hedberg's excellent account (1964a) and readers requiring more detail should consult this article.

Vegetation types

Ericaceous woodland and wooded grassland

Upland forest gives way to woodland of giant heaths and heath-like plants between 3000 and 4000 m. This ericaceous woodland is most vigorous and extensive on the wetter mountains such as the Virunga volcanoes and the Ruwenzori. On the remaining mountains, drought, fire, and sometimes grazing, have thinned the woodland and eliminated it from large areas, so that much of the ericaceous vegetation on these drier mountains consists of low-stature tussock grassland with scattered shrubs. The soils are acid and the soil profile usually has an upper dark, greasy, peaty layer which gives way abruptly to underlying material which may show little weathering. At lower altitudes the upper organic horizon fades gradually into the underlying soil. (Harrop, 1960).

The ericaceous woodland on the Ruwenzori is formed of twisted and gnarled trunks and branches reaching a height of about 13 m and draped with *Usnea barbatus*. Except on flat, boggy places the forest is dominated by *Philippia trimera* spp. *trimera*. Its trunks grow up from a thick layer of moss—including Sphagna—which conceals a tangle of fallen trunks. The only other trees present in this woodland are *Rapanea rhododendroides* and *Senecio erici-rosenii* (both mainly in valleys), *Hypericum keniense* and *H. bequaertii* 3600 m, and above this *Senecio adnivalis*. There are also the shrubs *Myrica meyeri-johannis* and *Myrsine africana*. *M. meyeri-johannis* grows up to a Fig. 3.6 Common algae of the phytoplankton.

numerous branches growing over rocks and, on Mt Kilimanjaro, Greenway (1965) states that it forms almost pure stands in very stony places. *Myrsine africana*, which reaches a height of 1·5 m, has small leathery, round leaves, scarlet berries, and ranges from 2700–3500 m. It is common also on rock outcrops and in the shade of trees on exposed mountain slopes. Where the ground flattens and becomes swampy, the woodland gives way at lower altitudes to *Erica bequaertii* and at higher altitudes to swamps with tussocks—up to a metre in height—of *Carex runsorroensis*. The undergrowth of the woodland varies with altitude. In some places it is mainly mosses. At lower altitudes it is sometimes a dense tangle of *Mimulopsis elliotii*, an acanthaceous plant with white flowers often growing mixed with species of *Rubus* and reaching nearly two metres in height. Other common plants here are several species of *Helichrysum* and *Peucedanum kerstenii*, an almost arborescent umbellifer. At higher altitudes the woodland undergrowth is sometimes dominated by the silvery-leaved *Alchemilla* which forms quite tall shrubs.

Above 4000 m, the giant heaths diminish in size, and at about 4200 m they are replaced by scrub of the straw-coloured 'everlasting' *Helichrysum stuhlmannii*. In some places this scrub is so thick that it is impossible to walk through it without cutting a path. It is best developed on steep, rocky slopes and the abundance of these on Ruwenzori probably accounts for its prevalence there between about 3600 and 4300 m.

In some circumstances and on some mountains there may be very little *Philippia* and *Erica* and their place is taken by *Stoebe kilimandscharica*, a member of the Compositaeae with an ericoid habit and a leaf very strongly resembling that of *Philippia*. There is much *Stoebe*, along with *Erica* and *Philippia* scattered over the tussock grasslands which occupy much of the ericaceous belt on Mt. Elgon. The ericaceous belt here is very different in general appearance from that on Ruwenzori and this may be due to human influence. The Musopisiek people of Sebei live permanently with their cattle between 2500 and 3350 m on Mt Elgon and they often burn the grass in the dry season (Thomas, 1963; Weatherby, 1962). The greater diversity of conditions thus created in the ericaceous belt on Mt Elgon is probably in part responsible for the richness of the flora compared with that of Ruwenzori (Table 5.2).

Dendrosenecio woodland and wooded grassland

Dendrosenecios, usually scattered through grassland,

Table 5.2 Plants of the ericaceous zone on Mount Elgon

This is a grass-dominated vegetation with scattered bushes of *Erica arborea*, *Stoebe kilimandscharica* and *Phillipia trimera* with many shrubs and herbs, often patchily distributed.

COMMON AND WIDESPREAD GRASSES

Agrostis gracilifolia
A. kilimandscharica var. *kilimandscharica*
A. producta
A. quinquiseta
A. volkensii
Andropogon abyssinicus
A. amethystinus
A. pratensis
Anthoxanthum nivale
Bromus afromontanus
B. pectinatus
Deschampsia caespitosa
Festuca abyssinica
F. pilgeri
Helictotrichon elongatum
Koeleria capensis
Pentaschistis borussica
P. imatongensis
Poa schimperana

SHRUBS AND HERBS

Alchemilla elgonensis	Rosaceae
A. ellenbeckii	,,
A. johnstonii	,,
Alepidia massaica	Umbelliferae
Anagallis serpens ssp. *meyeri-johannis*	Primulaceae
Anemone thomsonii	Ranunculaceae
Anthemis tigreensis	Compositeae
Anthospermum usambarense	Rubiaceae
Arabis alpina	Cruciferae
A. cuneifolia	Cruciferae
Artemesia afra	Compositeae
Bartsia decurva	Scrophulariaceae
B. macrophylla	,,
B. petetiana	,,
Callitriche stagnalis	Callitrichaceae
Campanula sp.	Campanulaceae
Carduus keniensis	Compositeae
Carex mannii	Cyperaceae
C. runssorensis	
Celsia scrophularaefolia	Scrophulariaceae
Cerastium afromontanum	Caryophyllaceae
Cineraria grandiflora	Compositeae
Coreopsis elgonensis	,,
Crepis ruepellii	,,
Delphinium macrocentrum	Ranunculaceae
Dierama pendula	Iridaceae
Dipsacus pinnatifidus	Dipsacaceae
Discopodium eremanthum	Solanaceae
Disa stairsii	Orchidaceae

146

Droguetia iners	Urticaceae
Eriocaulon abyssinica	Eriocaulonaceae
Euphorbia wellbyi	Euphorbiaceae
Euryops elgonensis	Compositeae
Galium glaciale	Rubiaceae
Geranium vagans	Geraniaceae
Haplocarpha rueppellii	Compositeae
Haplosciadium abyssinicum	Umbelliferae
Hebenstretia dentata	Scrophulariaceae
Helichrysum amblyphyllum	Compositeae
H. argyranthum	,,
H. cymosum	,,
H. formosissimum	,,
H. globosum	,,
H. meyeri-johannis	,,
H. nandense	,,
H. odoratissimum	,,
Hydrocotyle mannii	Umbelliferae
Hypericum bequaertii	Hypericaceae
H. kiboense	,,
H. peplidifolium	,,
Lathyrus hygrophilus	Leguminosae subfam. Papilionoideae
Leonotis molissima	Labiateae
Leucas calostachys	,,
Lobelia anceps	Campanulaceae
L. lindblomii	,,
L. telekii	,,
Lotus goetzii	Leguminosae subfam. Papilionoideae
Luzula abyssinica	Juncaceae
L. campestris var. *mannii*	,,
L. johnstonii	,,
Lysimachia ruhmerana	Primulaceae
Myosotis vestergrenii	Boraginaceae
Nepeta azurea	Labiateae
Peucedanum kerstenii	Umbelliferae
P. linderi	,,
Pimpinella kilimandscharica	,,
P. oreophylla	,,
Plectranthus sylvestris	Labiateae
Ranunculus oreophytus	Ranunculaceae
R. volkensii	,,
Romulea fischeri	Iridaceae
Rorippa indica	Cruciferae
Sanicula elata	Labiateae
Satureja abyssinica	,,
S. kilimandscharica	,,
S. punctata	,,
S. simensis	,,
S. uhligii	,,
Scabiosa columbaria	Dipsaceae
Scirpus setaceus	Cyperaceae
Sebaea brachyphylla	Gentianaceae
S. microphylla	,,
Sedum meyeri-johannis	Crassulaceae
Senecio elgonensis	Compositeae

S. snowdenii	`,,`
S. sotikensis	`,,`
Stachys lindblomiana	Labiateae
Swertia calycina	Gentianaceae
Swertia kilimandscharica	`,,`
S. crassiuscula	`,,`
Trifolium baccarinii	Leguminosae subfam. Papilio
T. cryptopodium	Leguminosae subfam. Papilionoideae
T. multinerve	Leguminosae subfam. Papilionoideae
T. usambarense	Leguminosae subfam. Papilionoideae
Umbilicus botryoides	Crassulaceae
Valeriana kilimandscharica subsp. *elgonensis*	Valerianaceae

Table 5.2a   Common lichens of upland habitats

COMMON LICHENS OF UPLAND HABITATATS (Dr C. D. T. Swinscow)

High montane rocks	*Alectoria fuscescens* *Cladina* spp. *Parmelia* spp. *(Xanthoparmelia)* *Umbilicaria africana* *U. vellea* *Umbilicaria* spp. *Stereocaulon ramulosum* *Stereocaulon* spp. *Usnea* spp.
Ericaceous zone	*Anaptychia boryi* *A. tremulans* *Cladina* spp. *Parmelia pseudonilgherrensis* *P. austrosinensis* *P. reticulata* *Stereocaulon ramulosum* *Stereocaulon* spp. *Usnea articulata* *Usnea* spp.
Upland forest	*Anaptychia boryi* *Cladonia balfourii* *Cladonia* spp. *Lobaria pulmonaria* *Parmelia andina* *P. austrosinensis* *P. pseudonilgherrensis* *P. reticulata* *P. tinctorum* *Pseudocyphellaria* spp. *Sticta* spp.

42 *Dendrosenecio* sp. Mt Elgon. (Morrison)

occur on all the mountains, but dense woodland occurs on the wetter mountains, namely, the Virungas and Ruwenzori, from about 3500 m to near the thermal limit of plant growth. On the drier mountains, such as Mt Elgon and Mt Kenya, the woodland is very open, (Plate 41), and it is best described as tussock grassland with scattered Dendrosenecios. On Mt Kilimanjaro woodland is limited to moist places such as stream sides and ravines, and Dendrosenecios are absent on Mt Meru which apparently is too dry. Many Dendrosenecios have a *lower* altitudinal limit around 3000 m, but in the lower part of this range, which lies usually in the ericaceous woodland and/or derived tussock grasslands, they are confined typically to moist grassland alongside streams. It is believed that they require much moisture; but their distribution might also be accounted for by temperature differences between the valley sides and valley bottoms. In some places these differences might be as great as the differ-

ences between neighbouring vegetation zones. The upper parts of the valley slopes, usually occupied by tree heaths, would tend to be warmer than the valley floors where cold air would collect and might favour the growth of the Dendrosenecios.

Most Dendrosenecios, which reach a height of about 8 m and may take a century to reach full height, have a few stout branches terminating in compact rosettes of large, simple leaves, often more than 80 cm in length (Plate 42). The exception to this growth-form is *Senecio brassica*, which has a prostrate trunk so that the rosette, seen from a distance, appears to be sessile on the ground. The leaves of many species are covered with a dense, soft, white tomentum giving a silvery appearance which perhaps reflects some harmful radiation. In all species the flower spikes are large, often more than 1 m in length, and they contain hundreds of yellow capitula; the lower altitude species, which have distinct ray florets, being the most handsome. Hedberg (1969a) considers that protection measures are urgently needed to preserve the Dendrosenecios from tourists who use them for camp fires, though it is usually the dead plants which are used as the living ones make poor fuel.

The field-layer of the *Dendrosenecio* woodland is usually of *Alchemilla johnstonii* though, on the Ruwenzori, grasses are common. In the dense woodlands of the Ruwenzori and the Virunga volcanoes there are many epiphytic and ground bryophytes. In the open woodlands of the drier mountains the *Alchemilla* field-layer is prominent, particularly on well-drained slopes, and large areas on Mt Elgon and Mt Kenya are covered by *Alchemilla* scrub.

Helichrysum scrub

This appears to be a stable vegetation which occurs, to a slight extent, in the upper part of the ericaceous belt, but is more common in the lower part of the paramos belt, especially on rocky ground and porous sandy soil. It is possible that rocky slopes may provide a favourable thermal habitat. During the day, especially on east-facing slopes, the heat capacity of rocks may be significant in keeping the slope warm, while at night, cool air will drain rapidly towards the valley bottom. *Helichrysum* is recorded from all the East African high mountains, but it is most abundant on the Ruwenzori, on the south-east slopes of Mt Kilimanjaro and on the Shira Plateau. Its abundance on these mountains is due perhaps to the large number of suitable habitats. On steep slopes the *Helichrysum* stems follow the slope of the ground for about 3 m, the ends of the branches curving upwards and bearing clusters of flowers. On the Ruwenzori the upper parts of the stems carry *Usnea*, and other lichens and bryophytes are common. Sometimes mosses form a thick mat below the bushes. In some places shoots of *Lobelia wollastonii*, with vertical flowering spikes at least 3 m tall, protrude above the *Helichrysum*. In the upper part of the paramos belt the *Helichrysum* scrub is low and open.

Helichrysum has about 350 species and a distribution covering Europe, Asia, Africa and Australia; about 150 of the species occur in South Africa, mainly on high-altitude grasslands. They are commonly called 'everlasting flowers' because the mature flowers, if gathered, do not disintegrate and the stiff, membranous floral parts retain their bright yellow, pink or white colour.

Tussock grassland or moorland (Table 5.3)

The tussock grasslands may be the ultimate re-

placement of the ericaceous and *Dendrosenecio* woodlands as the result of burning, or they may take their place naturally on the driest mountains. They are widespread on all the mountains, except the Virungas and the Ruwenzori, and are regarded as the most characteristic vegetation at the high altitudes. They are very well developed on Mt Elgon probably due to the influence of the Sebei mountain people.

Some thirty-four species of grasses are listed by Hedberg (1957) in the Afroalpine flora. The majority of these are within the tribes Festuceae, Aveneae and Agrosteae which is a contrast with the grasses of the lowland wooded grasslands which belong mainly to the Andropogoneae and Paniceae. Many of the Afroalpine grass species are endemic to the East African mountains though some have a distribution extending to the north temperate latitudes, namely, *Aira caryophyllea*, *Deschampsia flexuosa*, *Koeleria capensis*, *Poa annua* and *Vulpia bromoides*.

The most common grass of the high moorlands is *Festuca pilgeri*, which is a densely tufted perennial forming large tussocks. It is endemic on Mts Elgon, Kenya and in the Aberdare range. It is dominant over large areas of more or less sloping ground from the upper part of the bamboo vegetation into the paramos vegetation, that is from 3000–4250 m. On Mt Kilimanjaro it is replaced by sp. *supina* which occupies similar habitats from the upper part of the ericaceous vegetation into the paramos vegetation, that is from 3650–4400 m. However, on dry, stony and rocky slopes on Mt Kilimanjaro another grass, *Pentaschistis minor*, is dominant in this altitudinal range and forms, along with some heaths, a very open vegetation. Greenway (1965) notes that *Pentaschistis minor* forms very large tussocks, the central part of each tussock finally dying so as to leave a central mass of compacted hay surrounded by the living part of the tussock in the shape of a ring or crescent. He observed that this central portion on a number of the tussocks was much charred, while others which had not been burnt had large water drops, like large glass beads or marbles, resting on the dry centre and he suggested that perhaps these water drops acted like lenses which focused the intense sunlight and set the central part of the tussock alight. This could be one of the causes of the fires that break out on the upper slopes of Kilimanjaro and are not due always to the activities of honey hunters.

The many attractive flowers in these high altitude tussock grasslands include the red-hot poker *Kniphofia thomsonii*; *Anemone thomsonii*; *Scabiosa columbaria*; *Disa stairsii*, an orchid with a dense spike of small pink flowers; a thistle, *Carduus nyassanus*; *Selago thomsonii*; the beautiful *Dierama pendulua* of the

Table 5.3   Tussock grassland, west slope of Mt Kenya (3800–4000 m) (from Hedberg, 1957)

Field layer	*Alchemilla johnstonii*	} Dominants
	Festuca pilgeri	
	Arabidopsis thaliana	
	Cardamine obliqua	
	Carduus chamaecephalus	
	Galium glaciale	
	Lobelia telekii	
	Myosotis keniensis	
	Senecio keniodendron (seedlings)	
	S. purtschelleri	
	Valeriana kilimandscharica sp. *kilimandscharica*	
Bottom Layer		
Mosses	*Brachythecium ugandae*	
	Bryum argenteum	
	Didymodon papillinervis	
	Encalypta ciliata	
Lichens	*Peltigera canina*	

The principal tussock forming grasses on the East African mountains in the paramos belt are *Agrostis trachyphylla, A. volkensii, Andropogon amethystinus, Deschampsia caespitosa, Festuca abyssinica, Festuca pilgeri* sp. *pilgeri, F. pilgeri* ssp. *supina, F. kilimanjarica, Koeleria capensis, Pentaschistis* spp. and *Poa leptoclada.*

Iridaceae with long arching racemes of pink flowers; *Homoglossum watsonioides*, with erect spikes of showy crimson flowers; the clover *Trifolium cryptopodium*, with heads of small blue flowers; the pinkish-white flowers of *T. burchellianum* ssp. *johnstonii*; cushions of *Anagallis serpens;* and the wonderful blue flowers of the erect herb *Delphinium macrocentrum*. The small *Lobelia lindblomii* contrasts strongly with its giant neighbours.

High altitude acidic mires

Acidic mires, (*p*H 3·5–4·5) of tussock-forming *Carex* species occur on some of the high mountains and are one of the most striking and unpleasant features of the Ruwenzori vegetation (Osmaston, 1965). According to Hedberg (1964a), who describes such mires from several mountains, they are most common in the region of the ericaceous woodlands and derived vegetation, and they are rare at higher altitudes. They often occur in valley bottoms, sometimes behind lava dykes, and on the Ruwenzori they may even form blanket mire on steep slopes.

The two species of *Carex* involved are *Carex monostachya* and *C. runssoroensis* which are capable of forming peaty, rhizomatous tussocks up to a metre in height and nearly as much in diameter. The *Carex* rhizomes may extend to the base of the tussock and right through the underlying peat, which can be as much as 6 m in thickness and is generally suitable for statistical investigations of pollen, the pollen being well preserved; but the fibrous texture of the peat makes it very difficult to sample.

The tussocks are spaced about a metre apart and the commonest subdominant between the tussocks is *Alchemilla johnstonii*, while the wet hollows are often covered by a spongy layer composed mainly of species of *Sphagnum*. Other plants recorded from these mires are listed in Table 5.4, which is not a list from any *one* mire but includes plants which have been recorded from such habitats on various mountains by Greenway (1965), Hedberg (1964a) and Osmaston (1965).

High altitude desert

This has been described from the saddle between Mawenzi and Kibo peaks, at an altitude of 4350 m, on Mt Kilimanjaro. On this saddle, where the average

151

Table 5.4   Plants of the high altitude acidic mires (Data from Greenway, 1965; Hedberg, 1964a; Osmaston, 1965)

DOMINANT TUSSOCKS

	Carex monostachya	Cyperaceae
	C. runssoroensis	,,

SUB-DOMINANT

	Alchemilla johnstonii	Rosaceae
	Accompanied by:	
	Agrostis gracilifolia	Gramineae
	Anagallis serpens ssp. *meyeri-johannis*	Primulaceae
	Anthoxanthum nivale	Gramineae
	Cardamine obliqua	Cruciferae
	Carduus keniensis	Compositeae
	Cerastium octandrum	Caryophyllaceae
	Haplocarpha rueppellii	Compositeae
	Helichrysum stuhlmannii	,,
	Lobelia bequaertii	Campanulaceae
	L. deckenii	,,
	L. elgonensis	,,
	L. keniensis	,,
	L. sattimae	,,
	Luzula abyssinica	Juncaceae
	L. johnstonii	,,
	Montia fontana	Portulacaceae
	Poa schimperana	Gramineae
	Ranunculus oreophytus	Ranunculaceae
	Senecio cottonii	Compositeae
	S. transmarinus var. *sycephyllus*	,,
	Swertia crassiuscula	Gentianaceae
	Vernonia glandulosa	Compositeae
On the tussocks	*Antitrichia curtipendula*	Bryophyta
	Campylopus stramineus	,,
	Hylocomium splendens	,,
	Hypnum cupressiforme	,,
	Sematophyllum elgonense	,,
In the hollows between the tussocks	*Sphagnum davidii*	Bryophyta
	S. madagassum	,,
	S. pappeanum	,,
	Brachythecium spectabile	,,
	Breutelia stuhlmannii	,,
	B. subgnaphalea	,,
	Campylopus stramineus	,,
	Cyclodictyon borbonicum	,,

annual rainfall is only 150 mm, very few plants survive. Greenway (1965) records *Helichrysum newii*, *H. cymosum* ssp. *fruticosum*, and some herbaceous groundsels, such as *Senecio telekii* and, possibly *S. schweinfurthii*.

The thermal limit of plant growth

The thermal limit of plant growth is in the neighbourhood of 5000 m. The upper limit of the *Dendrosenecio* woodland is probably largely determined by available moisture, since these plants appear to need much moisture and can therefore go to higher altitudes on the wetter mountains, such as the Ruwenzori and Mt Kenya, where it is believed they get suitable conditions almost everywhere up to their thermal limits. Individual Senecios extend up to about 4500 m, but dense woodland of *Senecio adnivalis* with much *Lobelia wollastonii* ends at about 4000 m. Above this altitude these plants become rare, and there is, generally, little soil. *Helichrysum stulhmannii* may be common or occasional. From 4300 m upwards moss dominated communities, where vascular plants are absent or widely spaced, are common. Near the glaciers, the lower edge of which varies from 3960–4570 m the vegetation is very open; *Poa ruwenzoriense* is the commonest vascular plant, occurring mainly in rock crevices and damp places. Other plants recorded within 14 m of the glaciers are *Helichrysum stuhlmannii*, *Subularia monticola*, *Alchemilla subnivalis*, and more rarely *Senecio adnivalis* and *Lobelia wollastonii*. The rocks below the glaciers are first colonised by lichens, notably species of *Umbilicaria*. Zahlbruckner and Hauman (1936) state that on the highest parts of the Ruwenzori range there is almost continuous cover of *Umbilicaria* on the rocks. As well as species of *Umbilicaria* they mention the occurrence of the lichens *Caloplaca*, *Rhizocarpon*, and *Usnea*. Greenway (1965) mentions that *Helichrysum newii* is recorded as the highest known vascular plant on Mt Kilimanjaro, having been collected on Kibo at 5760 m near the eastern fumarole.

Part Two

Vegetation
and
Environment

Six

Climate and vegetation

CLIMATE AND VEGETATION

Introduction

Throughout the world, the distribution of natural vegetation is governed largely by two factors, climate and soil. The larger plant formations, such as forest, woodland, and desert are mainly related to climate; but within them, differences of soil often account for the presence of certain plant communities.

In cooler climates, the type of vegetation may change abruptly with a change in the basic rock from which the soil is derived. This kind of sudden change is unusual in tropical East Africa because the underlying rocks are very old and have been subject to the effects of climate over a long period, so that the soil is often of great depth. In this way, differences in the nature of the rock have been minimised while features of climate and topography have become all-important.

In the following chapter we have tried to outline the main ways in which climatic factors, especially rainfall and temperature, affect the distribution of vegetation in East Africa. Effective rainfall is the more important because at altitudes below about 1840 m the main daily temperature is relatively high and does not, by itself, restrict plant growth, though it may have indirect effects.

Rainfall

Distribution

Rainfall in East Africa seems to depend primarily on convergence between and within air streams. Two main low-level air streams (locally known as monsoons) are recognised; the south-easterlies which prevail from approximately May to October and the north-

easterlies which prevail from approximately November to April. The wet seasons are associated with the period of change in direction of the prevailing wind. Complete understanding of the influence of patterns of air flow at middle and upper levels of the troposphere is not yet possible although advances are being made (Johnson and North, 1960). Incursions of westerly winds are often associated with disturbed weather and rainfall although their origin and exact behaviour are not well understood (Nakamira, 1969).

Three main regions of seasonal rainfall have been recognised in East Africa (Fig. 6.1). Over much of Uganda and Kenya and the Usambaras in Tanzania rainfall is biseasonal coming from March to May and mid October to December, the first rains being the longest and heaviest. The dry seasons are sometimes not well defined and are broken by occasional showers. The wet seasons are also unreliable, and in some years the rains may come late or fail altogether. In the West Kenya Highlands the main rainy season continues from April to July or August. In the extreme north the rainy seasons merge and there is a single dry season. The counterpart to this northern regime is the southern regime which covers most of Tanzania south of a line from approximately 3°S. inland to 7°S. on the coast. The rains in this southern region begin from November onwards and may continue into April; thereafter the dry season is well marked, lasting for about five months.

The distribution of rainfall amount is largely controlled by relief. For example, the eastern faces of the plateau rim mountains of Tanzania are high rainfall areas, in some places having over 2000 mm per year. As a rule the mountains stand out as wet areas and they are usually wetter on their eastern and south-eastern faces. In the northern half of East Africa there are no rim mountains to catch the moisture of the incoming air streams. Conditions inland from the coast until the Eastern Rift highlands are reached are semi-arid, and in the north-east the very extensive low ground

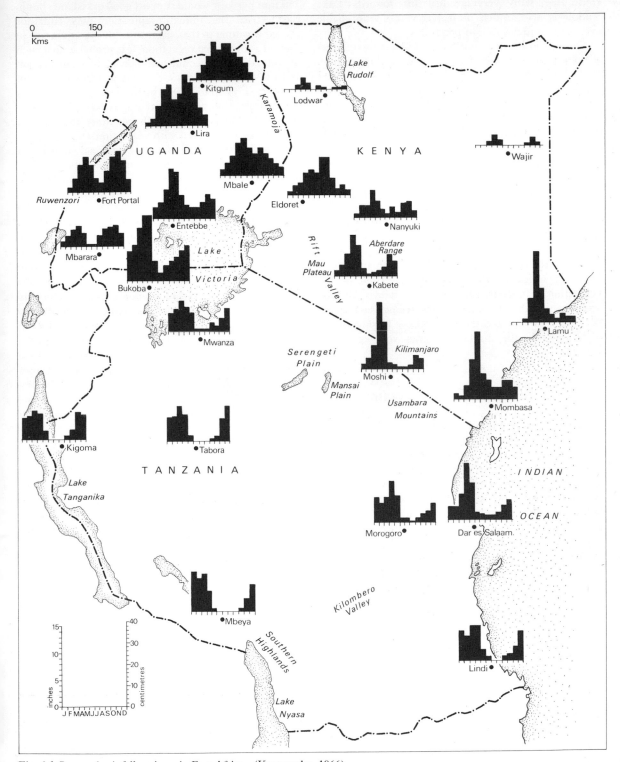

Fig. 6.1 Seasonal rainfall regimes in East Africa. (Kenworthy, 1966)

157

continuing with very gradual lift towards Lake Rudolf is arid, much of it receiving an unreliable average of under 250 mm per year.

The altitudinal distribution of rainfall is also of great ecological importance (Trapnell and Griffiths, 1960). It generally increases up to about the top of the forest belt and then decreases beyond that altitude, depending on the aspect and direction of prevailing rain-bearing winds. There are very few rainfall isohyets for the East African mountains, but, for example, on Mt Kenya, according to the East African Metero-logical Department, rainfall increases with altitude up to between 2740 and 3050 m and decreases up to the peak area, where it is about 762 mm per year. According to figures given by Hedberg (1964a) and the East African Meteorological Department the decline after the maximum is more marked on Mt Kilimanjaro. This altitudinal distribution of rainfall, as well as the decrease of temperature with altitude, is likely to influence the altitudinal zonation of the vegetation. The consequences of this rainfall/altitude relationship differ between various regions such as the coast, East of Rift and the West Kenya Highlands. At high altitudes the rainfall becomes very erratic or variable.

The estimation of available water

Observations on rainfall, temperature, radiation and wind which are available in meteorological records are not usually in a form very suitable for defining any ecological environment. As Gates (1962) observes, it is not surprising that their use leads to such statements as 'little correlation with plant growth was noted'. What needs to be measured is not just the amount of water that falls as rain, but the amount which is available to the plant after due attention has been paid to the water-holding capacity of the soil and the loss of water vapour to the atmosphere due to evaporation. This takes place from bare soil or a water surface and also by transpiration by plants. The combined water loss is known as evapotranspiration and is a key factor affecting the effectiveness of rainfall.

There are three main reasons why, in East Africa, measurements of mean annual rainfall alone are unable to give significant correlations with plant growth. First, the key factor for natural vegetation is the balance between rainfall and evaporative demand and the variation in evaporative demand is very large over East Africa. Annual evaporation rates for open water vary from just over 2450 mm inland from the Kenya coast to less than 1043 mm above 2740 m in the Aberdares. In temperate lands an annual rainfall of 750 mm per year would support good grassland; but in East Africa the annual evaporation may equal or exceed this total so that, even with full water storage in the soil and a stable vegetation, it is probable that such an annual rainfall could maintain plant transpiration and growth for not more than six or eight months of the year. This would often be inadequate to support a satisfactory plant cover when grazed and would result in a semi-arid rangeland with not more than 30 per cent of the soil covered with vegetation.

Second, the rainfall is highly seasonal in parts of East Africa and under these conditions the physical properties of the soil assume great importance both in relation to its depth and capacity to store water and the ability of the plant roots to exploit it.

Third, the amount of rainfall is very variable from year to year. Nairobi, for example, has an annual average of 889 mm and a variability which ranges from 483 mm to 1524 mm. Also the number of rainy days is very variable.

The knowledge that the average annual or monthly precipitation was insufficient to define the climate of a locality for ecological work led to various attempts to devise an index which would be a measure for the effectiveness of the available rainfall. Many of these indices were developed empirically and were refinements of the Lang rain factor, $I = P/T$ where P is the mean annual precipitation and T is the mean annual temperature (Richards, 1962). One of the most successful of these is Lauer's (1952):

$$I = \frac{12n}{T+10} \quad \text{where } n = \text{precipitation in mm and}$$
$$T = \text{temperature in degrees centigrade}$$

Lauer defines a *dry* month as one in which the index of aridity (I) is less than 20. With data from about 4000 places in Africa and America he has shown that there is good correlation between the type of natural vegetation and the number of dry months calculated in this way.

The success of this index depends, in part, on the assumption that daily temperature is a reliable guide to solar radiation and hence to the energy available for evaporation. But this relationship does not hold under the conditions of near constant daylength in tropical latitudes, particularly at high altitudes where the net radiation will be strongly influenced by cloud layers.

Sansom (1954) mapped East African climates using the classification developed by Thornthwaite (1948) according to the moisture index derived from a comparison of mean monthly rainfall and mean monthly potential evapotranspiration. Values for potential evapotranspiration were estimated from an empirical formula based on values for mean air temperatures and daylength. This method seriously underestimates evaporation in East Africa and more realistic values are obtained using the Penman equation (1948, 1956) which requires data for radiation, temperature, saturation deficit and wind speed, and has been modified by an altitude factor for use in East Africa (McCullock, 1965). Estimates of potential evaporation according to Penman have been mapped for the whole of East Africa (Rijks and Owen, 1965; Woodhead, 1968 a and b). Dagg, (1965) has adapted Thornthwaite's classification of climate using Penman values and this approach has been used in the classification of East African rangeland (Pratt, Greenway and Gwynne, 1966; Woodhead, 1970).

Although the development of the moisture index was an important advance in mapping climates for ecological work, it does not integrate all ecologically important features of the climate. For example, it fails to take account of the variability of the annual rainfall which is a major factor in East Africa. The moisture index is based only on the *average* values of potential evaporation and rainfall.

The most useful approach to the analysis of rainfall has been to calculate probability values. Glover, Robinson and Henderson (1954) mapped the probabilities of failing to get 500 mm and 750 mm of rainfall respectively. Probability maps in the National Atlases show the isochyets that are likely at defined probability levels, illustrating the minimum amounts that are likely to be received 80 per cent and 90 per cent of the time respectively. In other words there is a 20 per cent and 10 per cent likelihood of failure to obtain these amounts of rainfall. From these maps it is possible to calculate the percentage of the total area of East Africa receiving reliable amounts of the annual totals. These calculations show that only 4 per cent of the land surface can expect to receive, in four years out of five, as much as 1300 mm of rain per year. Only 15 per cent of Kenya's land surface receives a reliable 762 mm per year compared with 50 per cent of Tanzania and about 75 per cent for Uganda. Particularly detailed work for Uganda has been carried out by Manning (1956).

The effect of the structure of the root system on water availability

The readily available water in the soil is effectively restricted to the water in the immediate neighbourhood of the root system and not very far beyond. Soil water moves· very slowly, even when there is only a very small soil moisture deficit. For these reasons, the extent to which the root system is able to exploit the soil profile is extremely important (Gwynne, Taerum and Opile, 1966). Unfortunately very little is known about the rooting depths of plants in East Africa. Kerfoot's studies (1962 a and b, 1963) indicate that trees and woody shrubs are probably capable of exploiting the soil to a depth of at least 7 m and that some grasses, for example, *Chloris gayana, Cynodon dactylon, Pennisetum clandestinum* and *P. purpureum*, are capable of drying the soil profile to wilting point to a depth of 3 m, their roots probably going beyond this depth. Research on the physiology of the growth of tropical grasses, with particular reference to root depth and the effects of water stress, should be work of high priority. Without such information on different plants it is not possible to indicate precisely how soil depth and its water capacity will influence the distribution of different plants. In considering the effectiveness of root systems it should be borne in mind that a plant with only a few deep-ranging roots may be able to obtain most of its needs in periods when the upper soil layers have been dried out to wilting point. However, in the semi-arid regions of East Africa where rain showers often will not penetrate the soil to a depth of more than 30 cm, shallow, extensive root systems are probably more common than deeper-rooting systems. Glover, in his studies of roots in the dry areas of Somalia (Glover, 1950, 1951, 1952) examined trees, succulents and herbs. Among trees such as *Acacia etbaica, A. mellifera, A. reficiens* and *Balanites orbicularis*, he found most of the roots lying between 3 and 30 cm below the soil surface, while some extended to a depth of from 10 to 19 metres. Similar extensive and shallow-rooting systems were found to be characteristic of succulents such as *Adenium, Aloe, Caralluma, Euphorbia* and *Sansevieria*, and some of the roots grew upwards to within a few millimetres of the soil surface. Antigeotropic roots of this nature are ideally placed for absorbing water after light rain showers. In the herbs, the extensiveness of the root system in relation to the size of the aerial part of the plant was very marked.

The effect of soil texture on water availability

Various characteristics of the soil affect the availability of water to plants. Among these are the relative amounts of clay, silt and sand, the type of clay, the amount of organic matter and the structure and depth of the soil profile.

Soil texture is measured by the relative amounts of clay, silt and sand particles. Clay particles are the smallest in size, and have the largest pore space between them. The size of the pore space is critical to water penetration into a soil and to water storage and availability. Clay soils with their high percentage pore space have the greatest potential for water storage or 'field capacity', but the small size of the pores in clay soils may prevent a rapid flow of water into and through them.

The degree of infiltration of water into a soil is particularly important in tropical countries where water may be received in a heavy downpour and lost very rapidly by evaporation or surface run off. This is especially true on bare ground or in overgrazed areas where soils are compacted. Sandy soil absorbs water more rapidly than clay unless the clay particles are aggregated into crumbs which allow water to pass between them.

Once water has entered a soil, however, it may be held there in a form unavailable to plants. It is only the water held by capillary forces which tends to be readily available; if this is insufficient for the plants' needs, the soil is said to be at 'wilting point'. The water available to plants at a particular time is the difference between the actual water content of the soil and the water content at wilting point.

The water lost to the atmosphere by evapo-transpiration is probably greater in clay soils than in sandy soils, especially in semi-arid areas. In clay soils, where the water does not usually penetrate to great depth, it is lost by a capillary rise of soil moisture and by evaporative loss from cracks which expose a large soil-air surface.

The importance of soil texture has been stressed by Smith (1949) in the Sudan, where he claims that the soil texture greatly influences the effectiveness of rainfall. He cites examples of many species which are able to extend much further into drier areas on the sandy soils than on the clay soils. This applies to the northern limit of mixed deciduous forest and to several Acacias, such as *Acacia seyal*.

The effect of plant-form and the structure of the ecosystem on water availability

The overall structure of a plant community and the form of individual plants will affect the amount of rain water reaching the soil and its distribution there. In a forest, woodland, or grassland or indeed in any vegetation which has a uniform, closed canopy the rain falls on the top of the canopy much as it would on the ground and—on the whole—no plant obtains much additional water at the expense of another. But some rain water remains on the canopy and is soon evaporated, and in this way there is a real loss to the ground.

The amount of rainfall intercepted by the upper canopy of the forest and subsequently lost through direct evaporation varies according to the intensity and quantity of the rainstorm and the type of vegetation. Pereira (1953) found that rainfall interception by natural bamboo thicket and by planted *Cupressus* in the same area was about the same, between 20 and 25 per cent. This is of the same order of magnitude as recorded by Freise (1936) in the subtropical forests of Brazil. He found that 20 per cent was intercepted and evaporated directly from the canopy, 28 per cent reached the ground via the trunk, 34 per cent was throughfall, and the remaining 18 per cent was taken up by bark, hollow trees, epiphytes and evaporation.

Rainfall in a complex layered forest will also be intercepted below the main canopy. In one of the forests of the Lake Victoria belt in Uganda, Hopkins (1960) found that most interception took place between ground level and 9·2 m above the ground, apparently on a well-developed low understorey of *Dracaena steudneri*. Some 35 per cent of the rainfall was intercepted by this layer. However, most rainfall intercepted by an under storey monocotyledon with upright leaves like *Dracaena*, would soon flow down the stem and into the ground. The presence of forest is important in lessening the contrast between dry and wet season stream flow. In some parts of East Africa it has been necessary to replace natural forest by plantations, often of conifers, or to take over upland forest belts for tea growing. East African foresters have been anxious about the effect of such changes on the water economy and an important series of experiments was established in four different types of catchment area. These were (a) in tea estates in tall rain forest near Kericho in Kenya, (b) in softwood plantations in the bamboo zone of the Aberdare mountains in Kenya, (c) in peasant cultivation in deep stream source valleys near Mbeya in Tanzania and (d) in semi-arid range-land near Moroto in Uganda, where overgrazing and hardening of the soil surface had been increasing the loss of water by surface run off immediately following rain. Early results are published in the *E. A. Agricultural and Forestry Journal*, 1962.

If the rate of stem flow does not exceed the capacity of the soil to imbibe water quickly, then ground storage of water round the roots may be substantial. Thus, light showers which are normally neglected when discussing water available to plants in dry areas, may be very important for plant survival. Glover and Gwynne (1962) give details of the interception of light rainfall by maize at Muguga in Kenya, and show that the aerial parts of the plant can be considered as a catchment area which collects rain water and delivers it to the soil at the base of the plants where it accumulates in amounts considerably greater than those available from ground with no plant cover.

If there is no closed canopy and the vegetation consists of well-spaced plants either singly or in clumps, the plant surface exposed to the rain is quite different. With slanting rainfall the vegetation now obstructs the wind and the rain and intercepts additional rain at the expense of the part of the surface lying in the rain shadow of the plant or clump. The isolated plant presents a greater surface area and can make more effective use of light rains. In arid grassland especially if trampled, a surface crust may form and, with sparse vegetation, there is considerable run off in contrast to accumulation of water below individual plants.

In a study in Kenya Masailand, Glover, Glover and Gwynne (1962) were able to show that the depth of rainwater penetration below plants is approximately equal to the height of the plant plus the normal penetration of the rain shower into the bare soil. The sectional area of the wet soil beneath each vegetation clump is also approximately equal to the sectional area of the clump showing above the ground, plus an area corresponding to the amount which would have fallen there had there been no plant cover. Thus, heavily grazed, close-cropped and burnt grassland affords less chance of useful rain penetration than does more lightly grazed grassland. In the dry areas—for example in the semi-desert *Chrysopogon* grasslands of north and north-eastern Kenya—the growth habit of the grass is of importance, especially the tussock or bunch-grass habit which is frequent in the tropics. Another example is the tussock grassland at high altitudes in the ericaceous and alpine belts. There the tussock form protects growing points against very low temperatures, preserves moisture around them and also acts as we have just described. This full use of rain showers may be very important in some years at the higher altitudes, where rainfall variability is very great.

Temperature

Temperature and growth

Although East Africa lies in the tropics, very high daytime temperatures are reached only in limited areas because most of the plateaulands lie at about 1230 m. The mean annual temperature of the low-lying coastal belt and the islands is about 27°C with a mean annual range of approximately 4°C and an average diurnal range of 5–8°C. Inland, on the plateaux, the mean annual temperature is in the neighbourhood of 25°C (with a mean annual range of 4°C) and an average diurnal range of 8–11°C. Temperatures in excess of 38°C are very rare and an air temperature of 40°C has been recorded only once near Lake Magadi in Kenya. Ground frosts seldom occur below 2460 m.

Thus, until we reach the higher altitudes—which account for only a small percentage of the land surface —temperature by itself is probably not a factor directly limiting plant growth. However, there may be indirect effects in the form of lethal leaf temperatures, high night respiratory rates, and marked temperature fluctuations at the soil-air boundary. Plant physiologists, unlike human and animal physiologists, have paid little attention to the effects of high *internal* temperatures, although plant cells like animal cells are readily killed or inactivated by high temperatures. For example, although photosynthesis in many species is believed to be relatively constant between 10° and 30°C, at temperatures between 30° and 40°C the inactivation of enzymes prevents the photosynthetic machinery from working properly.

There is very little information on the internal temperatures experienced in leaves under tropical sunlit conditions, perhaps because it has been technically difficult to obtain satisfactory measurements. But, the introduction of the Stoll-Hardy radiometer, an instrument which measures surface temperatures (Gates 1962) has made estimation easier. The internal temperature of the leaf will be intermediate between the temperatures of the upper and lower surfaces.

Wind speeds in East Africa are low, but should be sufficient to prevent the development of lethal leaf temperatures during the day, certainly in open vegetation.

High internal leaf temperatures are most likely to arise in broad-leaved plants. Gates (1962) considers that needle-like leaves, such as those in many conifers, should seldom depart much from the ambient air temperatures, because of efficient convection cooling and heat exchange. It would seem obvious, too, that a finely divided leaf like that of an *Acacia* or *Albizia* is thermally at an advantage compared with a large, undivided leaf.

With large, relatively undivided leaves, the degree to which the lamina is tilted away from the horizontal plane may be very important, a leaf held almost horizontally tending to accumulate a pocket of warm air on the underside. The large leaves of some of the plants of semi-arid bushland, such as *Calotropis procera*, would appear to be disadvantageous for efficient sensible heat loss, the waxy cuticle and sunken stomata ruling out substantial cooling via latent heat transfer by transpiration. But it may be that the olaugus surface is highly efficient in reflecting incident energy. The same difficulties arise in attempting to comprehend the heat dissipating mechanism of some other plants of the same type of habitat, namely, the *Aloes, Cissus, Sansevieria, Caralluma, Jatropha, Euphorbia candelabrum* and *Kalanchoë*. All these are succulents with fairly large, undivided, thick leaves or photosynthetic shoots, provided with water storage tissue, mucilage, a strongly cutinized epidermis and, often, sunken stomata. It is known that these plants dry out very slowly and they appear to be excellently proofed against water loss; cooling is therefore unlikely to be achieved through transpiration.

Surprisingly low production values appear to characterise low and medium altitude broad-leaved forests in the tropics (Dawkins, 1964). Dawkins has conjectured that high night temperatures favouring high respiration rates may be responsible for low production, but it may well be due to a combination of relatively low photosynthetic rates, prompted by high leaf temperatures, and high respiration rates during day and night. It should be remembered that the main physical factor governing the rate of respiration is temperature, and an increase of temperature by 10°C is expected to double the respiration rate.

Temperature and distribution of vegetation

The upland vegetation of East Africa lies in broad altitudinal belts and it is frequently assumed that this zonation is the direct consequence of the gradual decrease of temperature with altitude. But the zonation of vegetation must be considered not only in relation to temperature but also to available water, net solar radiation, and competition. The altitudinal ranges of plants in tropical East Africa are determined by a variety of factors.

It is clearly established that temperature decreases steadily with increase in altitude. Thus the thermal regime at a particular altitude is readily calculated. For example, Griffiths (1962) gives the following formulae:
Mean annual maximum temperature $(F) = 93 - 3 \cdot 0\,A$.
Mean annual minimum temperature $(F) = 76 - 3 \cdot 8\,A$.
where A = altitude in thousands of feet.
Slightly different formulae are given by Glover and Kenworthy (1957). The different formulae are probably due to the scarcity of meteorological stations at the higher altitudes. This general relationship may,

however, be modified by local topography and cloud cover. Troll and Wein (1949), Salt (1954), Hedberg (1964) and Coetzee (1967) have described the regular daily formation of a cloud belt which coincides, more or less, with the uppermost part of the forest belt, and may be responsible for the high precipitation, and high atmospheric humidity. During the early morning the tops of the high mountains are entirely free of cloud, and they are subject, especially on their eastern slopes, to intense solar radiation. In the course of the morning warm air moving up the mountain slopes forms a cloud belt at the altitude of the forest belt and by midday this cloud belt has increased in depth to envelop most of the mountain top, so that incoming and outgoing solar radiation will be severely reduced. During the night this cloud cap begins to disperse, the air cooled by outgoing radiation flows down the mountain slopes forming a narrow belt of clouds just above and below the top of the forest belt. It is thought that this cloud formation at the level of the forest belt causes the maximum precipitation to occur there, and above that level there is a strong decrease of rainfall so that the ericaceous and paramos belts have a low and very variable mean annual rainfall. On some steep scarp faces the lapse rate of temperature with altitude may be substantially greater than that given by Griffiths' formulae. For example, Moreau (1935a) presented evidence to show that Amani at 840 m in the eastern Usambaras has a mean annual temperature of 16°C which is characteristic of conditions at 1400 m in the mean temperature/altitude gradient.

The temperatures given by the above formulae refer to screen temperatures and the temperature within stands of vegetation—certainly inside stands of forest—will be slightly different. Moreau's (1935a) observations on the Usambaras demonstrate that, among other things, plants in the forest undergrowth experience lower mean maximum temperatures than plants nearby in the open. Figure 6.2 shows that mean monthly maximum temperatures inside forest, near Amani at 920 m altitude, are consistently lower by 3°C to 4°C except in the very wet and cloudy month of May, when the difference was hardly more than 2°C. This 3°C reduction in the maxima agrees closely with that found by Philips (1931) in the extra-tropical Mt Knysna forest in South Africa. Thus plants in the forest undergrowth experience a mean maximum temperature equivalent to that 615 m higher in altitude.

Despite the considerable range of mean temperatures between sea level and the highlands of East Africa the temperature through most of the altitudinal range is relatively high and most regions are free from frost. It is to be expected that the altitude at which frosts are *occasionally* experienced will be critical for many tropical plants, the majority of which seem not to have bracts and/or stipules which would protect growing points from frost damage. In the ericaceous and alpine belts the nightly recurrence of temperatures at and below freezing has unquestionably prevented the establishment of many plants from lower altitudes and has favoured the development of growth-forms which afford protection to tissues susceptible to frost damage (see chapter 5).

An important outcome of these interacting factors is the increase in the quantity of available water for plant growth as we move to higher altitudes in the

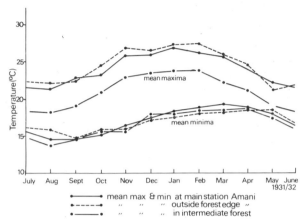

Fig. 6.2 Temperatures inside and outside forest at Amani, Tanzania. (Moreau, 1935a)

upland forest belt. The peak of water availability is probably reached at the top of the forest belt in the *Hagenia-Rapanea* woodland zone, though the view has sometimes been advanced that it coincides with the zone of bamboo thicket. At higher altitudes in the ericaceous and paramos or alpine belts water availability will in general be less than in the upland forest belt, and in some years there may even be acute shortage of water. The total annual rainfall is markedly lower than in the upland forest belt, it is highly erratic, and low temperatures may render the soil water less accessible to the vegetation. The water regime of natural bamboo thicket and of planted conifers both at about 2600 m in the Aberdares was studied over a number of years and no water stress was detected. (Pereira and Hosegood, 1962a).

The altitudinal range of a species is also influenced by its competition with other species. Many plants freed from competition can grow, flower and seed in climates substantially different from those under

which they usually occur. This is clearly demonstrated by the fact that a number of species from the ericaceous belt of Mt Elgon were successfully transplanted and grown in the University Botanical Garden in Kampala at an altitude 1200 m. In the same garden it has been possible to grow *Lobelia gibberoa* and *Rapanea rhododendroides* both taken from about 3230 m in the *Hagenia-Rapanea* zone of Mt Elgon. However, the slow growth rate of the *Rapanea* in Kampala leaves no doubt that if left to compete with the indigenous flora it would soon succumb from competition. *Hagenia* is obviously capable of growing through an altitudinal range of 2200–3400 m though, at the lower altitudes, it appears to be confined to forest edge and clearings while at the top of the forest belt it forms climax woodland. Again, *Podocarpus malanjianus*, which is usually seen most fully developed in the highlands at about 2000 m, also occurs plentifully in the seasonally swampy forests on the west shore of Lake Victoria. Slightly further north in broad-leaved angiosperm forest, on soils that are not swampy, it is rare or absent. Perhaps it is excluded from these mainly by competition.

These few examples, and those of gardens throughout the world, prove clearly that many plants do not necessarily show their optimum growth only in the climate under which they naturally occur. Nor should we assume that the best growth or maximum production of a species occurs inside its present range. Consider the luxuriant growth of *Salvinia auriculata* after its appearance in Lake Kariba or *Galinsoga parviflora* a native of South America and now abundant on cultivated land throughout East Africa and Europe. Brenan and Greenway (1949) includes numerous references to plants which are acclimatised successfully in Tanzania usually following their introduction at the former research station at Amani.

Soil temperature

Soil temperatures are important in controlling seed germination, leaf expansion rate, root growth, root permeability and evaporation. Unfortunately few data are available on soil temperatures in East Africa.

Banage and Visser (1967) measured temperatures on a bare soil surface and at 7·6 cm depth near Kampala, which is at an altitude of 1300 m and has a somewhat cloudy climate. Their results are summarised in Table 6.1 It will be noticed that there is a reduction in the range of soil temperature with depth. The figures given in Table 6.2 summarise information for stations representative of the coast, Kenya Highlands and the Lake Victoria area.

Table 6.1 Mean temperatures (°C) in soils at Kabanyolo, 21 km north of Kampala, Uganda. Altitude 1317 m. (From Banage and Visser, 1962)

	Soil surface		7·6 cm below soil surface	
	range	average	range	average
Daily mean	28–29	28·5	24–26	25·0
Daily mean, maximum	38–42	40·0	29–32	30·5
Daily mean, minimum	17–18	17·5	20–21	20·5
Diurnal range dry soil	30–38	34·0	14–17	15·5
wet soil	12–17	14·5	6–10	8·0

Table 6.2 Soil temperatures at selected places (after Griffiths, 1962)

Place	Altitude (m)	Mean air temperature °(C)	Mean soil temperature °(C)	Mean range at 15 cm 122 cm		Excess of mean soil temp. over mean air temp.
Dar es Salaam	0	26	29	7	4	3
Entebbe	1189	22	24	–	1	2
Kabete (Nairobi)	1829	18	22	5	2	4
Muguga (near Nairobi)	2103	16	21	6	2	5

We know of no measurements from the wooded grasslands in East Africa, but Lawson, Jenik and Armstrong-Mensah (1968) from West Africa have recorded soil temperatures at different depths from different parts of a soil catena under savanna vegetation (see Fig. 6.3). The highest ground in the area studied was not more than 275 m above sea level. The diurnal range of air temperatures at 10, 35, and 150 cm above the ground approached 20°C. The diurnal range of soil temperatures near the surface was about the same, but at 20 cm below the surface the range fell to about 7°C on the middle slope of the catena.

Although the magnitude of these diurnal soil temperature changes will differ according to the climate and the character of the overlying vegetation, it is clear that in many of the more open vegetation types, organisms such as bacteria, blue-green algae, and plants at the surface or with roots close to the surface, will be exposed to quite substantial fluctuations of temperature. But it must be remembered that as the temperature rises at the surface so also the amount of convection increases, which, in turn, has a regulating influence on the temperature.

A vegetation cover will strongly modify the diurnal temperature fluctuations in the soil, and its extent will depend on various characteristics, such as its height, colour and density. The greatest modification is under forest which favours very equable soil temperatures even quite close to the soil surface. Weber (1959) reports isothermal temperatures (26±0·3°C) in tropical forest in Panama at subsurface depths of 10, 20 and 30 cm, while in clearings in the forest, at the same depths, the range was 28·6–30·8°C. High altitude vegetation experiences great air and soil diurnal temperature changes as does the vegetation of semi-arid areas.

Harsh soil surface temperature regimes may be experienced also under humid grasslands after burning. Fire sweeping rapidly across the grassland may momentarily raise the temperature of the soil surface to high values, and subsequently there is a more lasting effect as a result of the removal of the blanket of vegetation and litter. During the day the dark ash on the soil surface will be a good absorber of incoming solar radiation and the soil surface temperatures will rise to higher values than under the previous grass cover. At night the black soil surface will emit radiation and the outflow will be unimpeded so that temperatures will probably be lower than recorded under the usual grass cover. The initial heat caused by the passage of the fire, and the subsequent stronger daily temperature fluctuations, may be important in triggering off the germination of seeds at the surface or just below it. Although some seeds germinate readily if maintained at one specific temperature, other examples are known where periodic alternation of temperature is required for germination.

In Queen Elizabeth National Park in Uganda, Lock (1967) investigated the effect of soil temperatures and grass fires on the germination of the 'seeds' of *Themeda triandra*. His observations were made on dry soils and with fire at least as vigorous as normally encountered. At a depth of 1 cm he recorded a maximum soil temperature of 45°C, the normal maximum under grass cover being 29°C. The degree of heating by the fire depends, of course, on the wetness of the soil, the initial soil temperature, and on the amount of fuel available to the fire. Lock found marked differences between the soil temperatures of burnt and unburnt areas of *Themeda* grassland (see Fig. 6.4). The maxima

at a depth of 1 cm differed by 8·9°C, a difference which, if accompanied by rain after what is effectively a period of dry storage, may be large enough to break the dormancy of some seeds.

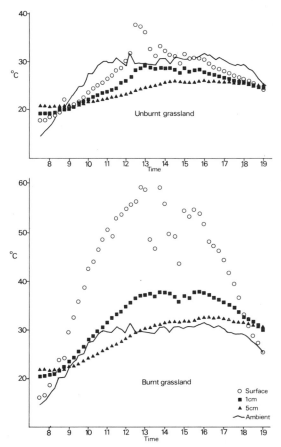

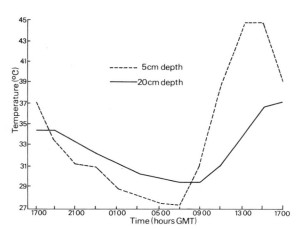

Fig. 6.3 Daily march of soil temperature at 5 cm and 20 cm depth in the middle-slope savanna near Lovi Camp, West Africa. (1–2 April 1967)

Fig. 6.4 Soil temperatures in burnt and unburnt *Themeda* grassland. (Lock, 1967)

Seven

Soils

Factors determining soil type

Parent material, climate, topography, age, vegetation and previous land use have interacted to produce a considerable variety of soil types in East Africa. For example, Anderson (1963) lists fourteen main soil types in Tanzania. It seems that no one physical factor can be said to have a directing influence on soil development in East Africa as a whole. Milne (1936) was of the opinion that if we apply each hypothesis in turn, supposing that climate, or parent material, or topography, provides the key to the distribution of soil types, then each gives encouraging results in some region but fails in others. Each of the major factors tends to govern soil development only towards the extreme of its range. Thus, regions with similar climates have similar soils only when their topography and drainage, their surface geology and their past history of land utilisation have much in common.

Climate

Climate is most important in determining the soil type in the wetter regions. Thus, those regions with an average annual precipitation of 1500 to 2000 mm or more, nearly always have red soils or *latosols* (Ellis, 1952). Given sufficient rainfall and free drainage through the soil, these red soils develop over a wide range of parent materials. They are known to have developed out of such diverse and dissimilar materials as granites, basalts, sandstones, limestones, shales, schists, dolerites and pegmatites, and they occur on steep mountain slopes and on gently undulating surfaces. The highly leached red earth overlying the rock is the end-product of weathering under generous rainfall and free through-drainage, and in East Africa could be regarded as a climatic type. Differences in rock formation and geological origin are largely masked and rendered subordinate by the dominating climatic influence.

There are in East Africa red soils occurring outside the limits of what may be regarded as the climatic domain of such leached soils. The semi-arid conditions under which they now occur—for example, over much of the Masai plain—would appear not to provide the necessary leaching conditions for their development. Thus, it is thought that these soils are relics of earlier wetter conditions; that in the fairly recent past the climate became drier, and they still carry the impress of former conditions.

Parent material

Parent material is a factor of increasing importance in determining the soil type as one passes from wetter to drier regions. We might expect the greatest diversity of soils overlying different sedimentary rocks because the chemistry and mineralogy of these is much more varied than that of the igneous rocks. For example, a sandstone may consist of up to 99 per cent SiO_2, whereas a pure limestone, such as we would find on the coastal corals may have as much as 56 per cent of its weight as CaO. Sedimentary rocks are common along the coast facing the Indian Ocean and a broad wedge of Cretaceous and Jurassic sandstones and limestones extends inland over south-eastern Tanzania.

Erosion and weathering (Plate 43)

It is important to remember that in many parts of East Africa the soils are not weathering directly out of the parent rock. As a consequence of the great age of the landscape, and the several cycles of erosion or planation that have passed over it, the soil is often forming out of already deeply-weathered and impoverished debris which began to accumulate on the

43 Erosion gulley now used as hippopotamus track, Murchison Park. (Osmaston)

basement rock complex in very remote geological time. Geologists believe that the land over much of Africa was tectonically quiescent for an immense stretch of time, and during this time the landscape was reduced by erosion to a relatively featureless plain, usually termed a peneplain. Beneath this plain, chemical rotting of the rock proceeded down to depths of some hundreds of metres, even where the rock was resistant quartzite. The junction between the weathered rock and the underlying surface was sharp but very uneven. This irregular rock surface is known as the 'basal surface of weathering' or simply as the 'basal surface' (see Fig. 7.1).

The *gross* effect of the renewed erosion following tectonic uplifts was to remove some of the rotted rock and to expose the most elevated parts of the basal surface as tors or rocky hills. However, although several cycles of erosion passed over the landscape, and cleared away as much as a hundred metres of rotted rock, the basal surface was not extensively exposed and in many places the tors are separated by countryside still underlain by deep layers of thoroughly weathered rock (see Fig. 7.1), and various types of laterite. We

should expect upper hillslope soils, weathered directly and relatively recently from the living rock, to be relatively rich in primary minerals, while red soils on the lower part of the hillslope should be very poor in primary minerals because they have been derived and weathered from material already weathered and degraded over an immense period of time. The parent

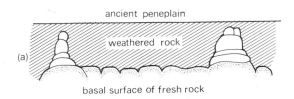

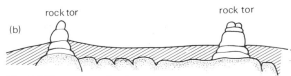

Fig. 7.1 Formation of rock tors.

materials of these lower hillslope soils will consist mainly of iron oxides, and various secondary minerals, principally kaolinite, together with only those primary minerals most resistant to weathering, namely, quartz, zircon, and tourmaline. The absence or scarcity of unweathered minerals, which release elements on weathering, renders these soils intrinsically base poor. The bulk of their nutrient supply resides in the humus complex in the topsoil; there it is held in an easily exchangeable form, and under moist conditions can be taken up by plant roots.

Topography

For the most part, in East Africa, we are not concerned with uninterrupted stretches of one soil type determined by climate or parent materials, but rather with soil complexes in which several soil types are in close spatial association. Milne (1935) drew attention to the importance of topography in giving rise to these compound soil units. He illustrated this feature by reference to southern Uganda where considerable areas of the landscape consist of little else but a repetition of crests and hollows. The crests and slopes carry well-drained red loams; the hollows are swampy and have black or grey clays; and murram soils occur towards the foot of the slope. A traverse from a point on a crest to a point in a hollow takes us over a range of differences in soil profiles that cannot be included in any grouping that depends on morphological similarity. The range is really continuous although, within it, it is possible to pick out three or four distinctive soil types. Milne considered that this type of soil sequence was of such common occurrence that it should be designated by a distinctive word. He proposed the word 'catena' (Latin, a chain). He believed that this term would help to indicate that the soils so grouped are linked by their topographic relationship. Milne also envisaged that the catena concept might find a wider application, as indeed it has, for it is now applied to the sequence of vegetation zones which occur corresponding to the soils. It could also be applied—on a grand scale—to describe the sequence of climatic and vegetation belts which surround mountains like Mt Kenya and Mt Kilimanjaro. As a mapping unit it has great utility; it means that we need not represent the soils of a region of undulating topography solely by one soil type. Prior to Milne's concept the well-drained soil of the hillside would have been singled out as the zonal type and represented on a soil map, and the ill-drained soils of the valley floors would have been relegated to intrazonal status and suppressed.

The soil catena

On account of the prominence of catenary soil associations in eastern Africa it will be well now to give examples and subsequently to consider more fully the processes involved in their development.

In describing these soil catenas we have not attempted to name the soil types according to any of the more elaborate or developed schemes of classification which have been proposed. There is as yet no generally accepted scheme of soil classification for East Africa, or indeed for tropical soils as a whole. Moreover, there are many views even on the correct naming of some of the more important soil types. The difficulties of the various major systems have been discussed in some detail by Chenery (1960). Anderson (1963) in Tanzania, used a very simple scheme; Bellis (1964) outlined a more elaborate scheme for Kenyan soils. Readers interested in problems and concepts of soil classification should refer to Chenery (1960) and to some of the following papers on this topic: Milne (1935); Kellogg (1949); Leeper (1956); Kubiena (1958); Manil (1959); Basinski (1959); Prescott and Pendleton (1952); D'Hoore (1960); Wambeka (1962). Soils in the following account are designated by some very obvious characteristic feature, such as red soils, brown soils, grey soils. In doing this we are, of course, well aware that not all red soils have been formed from the same parent material nor by the same soil-forming process. It is not primarily by their colours, but by the characters of their profiles that soils should be distinguished from one another.

Buwekula catena (Uganda)

Since Uganda is the type locality for soil catenas (Milne, 1935, 1936) we begin with a description of the Buwekula catena (Fig. 7.2). This was studied in some detail by Radwanski and Ollier (1959). It lies about 97 km west of Kampala and close to Mubende. The annual rainfall of the area is between 900–1126 mm. The basal rock surface in the area is overlain by deeply-weathered material and many of the hilltops are crowned with rocky tors. It is the kind of situation represented in Fig. 7.1(b).

Five soil types were recognised. From the hilltop downwards we find (a) shallow skeletal soil; (b) red earth; (c) brown earth; (d) yellow-brown soil; and (e) grey soil; the last two soils being alluvial in origin. The vegetation consisted of *Themeda triandra* grassland over the narrow zone of skeletal soil surrounding the hilltop tor. The remainder of the hillside had grassland dominated by *Cymbopogon excavatus*, *Brachiaria soluta* and *Brachiaria brizantha*, with shrubs of *Vernonia amygdalina* and *V. smithiana*. According to Radwanski and Ollier (1959) the vegetation changed very little on moving downslope. Permanently water-logged soils in wider valleys of the same area carry swamps of papyrus and/or *Miscanthidium*.

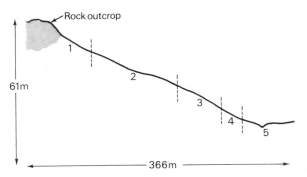

Fig. 7.2 The Buwekula catena. (1) Skeletal soils. (2) Red earth zone. (3) Brown earth zone. (4) Yellow-brown soil zone. (5) Grey soil zone. (Radwanski and Ollier, 1959)

Skeletal, shallow soil zone

The bare rock on the hilltop is exposed to alternate heating during the day and cooling at night. This slowly breaks up the rock into small fragments which collect at the foot of the outcrop. This gritty material forms a narrow soil zone. The soil is usually shallow but, in pockets, may be as much as 60 cm deep with boulders reaching the surface alongside them. The rock itself contains many cracks and fissures and acts as a water-collecting surface, rainwater penetrating these fissures and emerging at the floor of the rock as springs. Thus the skeletal soil surrounding the tor is subject to substantial eluviation. This accounts for the high content of residual quartz in its upper horizons.

Red earth zone

This occurs on the upper middle and mid slopes of the hill and is the most extensive soil of the catena. It could be regarded as the zonal soil of the area. Unlike the shallow skeletal soil of the preceding zone, its character is not dominated by the nature of the basement complex rock. The parent material of this soil may have undergone more than one cycle of weathering, as indicated in Fig. 7.7. Consequently, the primary minerals have been thoroughly decomposed,

apart from the most resistant, and the soil is deeply and thoroughly weathered. This is reflected in low pH values throughout the soil profile. It is a *latosol* type of soil. The loss of clay by eluviation from the upper part of the soil profile permits relatively rapid internal drainage while the underlying layers are only slowly permeable. Due to the steepness of the slope, movement of water downslope is as important as down-profile movement. This prevents the formation of a true perched water table above the slowly permeable horizon, although the latter becomes completely saturated in the rainy season. Thus, the mottling in the lower part of the profile and the precipitation of laterite takes place under alternate wetting and drying.

The whole profile is in equilibrium with the present environment and the development of all its horizons is keeping pace with all the processes responsible for their removal downslope (see Fig. 7.7). Thus, although the profile as a whole is lowered the profile morphology remains unaltered.

The brown soil zone

This soil differs in colour and texture from the red soil. It has a higher quantity of quartz gravel in its upper layers, indicating a more intensive eluviation of clay. Ground water is moving laterally, but, owing to the lower position of the brown soil zone on the topography, a lot of water derived from the upper part of the hill passes through the brown soil profile. It therefore undergoes considerable eluviation and this process is repeated each rainy season. The lower horizons of the brown soil remain moist longer after the rains than those of the red soil, although the upper light-textured part of the profile drains quickly.

The yellow-brown soil and grey soil zones

Because of the relatively steep topography, the erosion products from upslope are moved straight out onto the valley bottom and subsequently carried away by running water in the wet season. They do not therefore, as in some catenas on less steep hills, accumulate to give a zone of hillwash soils at the foot of the slope.

It is envisaged that initially the grey type of soil covered most of the valley floor. As the stream cut down more deeply, the upper layers of the grey soil emerged above the reach of the ground water. Due to the improved drainage and aeration these upper layers changed in colour from grey to yellow-brown.

In the wet season the grey soil, which is only a few feet above the level of the stream, is completely submerged. In larger, flat-bottomed valleys such soils are permanently waterlogged and remain under swamp vegetation of *Cyperus papyrus* and/or *Miscanthidium*.

Other Uganda catenas

Thomas (1945–46) has given a very good impression of the vegetation in different parts of southern Uganda by describing some catenas in different places ranging from the Sese Islands in the south to Seuajongo on the Kafu river about 400 km farther north. On many of these hillsides the human factor has strongly modified the vegetation and the soils, but the catena concept is still very relevant. All parts of the slope from hilltop to valley bottom make special contributions to the life of the people and land is divided up to allow access to all the catenary zones, the boundaries between villages usually lying along the centres of the valleys.

Lang Brown and Harrop (1962) have described a soil-vegetation catena from the Kibale grasslands in the western province of Uganda (Fig. 7.3). The distribution of vegetation and bases on this catena is interesting since it shows an increase of bases as one proceeds upslope out of the forest and into the grassland, and an unusually high concentration of phosphorus on the hilltop. Thomas (1945–46) suspected that the upslope soils of many Ugandan catenas are richer than the soils of the lower-lying zones. He attributed this to the fact that cattle are kraaled at night upslope and around the houses, the latter being placed there to avoid the mosquitoes of the valley swamp. Nutrients exported upslope via the dung would in the

course of time substantially enrich the hillslope red earths. However, Lang Brown and Harrop considered that this explanation might not apply to the Kibale areas because the Batoro, who inhabit the area, do not build on the hilltops although they graze their cattle there. They considered that the upslope bases were derived from weathered hillside rocks and shelves of laterite. They conceded, however, that the high quantities of phosphate on the hilltops themselves might have been due to settlement there by some tribe other than the Batoro. But they advanced another possibility, namely, a volcanic source. They pointed out that there have been volcanic eruptions from explosion craters to the immediate west of Kibale and in a number of places further to the south. These occurred in relatively recent times and layers of tuff—which has a high phosphate content there—lie on the surface of the ground over red earths for a radius of several kilometres. On the non-eroding hilltops some of the tuff might have remained longer than on the hillsides and contributed to the phosphate content of the hilltop soils.

Catenas (northern Tanzania)

In Tanzania, Milne (1947), Burtt (1942), and Calton (1963) have described the appearance of soil and vegetation catenas. Many of the catenas developed on long, gentle slopes under medium rainfall in northern Tanzania show a very clear sequence of soil and vegetation types. Figures 7.4 and 7.5 illustrate some catenas described from northern Tanzania by Milne

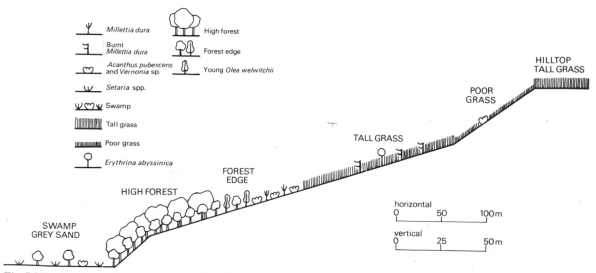

Fig. 7.3 Profile of a typical transect through Kibale grasslands. (Lang-Brown and Harrap, 1962)

and Burtt (Milne, 1947).

Generally, these northern Tanzania catenas include about six main soil types. Skeletal soils surround the hilltop rock outcrops and the sequence downslope passes through red earths, pale red earths, brown earths, hardpan soils, and ends in grey and black clays on the valley floor. The pale red earths which cover the broadest zone of the hillslope are the most weathered and leached of the soils and, over a long course of time, this zone expands upslope and downslope so that ultimately the rock outcrop, skeletal soils, and red earths will be succeeded by pale red earths. To convey more fully the nature of such a catena we will describe its soil types in some detail and indicate the type of vegetation which each soil supports. Imagine a hilltop in northern Tanzania crowned with outcrops of granite.

Skeletal soil zone

Physical weathering, and to some extent the disintegrating action of lichens and small plants, breaks up the rock outcrop slowly to produce a narrow belt of coarse rock debris at the foot of the rock; analogous debris occurs below cliffs and escarpments. These initial soils, in which rock fragments predominate, occupy the highest zone of the catena. The rock fragments usually overlie a thin layer of dark brown or black gritty loam. Such soils are free draining, but they may have favourable moisture relations for plant growth because of water draining off nearby rock faces. In desert areas like the Turkana and Chalbi deserts of northern Kenya these skeletal soils support

a fringe of thorn trees around the rock outcrops; the commonest trees being *Acacia senegal*, *A. mellifera*, *Boswellia hildebrandtii* and *Commiphora* spp. In Tanzania, under slightly moister conditions, they are reported to carry (Milne, 1947; Burtt, 1942) thickets of *Euphorbia* spp., *Sansevieria* spp., other succulents and the trees *Fagara* and *Pterocarpus*. With enough rainfall or catchment from the nearby rocks, their high mineral content may render them desirable for cultivation. However, this is liable to lead to erosion since they are often on the steepest part of the slope. Anderson (1963) maintains that they are better looked upon as regulators of the run off from the bare rock outcrop and should be kept under bush.

The red soil zone

The skeletal soils, especially if derived from acidic rocks, such as gneisses, granites and sandstones, weather under humid climates to give some form of red earth or *latosol*. They are characterised by a low base-exchange capacity, a usually low silica sesquioxide of iron ratio of the clay fraction, low activity of the clay, low content of most primary minerals, and of soluble constituents, and a high degree of crumb or aggregate structure.

In many of the Tanzania catenas the bright red earths—the colour incidentally refers to the subsoil not to the topsoil—occur on only the uppermost part of the hillslope, forming a relatively narrow zone beyond the skeletal soils. They grade downslope into paler red earths.

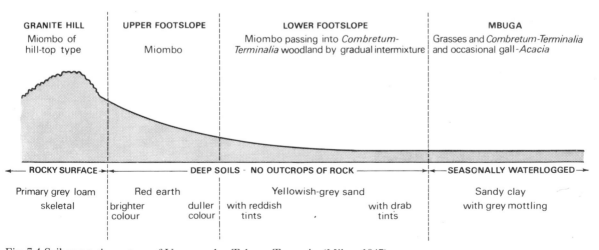

Fig. 7.4 Soil-vegetation catena of Unyanyanbe, Tabora, Tanzania. (Milne, 1947)

The pale red soil zone

This is the common soil type on hill-slopes in the region of Tanzania between Tabora and the Rungwa river. It is known there as *lusangu* or *lusenye*. The topsoil (pH 5·5–5·8) is grey-brown to dull yellow in colour, and becomes progressively paler with depth, developing slight rusty flakes and mottlings. The base of the soil has dull brown murram (or pisolitic laterite) overlying the weathered rock.

Although very sandy it is possible that some of these pale red earths have a better water supply during the dry season than the more clayey red soil upslope. They are not so high above the water table and they probably receive some water by lateral seepage from higher ground, and this will bring in plant nutrients. Thus the soils themselves, although intrinsically infertile and strongly leached, may be relatively productive on account of seepage.

Both the red soils and the pale red soils or yellowish soils on the upper part of the catena would be classified as *latosols* or *laterosols*. According to Buckman and Brady (1969), in the genesis of latosols through latosolization, the soluble bases such as calcium, magnesium, potassium and sodium are quickly released maintaining a near neutral soil reaction. The solubility of silica, at low acidity, is high relative to the solubility of iron and aluminium and, if drainage is free, soils high in sesquioxides of iron and aluminium and relatively low in silica are the end-product of the weathering process. The iron oxides are responsible for the red and yellow hues.

Brown earths of high base status

Being low in the catenary sequence the brown earths are not subject to through vertical leaching; they receive considerable water by lateral seepage. This water will be rich in silica and bases. Under these conditions, though we do not know precisely how, montmorillonitic clays are formed and these become saturated or nearly saturated with calcium, excess calcium giving rise to concretions in the subsoil. These brown earths, of fairly heavy texture, are highly fertile but unfortunately the conditions under which they form occur usually in only a narrow zone of the catena. In parts of Tanzania it is these earths which are used for cultivation.

Hardpan soil zone

The sorting of the soil particles as they move downhill by creep and erosion often results, on long gentle slopes, in a narrow zone very rich in coarse sand just above the valley floor. The illuvial clay in this zone is in a deflocculated state on account of the presence of much magnesium and sodium, brought in by drainage. Consequently the clay is readily eluviated down the soil profile where it fills up the pores between the sand grains and creates a hard pan. Sometimes the topsoil is a loose sand and in wet weather it may be possible to dig a short depth below this, but in dry weather the underlying clay and coarse sand set with cement-like hardness.

These hardpan soils do not shrink much on drying, and rainwater penetrates only the top few centimetres of the profile, the lower soil layers remaining hard and dry. Since these hardpan soils occur at the foot of slopes and slightly above the valley bottom, any rain water, in excess of that absorbed, runs off onto lower ground. Thus the moisture relations in the topsoils are unfavourable for plant growth. The vegetation they

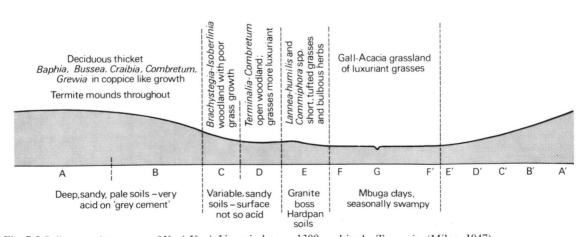

Fig. 7.5 Soil-vegetation catena of Kazi-Kazi, Uyansi plateau, 1300 m altitude, Tanzania. (Milne, 1947)

support is often of a more arid type than that on neighbouring soils. Generally, it is thornbush with clumps of shrubs and/or small trees separated by wider spaces with a thin cover of ephemeral grasses and herbs. The grasses and herbs live on the moisture in the surface horizons of the soil while the roots of some of the shrubs and trees penetrate the hardpan, though many may rely on a wide spread of roots at a shallow depth for recruiting water.

Burtt (1942) has described and illustrated the following kinds of thornbush on hardpan soils in the Lake and Central Provinces of Tanzania: (a) *Commiphora schimperi–Lannea humilis;* (b) *Acacia drepanolobium;* (c) *A. fischeri;* (d) *A. clavigera* sp. *usambarensis;* (e) *A. tanganyikensis;* (f) *Commiphora campestris.*

Scattered over the hardpan soils there are conical termite mounds which are often surmounted by bush species not occurring on the flat. These mounds usually contain carbonates, and the bicarbonate-rich rain water shed from them is a further factor to be related to the genesis of the hardpan soils.

Black and grey clays of the valley floor

Some of the clay which is transported downslope by creep and erosion is carried away by flood water, but some remains and builds up the black and grey clays or earths of the valley floor. The vegetation on these dark clays is nearly always grassland with a few scattered bushes and trees, the latter often confined to termite mounds.

The drainage water reaching the valley floor is relatively rich in bases and some of the calcium which is absorbed onto the clay and humus colloids gives rise to a dark-coloured complex which is responsible for the colour of the valley soils (D'Hoore, 1955; Dudal, 1963; Singh, 1956). The dark colour is only rarely due to the persistence of dark-coloured primary minerals or to large amounts of organic matter. For example, the black clays of Karamoja, Uganda, have—according to Wilson (1960)—never more than 1 per cent organic carbon and often have less. Calcium, in excess of that absorbed onto the clay-humus complex usually precipitates in the soil to give concretions of calcium carbonate.

The predominance of montmorillonite and/or illite clays in the black earths exaggerates the effects of drought, because these clays swell and become nearly impervious when wet, thus increasing run off, and their marked shrinkage on drying results in broad, deep cracks which thoroughly dessicate the soil; the more the clay, the wider the cracks. The type of clay will also have some influence; illites, on the whole, undergo

much less expansion with wetting than do montmorillonites. When dry, the black clays are so hard that a pickaxe often makes little impression. The alternation of shrinking and swelling, together with loose material falling into the cracks, makes for uniformity with depth and sometimes creates a hummocky soil surface with depressions 2 to 3 m across. This is known as 'gilgai' relief and has been described for parts of Kenya by Stephen, Bellis and Muir (1956).

The amount of calcium carbonate in the soil and its distribution varies. Not all black clays are calcareous; sometimes, non-calcareous clays or more accurately clays with very little calcium and tending to grey in colour, are found where incoming and outgoing calcium balance. In Tanzania calcareous black clays commonly occur in the extensive valley floors or mbugas of Central, Lake, and Northern Provinces. In Kenya continuous areas are reported over level landscape at about 1900 m throughout Laikipia and North Nyeri, extending northwards from the foothills of the Aberdare Mountains and Mount Kenya towards the Uaso Nyiro river, and also on the Athi and Kapiti plains to the south of Nairobi. In Uganda, calcareous black clays form extensive sheets over large parts of the Northern and Eastern Provinces, around Lake Kyoga, and over considerable areas of the floor of the Western Rift Valley adjoining Lakes Edward, George and Albert. An extensive area of non-calcareous black clays is reported for the Kinangop plateau in Kenya extending some 50 km to the north-north-west just above Kijabe on the eastern side of the Eastern Rift Valley.

Many of the black clays, calcareous and non-calcareous, carry grassland with a few *Acacia* trees and they are often used for grazing. Under annual rainfall of 380–635 mm, the black clays often have *Acacia mellifera* wooded grassland which, with heavy grazing gives *A. mellifera* bushland. With increasing rainfall *A. seyal* becomes more important with occasional specimens of *A. drepanolobium*. Above an annual rainfall of about 760 mm *A. drepanolobium* often occurs scattered through grassland with much *Setaria incrassata.*

Some sites with black clays are waterlogged for several months of the year and they bear a characteristic grassland, often with *Phoenix* palms on termite mounds.

On the whole, the black clays are difficult to cultivate and they are not usually—as their common name black cotton soils suggests—suitable for cotton, but they are suitable for maize and sorghum. The structure of the topsoil often improves after two or three years of cultivation. Sites prone to waterlogging can be subdivided into compounds and used, as in parts of Tanzania, for growing rice.

Dark-coloured soils are not confined to valleys, plains and depressions. Such soils occur also, especially over volcanic deposits, on the lower slopes of some of the mountains, for example, Mt Elgon. Some of the dark colour of these soils may be due, as Ollier and Harrop (1959) suggest, to ferromagnesian minerals in the underlying rock and not alone to the calcium-clay-humus complex. Prolonged weathering would, over a very long period of time, presumably convert these soils by laterisation to red earths. Some of these dark mountain soils contain montmorillonitic clays and, although these can absorb a high quantity of exchangeable bases, they have often become sufficiently desaturated to render the soils acid, thus a low pH and a quite high base content are typical of some of them.

Halomorphic soils

In hot places with a high evaporative demand the valley soils of the catena are characterised sometimes by very high concentrations of salts in their upper horizons. Because of the effect of these salts on their structure the soils are termed 'halomorphic'. If there is very much exchangeable sodium in the soil enough may go into the soil solution to react there with carbonate and bicarbonate ions to give a solution of sodium bicarbonate which raises the soil pH to 9 or more. These soils are then referred to as saline-alkaline soils. If, however, the percentage of sodium is less than 15 per cent then the soil is termed saline and the pH is sometimes less than 8·5. The soluble salts present are mostly neutral, and, as they are dominant, there is only a small amount of exchangeable sodium present.

Many of the flat valleys in the drier parts of Tanzania have alkaline soils. The alkalinity causes a thorough dispersal of the clay particles so that permeability is low and drainage poor. Much of the rain is lost by run off and by evaporation. The vegetation of these alkaline soils is typical of land receiving a lower rainfall.

Anderson (1963) reports that alkaline soils on the Usangu plains in Tanzania support a bushland of *Commiphora* spp., *Acacia kirkii* and *A. stuhlmannii*, while adjacent, more permeable soils, have *A. tortilis* ssp. *spirocarpa* woodland. There are large areas of saline and alkaline soils on the flood plain of the Pangani and Mkomazi rivers in Tanzania; in places the

vegetation is tall grasses and sedges with the shrubs *Pluchea discoridis* and *Sesbania sesban* and the tree *Acacia xanthophloea*. The head waters of the Pangani river lie on volcanic deposits of Mts Meru and Kilimanjaro, and these deposits weather rapidly releasing a relatively large amount of salts into the drainage water. After leaving the mountains the river crosses a flattish area of low rainfall which becomes flooded during the wet season, but the flood water as it evaporates leaves its salts behind in and on the soils. These saline soils are sometimes detectable by a white efflorescence of salt crystals on the sides of trenches and erosion gulleys. Some plants such as: *Salvadora persica*, *Sporobolus robustus*, *Suaeda monoica*, *Triplocephalum holstii* and *Volkensia prostrata* are considered to be salt indicators. Minor areas of saline soils occur on sisal estates along the Ngerengere river and close to the Uluguru mountains in Morogoro district, Tanzania. In Kenya there are saline soils at the seaward end of the Tana river. Other saline soils occur around some of the lakes such as Lake Manyara and around the salt lakes in Uganda and Kenya where the chief plants are *Cyperus laevigata*, *Sporobolus spicatus* and *Dictyotaenium* sp.

Bogdan (1958) has described how varying degrees of soil alkalinity affect the composition of grassland under *Acacia xanthophloea* woodland in Kenya. He refers to flats of seasonally waterlogged, black soils close to the Kiboko river, 164 km from Nairobi on the main Nairobi–Mombasa road. The flats are about 900 m above sea level and the region is arid receiving a mean annual rainfall of 603 mm, biseasonally distributed. Bogdan states that:

Under 'normal' soil conditions, i.e. when the soil is only slightly waterlogged and slightly alkaline, the grass cover is formed mainly of *Cenchrus ciliaris*, in almost pure stand, or together with *Sporobolus fimbriatus* and other species. When the soil becomes slightly more alkaline *Chloris gayana* appears and forms either pure stands or is mixed with *Cenchrus ciliaris* and often also with *Sporobolus fimbriatus*. In places, where waterlogging occurs for longer periods, *Echinochloa haploclada* appears, usually in a mixture with *Chloris gayana* or some other species. With increasing moisture and alkalinity *Sporobolus robustus* comes into the grass cover, often as a dominant species. In fairly moist and alkaline habitats but with more shallow soil, *Cynodon dactylon* forms a low, dense, almost pure stand. This is a small and fine type of *Cynodon dactylon* with very numerous fine and wiry stems and numerous short leaves, a distinct ecotype which occurs also in similar arid habitats in other arid areas of Kenya. It differs distinctly from a larger type which, in the experimental (Kiboko) area occurs in woodland on red sandy loam or on volcanic ash. Finally, in places where the soil is more alkaline (pH=8), shallow and completely waterlogged in the rainy season, *Sporobolus spicatus*, a low stoloniferous grass, forms dense growth. The following scheme illustrates the relationships between associations of wooded grassland with *Acacia xanthophloea*.

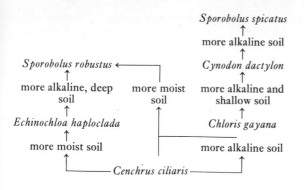

```
                              Sporobolus spicatus
                                      ↑
                               more alkaline soil
                                      ↑
Sporobolus robustus ←─────────┐  Cynodon dactylon
        ↑                     │        ↑
more alkaline, deep    more moist  more alkaline and
     soil               soil        shallow soil
        ↑                  ↑            ↑
Echinochloa haploclada     │        Chloris gayana
        ↑                  │            ↑
  more moist soil          │       more alkaline soil
        └──────────── Cenchrus ciliaris ──────────┘
```

Valley peats

In contrast to the halomorphic soils which form in valleys in hot, dry areas, peats accumulate in some of the cool valleys of the highlands. These peats are found mainly above 1800 m. They have been studied in south-west Uganda (Lind, 1956a; Morrison, 1961, 1966). Many of these highland peats develop beneath swamps of papyrus, *Cyperus latifolius, Cladium mariscus* or *Miscanthidium* and *Sphagnum*. Below 1800 m there are also very extensive swamps of papyrus and *Miscanthidium*, especially around the northern end of Lake Victoria. However, there is very little accumulation of peat under these; it is usually less than a metre in thickness. Presumably the absence of peat at low and medium altitudes is due to the greater activity of the bacteria responsible for decomposition of plant debris.

It is interesting to observe that deep organic deposits *do* accumulate at the base of lakes at low and medium altitudes. Moreover, this lake mud if brought to the surface and left under aerobic conditions shows little tendency to decompose. If boiled it decomposes rapidly. Beauchamp (1953) states that no satisfactory explanation for these facts has yet been reached. A tentative explanation is that antibiotic substances, originating in freshly precipitated plankton, inhibit the growth of the bacteria responsible for decomposition. Under tropical conditions, one might expect decomposition to proceed rapidly, but under a more or less continuous rain of plankton, carrying with it some antibiotic substance, the bacterial flora may be unable to flourish.

Large areas of the upland papyrus swamps, close to Kabale and Lake Mutanda in Uganda, have been reclaimed for cultivation. The peat, on drying, decomposes and releases elements. These are sufficient to support good crops of sweet potatoes, sorghum, and maize, when these are planted on cambered beds

about two metres wide. Useful grazing is possible on the reclaimed swamps where the water table is not allowed to get closer than about 15 cm from the surface.

In some places cultivation and drainage of the upland papyrus swamps led to complete sterility because of a build-up of excess acidity (pH $2 \cdot 4$–$2 \cdot 7$). Chenery (1952, 1954), who investigated these peats in some detail, showed that this extreme acidity could be due to the oxidation of sulphates to sulphuric acid. When the water table remains close to the surface an accumulation of sulphates, sulphuric acid and ferric iron occurs there by capillarity and by hydrolysis.

The sterile peats were reclaimed by lowering the water table and maintaining it about 30 cm below the surface. Also, dressings of lime were given (10 ton/ha $Ca(OH)_2$) and triple superphosphate (192 kg/hectare).

Further details of these interesting peats are given in Chenery's papers and in Harrop's memoir (1960).

Laterite

The term *laterite* was coined by Buchanan in 1807 to describe material which is quarried and trimmed into large building bricks in southern India. The derivation of the word is from the Latin *later*: a brick. The term thus refers to its use as a building material and *not* to its red colour. It is important to be clear how the term was used originally because the term *lateritic* is often misapplied to tropical soils on no better evidence than their red colour.

Pendleton and Sharasuvina (1946) believe that the term laterite should apply only to ferruginous rocks which are of secondary origin and are hard to the extent that they cannot be broken in the hand. The southern Indian material is fairly soft when first quarried but hardens after exposure to the atmosphere. Chenery (1960) states that in Uganda the term laterite is often synonymous with sheet ironstone or vermiform laterite. Semi-hard, indurated and other mottled material is sometimes referred to as 'plinthite'. Pea-like or pisolitic concretions of iron and aluminium oxides are commonly referred to as 'murram' or 'murrum'. These various types of laterite are related in space and time and it is perhaps best to allow the term 'laterite' to cover three main types recognised by McFarlane (1969). These are spaced pisolitic laterite or murram, massive vermiform laterite and packed pisolitic laterite.

In the highly weathered substratum of a maturing peneplain under moist, forested conditions, iron be-

Table 7.1 The average composition of laterites from Southern Uganda.
(From McFarlane, 1969)

	SiO_3 %	Ai_2O_3 %	Fe_2O_3 %	$Ai_2O_3+Fe_2O_3$ %
Vermiform laterite	22·6	24·5	37·6	62·1
Spaced pisolitic laterite	22·6	22·9	41·3	64·2
Packed pisolitic laterite	22·8	24·8	42·7	67·5
Pedogenetic laterite	13·3	16·8	57·9	74·7
Average	20·3	22·2	46·4	68·3

comes mobilised in the zone of water-table fluctuation. It precipitates in the kaolin-rich matrix in the form of hard pisoliths which enclose reddish saprolitic material. They are widely spaced in a matrix in which kaolin, goethite, and quartz predominate. The deposit as a whole is soft and *permeable* to water and does not harden en masse if exposed to the atmosphere. Under more hydrated conditions vermiform laterite may develop. This is characterised by narrow (20–30 mm) more or less vertical pipes, the walls of the pipes consisting of paper-thin, concentric layers of yellow and brown goethite. The pipes are usually filled with almost pure kaolin. The unexposed deposit is soft and *permeable* to water, but hardens on exposure to the atmosphere and becomes *impermeable* to water. The hard material is often still somewhat fragile, but it seems to be the nearest approach in East Africa to Buchanan's laterite. By erosion this vermiform laterite breaks down and, on hill slopes, gives rise to layers of packed pisolitic laterite, though packed pisolitic laterite can also arise from the downwasting of horizons of spaced pisolitic laterite. Packed pisolitic laterite is relatively hard and *impermeable* to water even before exposure to the atmosphere.

Chemically these various forms of laterite (see Table 7.1) are very similar, but there are consistent differences in the degree of hydration. Thus, pisolitic laterite contains a high proportion of haematite, while the more hydrated vermiform laterite is richer in goethite. Pedogenetic laterite is an iron-rich form of laterite which develops on the surface of platforms of vermiform and pisolitic laterite. Its development could be described as a lateritisation of laterite.

The development of spaced pisolitic and vermiform laterites may be regarded as precipitation features characteristic of the fully-developed latosol profile under prolonged, moist, forested conditions. Such precipitations may also occur under less moist climates where, for example, rock resistant to precipitation occurs at no great depth and results in temporary waterlogging of the upper weathering zone during the rainy season.

Figure 7.6a–d shows the possible mode of formation of various forms of laterite during the evolution of the peneplain. In the earlier stages, iron became concentrated locally in the kaolin-rich, weathered rock resulting in a layer of spaced pisolitic laterite (7.6a). Continued down-wasting at the surface and the encroachment of the soil on the underlying pisolitic-bearing stratum then led to the development of a sheet of packed pisolitic laterite towards the base of the soil (7.6b). On account of the close packing of the pisoliths and the cementing of these together by iron oxides and silica, this sheet of packed pisolitic laterite was relatively impermeable to water. This restriction on water movement, perhaps aggravated by a less freely draining surface on the mature peneplain, created conditions more favourable than previously to the mobilisation of iron. Consequently, re-solution of the packed laterite sheet took place and it was changed slowly into a horizon of vermiform laterite (7.6c) which is permeable to water so long as it remains unexposed to the atmosphere. The existence of such horizons of vermiform laterite had far-reaching effects on the further evolution of the landscape, soils and vegetation when further erosion cycles were initiated by tectonic uplifts. These resulted in the removal and redistribution of much of the rotted rock and laterite and the formation of the flat-topped mesas so characteristic of the Buganda landscape (Figs. 7.6e and f) and of some very complex hillside catenas.

Under prolonged, moist, forested conditions the entire latosol-laterite profile may be very deep. The vermiform laterite at the base of the latosol will remain permeable to water and this downward moving water, over very long periods of time, will create a deep *pallid zone* extending downwards for 30 m or more.

Many people assume that the same process is involved in the formation of laterite and lateritic soils or latosols. This assumption is not always correct. Although latosols may contain horizons of laterite or be underlain by various forms of laterite—often in the form of pisolitic or colitic concretions—the mode of formation of the latosol is not necessarily quite the same as that of laterite.

In many parts of East Africa soil formation and catenary development will only be correctly comprehended when it is recalled that weathering has been operating over *millions* of years, and within this immense span of time several major erosion cycles have moved across the landscape.

Laterite-modified catenas

We have described already how an integral part of the end-product of peneplanation in moist, forested areas was the development of horizons of laterite at the base of the soil profile. Where such peneplains, underlain by laterite, were incised by later cycles of erosion—initiated by successive tectonic uplifts—then extremely complex hillside catenas resulted (Fig. 7.7). This type of catena is very common in southern Uganda (McFarlane, 1969) and in western Kenya (Makin, 1969a). In such a catena the benches and platforms of laterite, with their different degrees of permeability to water, profoundly affect the development of soils and of vegetation. However, the full extent of the complex interrelationships of soil, vegetation and water movement are not yet completely understood.

The stages through which such a catena probably develops are depicted diagramatically in Fig. 7.6d–f and may be described as follows. The vermiform laterite (d) developed on the ancient peneplain is cut into by a new erosion cycle and gives rise downslope to sheets of packed pisolitic laterite (e). Where the erosion has produced gentle slopes, as in tributary valleys, in contrast to the steep slopes in main valleys, these sheets of packed pisolitic laterite are continuous and will often link together the higher-lying vermiform laterite. The packed pisolitic laterite is relatively

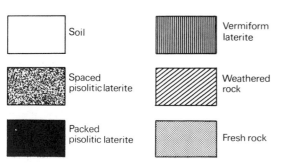

☐ Soil	▦ Vermiform laterite
▦ Spaced pisolitic laterite	▨ Weathered rock
■ Packed pisolitic laterite	▨ Fresh rock

Fig. 7.6 Evolution of land surface towards a peneplain, and effect of tectonic uplift. (After McFarlane, 1969)

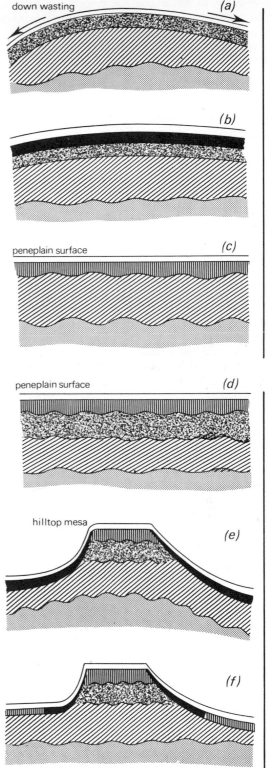

(a) down wasting

(b)

(c) peneplain surface

Evolution of land surface towards a peneplain

(d) peneplain surface

(e) hilltop mesa

(f)

Tectonic uplift - renewed erosion - water table lowering

impermeable to water and the restricted drainage, especially on the gentlest parts of the hillslope, leads to increased mobilisation of iron so that the packed pisolitic laterite slowly gives way to vermiform laterite. This process will be repeated with each tectonic uplift and the hillslope may eventually assume a gently stepped profile with a complex disposition of vermiform and pisolitic laterites (see Fig. 7.7).

Under relatively undisturbed forest, these various laterite platforms will probably remain buried under soil, but they will affect the movement of water in the hillslopes. McFarlane (1969) claims that unexposed vermiform laterite is permeable to water but unexposed pisolitic laterite is impermeable. Thus rain water from the upper part of the hillslope will, lower on the slope, flow over the surface of the packed pisolitic laterite and emerge, perhaps, as a line of ferruginous springs, giving rise downslope to a zone of wet, eluviated pediment.

With forest clearance, cultivation, grazing and burning the laterite platforms often lose their soil cover. Then, exposed to the atmosphere, even the vermiform laterite after hardening becomes impermeable to water. Water moving over these surfaces retards the development of deep soils and the surface remains free of soil for long periods. Large soil-free areas may be colonised by lichens and mosses. Shallow drifts of soil in hollows may encourage the growth of broad patches of the spreading herbs, such as *Aeolanthus* and *Cyanotis* and *Commelina*; species of *Borreria*, *Ipomoea*, *Indigofera* and *Murdannia* may root in crevices; and what appear like solution-hollows, commonly towards the edges of laterite platforms, may hold water for long periods. This gives rise to small, local hydroseres. The central part of the pool may have species of *Marsilea* and *Rotula*, while around the edges there will be species of *Ammannia*, *Micrargeria*, *Murdannia* and various sedges. Here and there shallow soils, often only about 10 cm in depth, cover the laterite and support short grassland of *Loudetia*

kagerensis and *Microchloa kunthii* and a very characteristic assemblage of associated species (see chapter 2). With deeper soils the short grassland gives way to taller grassland with species of *Hyparrhenia* and occasional low trees or bushes, such as *Acacia hockii*, *Combretum* spp. and *Hymenocardia acida*.

There is evidence that the thin soil on the laterite platforms is periodically lost through erosion. The eroding edge of the soil often exposes *mweso* grids— *mweso* being a game played by two people in a grid of usually 32 small hollows (8 × 4).

Podsols

Another soil type which, like vermiform and pisolitic laterite, results from the mobilisation and deposition of iron, is the podsol. This occurs on thoroughly leached soils. Characteristically this leaching produces an upper horizon *grey* in colour. Some of the colloidal humus and iron and aluminium leached from this upper horizon is usually deposited lower in the soil profile and the iron oxides often become cemented together to give an iron pan. This iron pan may be sufficiently well developed to impede drainage and root development.

At one time it was believed that the characteristic podsol could only form under a well-developed mat of surface organic debris, in other words, under the *mor* layer which is a feature of the temperate coniferous forests. However, Stephens (1950) drew attention to the almost complete absence of such a layer from Australian podsols. It is now clear that polyphenols leached by rain off overlying vegetation, and acids such as fulvic acid, produced in the decaying litter and organic matter of the topsoil, are the effective agents. These substances are active chelating compounds giving rise to mobile aluminium and iron organic

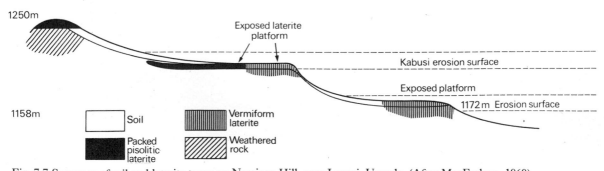

Fig. 7.7 Sequence of soil and laterite types on Naminya Hill, near Lugasi, Uganda. (After MacFarlane, 1969)

complexes which migrate in the soil profile.

It is not yet clear how commonly podsols occur in East Africa. Anderson (1963) makes no reference to their occurrence. However, they have been reported under wet *Ocotea-Podocarpus-Ficalhoa* forest at 2000 m on the Usambaras. Presumably they occur also in Kenya, for Bellis (1964) makes allowance for them in his proposed classification for Kenya soils. In Uganda they have been reported (Thomas 1942b; Radwanski, 1960) as low as 1700 m on the western side of Lake Victoria. They occur there under *Loudetia kagerensis-Eragrostis racemosa* grassland established over coarse sands in a rainfall between 1000 and 1500 mm. The typical profile consists of grey, coarse sand several feet thick and faintly to strongly mottled. The upper horizons frequently show signs of podsolisation. With the exception of phosphorus these soils have an acute shortage of all the major plant nutrients. Podsols are also reported in Uganda (Harrop, 1960) at higher altitudes, for example, the Lubare ridge in Ankole at 1800 m where they are developed over quartzites and under a grassland with bushes of *Philippia Protea* and *Pteridium aquilinum*.

At the highest altitudes, good examples of podsols seem to be rare. Their scarcity on the upper parts of the mountains may be due to extensive percolation of soil water on the sloping ground so that the surface soil layers never become markedly leached of iron and consequently retain a brown or reddish colour. It has also been suggested that soil animals, particularly earthworms, are so active at the higher altitudes that the chances of obtaining a well-stratified profile of the podsol type are low. However, there is as yet no evidence to show that the earthworm fauna is more plentiful and active at the higher altitudes.

It would be very interesting to know what factors, in the tropics, direct soil development on the one hand towards the podsol and on the other hand towards the latosol. *Broadly* speaking the genesis of both appear to be similar and to depend on mobilisation and precipitation of iron compounds. The prevalence of podsols, rather than latosols, in the wet, temperate regions of the world is suggestive that temperature is an important differentiating factor, but in the tropics and subtropics podsols and latosols occur within the same climatic regions. The *grey* upper horizon of the podsol is clear evidence of *very thorough* eluviation of iron compounds, and although this may not require the agency of the classic *mor* surface it may be that such extreme eluviation of iron occurs more readily under some vegetation types than others, some producing very large amounts of active, iron-chelating compounds. For further information on the factors involved in the formation of podsols readers are referred to Broomfield (1955), Martin and Reeve (1960).

Soil forming processes within the catena

Soil profiles and soil creep

Since the lower slope soils of most catenas are formed largely from material which has been transported downhill by creep and erosion, a complete understanding of the development of the catena and the processes involved requires a study of a soil profile from the upper part of the slope, an area which is rich in information about soil-forming processes.

Soil profiles are often investigated in terms of three major horizons. The upper A (eluviated) horizon being that from which, by the action of percolating water, substances are removed and carried downwards by the process known as 'eluviation', to be deposited in a second B (illuvial horizon). Below these two horizons lies the C horizon consisting of the substratum. The various horizons may show stratification and the subhorizons are indicated by the symbols A_0, A_1, A_2, B_1, B_2, and so on. This nomenclature, and way of looking at the soil profile, is often used in the tropics, but the system will not always reflect the most important processes operating in the soil. Nye (1954–55) has proposed a different system which focuses more attention on the importance of soil creep and on the activity of the soil fauna, especially ants, termites and earthworms. He maintains that the soil profile in the catena can often be divided into two major horizons: Cr, which is the horizon of soil creep; and S, which is the sedentary horizon.

Soil creep occurs on slopes of as little as 5 per cent and Nye believes that the importance of soil creep has been much underrated by soil scientists. He considers that many apparently sedentary profiles have derived their topsoils not from material directly below, but from soil higher up the slope. Probably material will also be moved downslope by storm rain water travelling over the soil surface. It is not, however, clear how commonly water moves in this way over a vegetated surface. Doubtless the extent to which water penetrates the topsoil rather than moves over its surface will be determined by the structure of the vegetation and the stability of the crumb structure in the topsoil. Nye (1954–55) observed good vertical drainage, without surface run-off under heavy rainfall on upper hillslope red earths in West Africa. It seems that red earths in general have fairly good drainage even though their clay content may reach as high as 70 per cent. The coating of the kaolinite clay particles by iron oxides (Greenland *et al.* 1968) may be an important factor in maintaining a stable texture. Other soil factors are undoubtedly involved in the maintenance of texture (Emerson 1959; Rose 1961, 1962).

Activity of soil fauna

Although the creep horizon as a whole may be moving slowly downslope, it must be appreciated that all the time there will also be an *upward*, vertical transport of material into the creep horizon from the sedentary horizon (Fig. 7.8). This material, taken upwards by the soil fauna, will, in turn, be sorted and further transported by them so that usually three subhorizons can be distinguished, namely: CrW, which is the horizon formed mainly of worm casts; CrT, the horizon formed mainly of material transported upwards from lower levels by ants and termites; and CrG, the horizon of gravel accumulation.

In East Africa the CrW horizon may be less well developed than in West Africa where casts of the earthworm *Hippopera* deposit humus and fine sandy

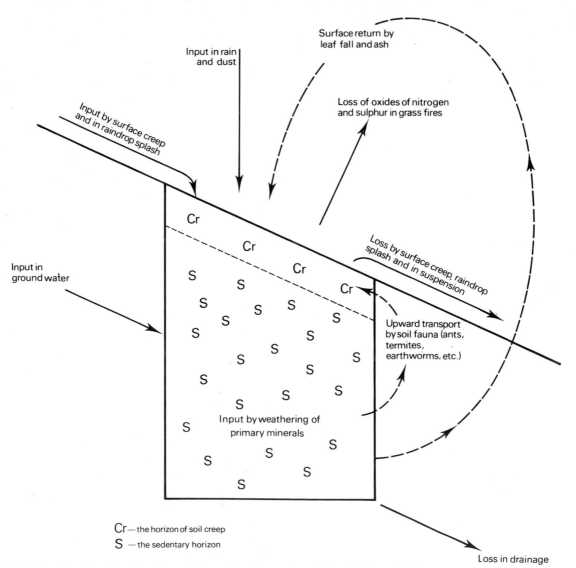

Cr — the horizon of soil creep

S — the sedentary horizon

Fig. 7.8 The dynamics of profile differentiation of a soil on a slope. (After Hallsworth, 1965)

particles on the surface. Earthworms occur in East African soils (Block and Banage, 1968), but many of them appear to cast *within* the soil and not on the soil surface. The activity of ants and termites is apparently great and the CrT and CrG horizons are often very clearly marked. The CrT horizon consists of relatively fine material, since the ants and termites do not usually carry upwards particles in excess of 4 mm diameter. This means that coarser material is left behind and accumulates gradually to form a distinct gravel layer.

The amount of material moved upwards by the soil fauna, presumably much of it from the sedentary horizon, can be very great and the importance of this process is vividly illustrated in some areas to the north and west of Lake Victoria. Here, in places, ancient lake beaches have been stranded above lake level by upward land movement. These beaches consist mainly of coarse, rounded pebbles and gravel. However, in the course of time—a matter of some thousands of years—the beach terraces have become buried under several metres of fine soil. In these areas where the terrace is virtually horizontal, and soil creep and erosion would be minimal, it is apparent that this overburden of soil has been moved into place by the soil fauna.

Sometimes on very steep slopes in dry areas and in arid regions swept by strong winds, as in the Turkana Desert, the soil profile becomes truncated and the gravel horizon is exposed at the surface. However, apart from these places, the CrT horizon is usually well developed and is an important feature contributing to the ease with which the soil can be cultivated.

Movement of water

The colours of the soil horizons may provide important visual evidence of the movement of soil water, of the existence and progress of leaching and accumulation, and the presence and absence of aerated conditions in the soil. Permanently waterlogged soil acquires a blue-grey appearance due to the presence of ferrous iron compounds, and this is in noticeable contrast to the reddish and brown colours of ferric salts in well-aerated soils. The blue-grey colours of the waterlogged soils may sometimes be masked by the dark complexes formed by humus and montmorillonite in the valley soils so that chemical tests or measurement of redox potential (Eh) are required to extend and confirm visual signs.

It seems likely that rain water drains vertically through many red earths on the upper slope of the catena; it may begin to move laterally only when it reaches the underlying rock or layers of packed pisolitic laterite. If these surfaces are resistant to percolation then a waterlogged zone will form in the lower part of the soil profile in the rainy season. Iron will go into solution and will be precipitated as concretionary nodules when the water drains off. On lower slopes, seepage water from above may lead to the accumulation of sheets of vermiform laterite in the subsoil.

If the rock is thoroughly weathered in its upper part it may allow relatively free entry of rain water. There may therefore be rarely any waterlogging in the subsoil and, accordingly, little or no mottling, but widely spaced iron pisoliths may still form deep in the weathered rock.

The junction of the CrG and S horizons is often a horizon of lateral seepage with grey colours indicating reducing or waterlogged conditions. On the lower slope of the catena the reduced zone may extend sufficiently upwards to embrace the lower part of the gravel horizon. In the soil profiles in the valley, seepage may occur at many points in the CrG horizon and a substantial proportion of the soil profile may have blue-grey colours due to ferrous salts. Close to the surface, however, the valley soils are usually grey or black.

Formation of secondary minerals or clays

Rain water percolating through the soil, enriched by chemicals dissolved in it, decomposes or hydrolyses many of the primary minerals in the sedentary horizons and their elements pass into solution. Some of the elements, such as sodium, are very mobile and find their way eventually into the drainage and accumulate in the valleys, in inland drainage basins, or in the ocean. Other elements, notably silica and aluminium, are unstable in solution and some of their ions aggregate into crystalline layers and come out of solution as stable, flattish, colloidal sized particles. These stable, secondarily-formed minerals constitute the important clay minerals which—together with organic matter accumulating from plant and animal remains—have a profound effect on the chemical and physical properties of the soil. Possibly some of the primary minerals do not pass into solution but may pass directly into clay minerals by ionic substitution in their primary crystalline lattice. Other of the primary minerals, such as quartz (SiO_2), zircon ($ZrSiO_4$) and the tourmaline (a complex borosilicate of aluminium), are extremely resistant to decay and accumulate in the weathered residue. Much of the soil fabric is made up of relatively inert grains of quartz. The bulk of the sand in a soil is quartz and this often constitutes more than 50 per cent of the weight of the dry soil.

In any soil type, usually one type of clay predominates, but often others are present in smaller amounts. Kaolinite is the commonest clay in the hillside red earths while montmorillonite is common in the valley deposits. The striking differences of chemistry and texture between the hillside and valley soils arises largely from the properties of their dominant clays. The properties of the clays are best understood in terms of their crystalline structure, of which detailed accounts are provided by Grim (1962) and van Olphen (1963).

The factors controlling the type of clay synthesised in the soil are not yet fully understood; Mason (1958) believes that the primary determining factors are, first, the chemical nature of the parent material and, second, the physiochemical environment in which alteration and hydrolysis of the primary minerals takes place. The structure of kaolinite does not accommodate cations other than silicon and aluminium and its formation seems to be favoured by an acid environment and a poor supply of bases. Kaolinite has the highest Al:Si ratio of the clay minerals, and its formation is probably promoted by weathering processes and free drainage which tend to remove silica in solution. Under acidic conditions, of about pH 4, the solubilities of silicon and aluminium are such that relatively more aluminium and relatively little silica is present (Millot, 1970). Montmorillonite has an Al:Si ratio of about 1:2 and other cations, such as magnesium and iron, are probably essential to its formation. It appears to form most readily in a slightly alkaline environment with a good supply of bases and especially from parent materials rich in ferromagnesian minerals, calcic felspars, and volcanic ash. However, with sufficiently prolonged destructive weathering or latosolization, even volcanic-derived soils will eventually contain mainly kaolinite. Presumably such prolonged weathering has been responsible for the friable red soils which overlie volcanic rocks in the Kikuyu area of Kenya. Here under a rainfall of 1000–1250 mm, and with free drainage, the porous volcanic tuffs have weathered deeply and are overlain with about 5 m of a bright red, very porous, and friable loam.

Montmorillonite is a much more active clay than kaolinite and this follows from its atomic structure as the layers of the montmorillonite clay particle are relatively loosely held together. Water and other polar molecules can easily penetrate between them. Conse-

quently montmorillonite undergoes much expansion and shrinkage as water is added or taken away. This expansion of the particle enables exchangeable cations to be held on the internal surfaces, as well as on the edges of the platelet. The cation exchange capacity of montmorillonite is further increased by substitutions in the crystalline lattice. These substitutions leave valencies unsatisfied and the resulting negative charge is balanced by exchange cations. These properties help to explain why the cation exchange capacity of montmorillonite is of the order of 100 milliequivalents per 100 gm soil, while that of kaolinite is about 8. Since this is so low we can understand the importance of the colloidal humus fraction in the kaolinitic red earths. This humus fraction has a cation exchange capacity of about 200 meq/100 gm which compensates for the cation poverty of the clay.

Finally, it should be mentioned that although some of the silica mobilized in the soil is transformed into clays, and some is lost in the drainage, a considerable amount is cycled through the ecosystem. Rodin and Bazilevich (1968) estimate that of the 2000 kg/ha of chemicals taken up by tropical forest some 780 kg is silica. Again, being a predominant constituent of grasses, there will be a large amount of silica cycled in the grassland ecosystems. Bamboo thicket is also a high utilizer of silica.

Importance of organic matter and humus

Prolonged weathering of the soils in the catena will result in very infertile red and yellow soils almost devoid of adsorptive properties and very poor in all the essential elements of plant nutrition. The fertility of such latosols depends very much on the colloidal fraction of the organic matter in the topsoil. The cation exchange capacity of this is high and much of the fertility of the latosol resides in its upper 30 cm. When this topsoil dries out the nutrient supplying capacity of these leached red and yellow soils must be very low.

The total amount of organic matter in the topsoil is usually small. Bates (1960) gives data on the distribution of organic matter under Nigerian forest. Additional information is given by Jenny et al. (1949), Russell (1960). Birch and Friend (1956a) studied the organic matter of more than 400 soils in Kenya occurring from sea level to just over 3000 m. They discovered that increasing rainfall is more favourable to vegetative growth and organic matter production than increasing temperature. The favourable effect of temperature on plant growth and organic matter production is counterbalanced by increasing microbial activity resulting in organic matter destruction—certainly within the range of temperatures encountered up to 3000 m in Kenya. They estimated that each 300 m increase of altitude involves a 0·08 per cent increase in organic matter in the topsoil. It is unlikely that the above relationships hold at higher altitudes since further increase in altitude is often accompanied by a decrease in annual rainfall and by very great variability in the annual totals.

Hedberg (1964a) believes that within the alpine belt the extent of organic matter accumulation decreases with increasing altitude and this can be attributed to the decrease in rainfall and in temperature which both lower the plant production. He observed that in the lower part of the alpine belt on the Ruwenzori the dark-coloured humus layer of the almost permanently moist soil was often a few decimetres thick, but in the more sparsely vegetated soils of the upper part of the alpine belt, on the same mountain, the humus content appeared to be insignificant, especially on the well-drained soils on talus slopes and moraines. He considered that similar conditions prevailed on Mt Kenya. The soils of this mountain have been described briefly by Coe (1967) who records carbon contents between 7 and 10 per cent in its alpine belt. The amount of organic carbon in the soils at any altitude is probably determined very closely by the soil water supply. Thus, Coe found up to 13 per cent carbon in soils close to valley streams at about 4000 m, while close to the base of the valley values nearer 9 per cent were recorded. Hedberg illustrates the importance of water supply more vividly by drawing attention to the *Carex runssorensis* bogs in the alpine belt, where peat accumulates to a depth of a metre or more, while on more well-drained soils nearby, humus accumulation is slight.

Other factors affecting the organic matter content of the topsoil are fire and termites. It would seem obvious that, in wooded grasslands and grasslands, the organic matter content would be adversely affected by the periodic destruction of surface plant debris by fires. However, firing appears sometimes to be beneficial, perhaps because growth of accumulated litter inhibits grass growth and firing stimulates growth and root development. Naturally the timing and the intensity of the burn are important. Moore (1960) in derived wooded grasslands in Nigeria found that the organic matter content of the soil was 30 per cent higher in the early, lightly burned plot than in the late, fiercely burned plot, and higher than in the protected plot, after thirty years of treatment.

Termites may tend to lower the organic matter of the topsoil because they devour a great deal of the dead plant material which would otherwise enrich it with humifiable material.

The organic matter reaching the soil is broken down and transformed by ants, earthworms, termites, nematodes, fungi and bacteria. Humus, on account of its important physical and chemical properties, is one of the most important end products. The term refers usually to the amorphous dark brown to black organic substance, as distinct from partially decomposed plant and animal remains. Dark colour is not always regarded as a characteristic, and some workers include colourless mucilaginous polyuronides which are products of bacterial and fungal synthesis. Humus itself is a complex substance. It is widely accepted that it contains four main compounds, namely, humin, humic acid, fulvic acid and hymatomelanic acid; all these occur in varying proportions.

The nature of humic acid is uncertain. However, there are reasons (Burges, 1960) for believing that it is primarily a non-nitrogenous substance which occurs in the soil as a metallic humic acid complex, the metal varying with soil conditions. Acidic soils, such as many of the African leached red and yellow soils, possibly have the humic acid present mainly as iron and perhaps aluminium humate, while in soils with a higher calcium content it exists probably as calcium humate. The type of metallic cation alters the colour and physical properties of the humic acid. Thus, the dark colour of the black cotton soils is believed to be caused by some form of calcium humate, though some of the dark colour could be due to the presence of free carbon from grass fires and to the persistence of certain dark-coloured primary minerals, for example, magnetite.

The importance of humic acid lies in its high capacity for cation adsorption, and for absorbing water, and its beneficial effects on the aggregation or crumb structure of the soil, which promotes good water acceptance and good aeration.

The cation exchange capacity of humus is somewhere in excess of 100 meq per 100 gm of soil while for comparison, montmorillonite has 80–100, illite 15–40, and kaolinite and iron oxides 3–15 (Buckman and Brady, 1969). Humus usually absorbs up to 90 per cent of its weight in water, whereas most clays will usually take up not more than 20 per cent.

The stability of the soil crumb structure is influenced both by the amount of humus present and by the dominant cation of the clay exchange complex. Rose (1962) found that calcium-saturated kaolinite crumbs broke down rapidly under the impact of rain drops, but almost no breakdown was observed for calcium-saturated montmorillonite or illite crumbs.

Influence of soil nutrients on plant growth

In most parts of East Africa the soil probably exerts its strongest influence on vegetation through its water-absorbing and water-supplying capacities and we have described how these depend on the depth of the soil profile, the amount of organic matter and humus in the topsoil, the dominant cations of the exchange complex, and the dominant type of clay present.

The quantity of rainfall absorbed or leaching through the soil in turn affects the quantity of nutrients retained on the colloidal exchange complex. According to Scott (1962), in low rainfall areas the soil clays and humus colloids remain almost fully saturated with bases such as calcium, magnesium, sodium and potassium, but increasing rainfall leaches these off the exchange complex and they may be absorbed deeper in the soil or lost in the drainage. Analyses of drainage waters could give a valuable indication of the rate of removal of elements from the ecosystem but, unfortunately, very little effort has been made in the tropics to study the chemistry of drainage waters (Perrin, 1965). With 760 mm rain the clays remain about 85 per cent saturated with exchangeable cations; with 1140 mm of rain they are about 50 per cent saturated; and with 1900 mm the saturation level is below 10 per cent and little is left on the clay colloids except hydrogen and aluminium, so that the soils are invariably acid in reaction. Scott believes that below a mean annual rainfall of 760 mm leaching follows a simple physical pattern, but as the rainfall increases above 760 mm the much better plant growth which the soil can support allows a more active recirculation of bases resulting in a rise of the base saturation. With further increase in rainfall above 1140 mm, leaching again dominates leading to a fall in base saturation.

In East Africa it is not clear how far the vegetation type and/or its physiognomy are influenced by the overall nutrient status of the soil. Beadle (1966) has claimed this type of correlation in Australia where he believes that low soil phosphate levels have been a major factor in inducing short-stature, sclerophyllous, xeromorphic-type of vegetation. It would appear, however, that in East Africa, given sufficient rainfall, even the most complex ecosystem, namely, the forest, can develop over very oligotrophic soils. Apparently a good example of this is the *Piptadeniastrum*-dominated Jubiya forest which developed, perhaps very slowly, over ancient lake sands near Lake Nabugabo, Uganda. Analyses have shown these sands to have an acute shortage of all the essential plant nutrients. The upward succession towards forest on such very poor sites may, of course, be highly dependent on a low degree of interference. Doubtless very little disturbance in the form of grazing and burning will halt the succession at *Loudetia kagerensis* grassland. Again an important factor accelerating the upward succession towards forest is the construction of termite mounds on the grassland, since these become centres of seed accumulation and provide some advantage for the vegetation growing on them, perhaps slightly better water supply and some protection from fire (Jackson and Gartlan, 1965).

Another major factor in the upward succession of forest must be the efficiency of nutrient cycling within the ecosystem. It is well established that this is very efficient in the forest (Fig. 7.9). It may be that under some conditions of drainage, even on very impoverished soils, the input of nutrients via rain, dust, and nitrogen fixation just slightly exceeds the various losses. Thus, in the course of time, forest with its vast reservoir of chemicals in the tissues of the trees can become established. Very little work has been done on the chemical composition of plants in the tropics and subtropics (Rodin and Bazilevich, 1966), but the values given in Fig. 7.9 are impressive and indicate that the chemical store, in the living and dead tissues of the trees, amounts to approximately 12 000 kg/ha.

In view of this vast reservoir of chemicals in the trees it is understandable why great care is required in converting forest land to some other form of land use. The removal of the trees as timber means a tremendous depletion of the nutrient reserve of the area. Moreover, even if the trees are allowed to decay *in situ*, the post-forest crop may not, at least in its earlier stages, have a sufficiently developed root system to recover all the nutrients as they are mineralised and released from the decaying tissues. Milne (1937 b) believed that imperfect mineral cycling accounted for the failure of coffee plantations on ground which had previously carried luxuriant moist forest on the Usambaras in Tanzania. Such luxuriance often leads to grave misconceptions about the nutrient reservoir in the soil.

It seems likely that with minimal disturbance the vegetation *type* in East Africa is determined primarily by availability of water. Nutrient differences between the soils of one site and another are reflected mainly in the details of the floristic composition of the vegetation, in the relative abundance of its component species, and in the level of productivity.

In East Africa, indeed in tropical Africa, one of the best documented examples of the effect of soil nutrients on plant growth is the limit on productivity determined, in part, by the shortage of nitrogen, and to a lesser extent by phosphorus and perhaps boron.

Various aspects of this problem are considered in *Bulletins* 45 and 47 of the Commonwealth Bureau of Pastures and Field Crops. See also Hesse, 1955; Keen and Dothie, 1953; Lemare, 1949; Meiklejohn, 1954, 1957; Moore, 1963.

Nitrogen deficiency is especially marked in the red and yellowish latosols and as with soils all over the world—the nitrogen content of the soil varies directly with the content of organic matter. The humus horizon of these soils often does not extend much beyond a depth of 15 cm and the organic content often does not exceed about 2 per cent.

Undoubtedly one of the major drains on the organic matter and nitrogen contents of these soils, which are common beneath the grasslands and wooded grasslands, is the annual burning of the grass. Probably this creates the greatest nitrogen deficiences in areas with a fairly high rainfall and a severe dry season.

According to Nye and Greenland (1960), tall Andropogoneae grassland produces about 9000 kg/ha dry aerial matter per season, but nearly all the carbon, nitrogen, and sulphur contained in this are lost as gaseous oxides in the burn. This means that the input of nitrogen from dead plant material is mainly from the root systems. The production of the root system is claimed to be approximately one-third of the aerial matter production—it might often be higher—and, using rough conversion factors Nye and Greenland estimate that about 17 kg/ha nitrogen are lost in the burn but about 11 kg/ha will be conveyed to the topsoil via mineralization of old root matter.

Under grasses other than the Andropogoneae nitrogen levels may be quite satisfactory. In fact Uganda soils, under elephant grass (*Pennisetum purpureum*), were among the first in the world to be shown to be very rich in nitrate (ap Griffith, 1949; ap Griffith and

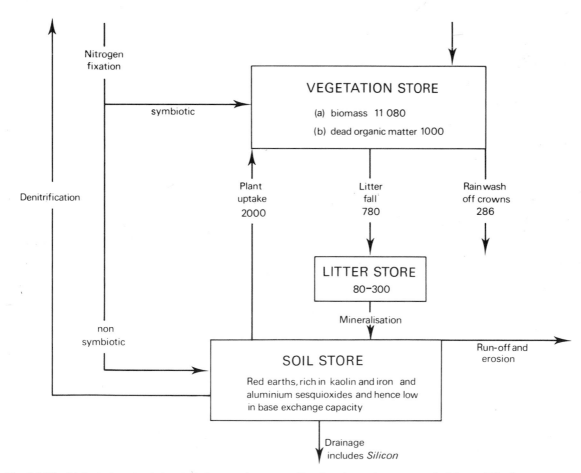

Fig. 7.9 The biological cycle of chemical elements between soil and moist angiosperm tropical forest. The figures represent the approximate quantities of chemical elements in kg/ha. (Data from Rodin and Bazilevich, 1966 and 1968)

185

Manning, 1949, 1950). This is a seasonal phenomenon controlled by the frequency of wetting and drying of the soil (H. F. Birch, 1958, 1959, Birch and Friend 1956 b; Calder, 1957; Greenland, 1958; Simpson, 1960; Weselaar, 1961; Stephens, 1962; Robinson and Gacoka, 1962). It appears that nitrification is more thorough in soils subjected to wet and dry periods than in soils which experience more steady conditions. It was at first believed that wetting and drying in some way liberated or exposed nitrogenous matter which was absorbed onto or shielded by the clay minerals. However, this cannot be the whole explanation because flushes of nitrification have been followed in systems free of soil particles.

Since satisfactory nitrogen levels are found in soils under species of *Panicum* and *Pennisetum*, but generally not under species of *Andropogon*, *Hyparrhenia* and *Imperata*, it has been suggested that the *Andropogoneae* emit substances which inhibit the growth of nitrogen-fixing bacteria. However, as yet no satisfactory proof of such inhibitory substances has been obtained.

Low nitrogen levels must, in part, be due to the scarcity of nodulating legumes in the rangeland ecosystems. Leguminous trees are, of course, common in the miombo, but generally herbaceous and shrubby legumes are rare. It is not clear why this should be so. Inability to compete with the tall-growing grasses may be one factor, another may be low levels of soil calcium, molybdenum, phosphorus and sulphur, all of which are essential to vigorous growth and nodulation of legumes (Norris, 1962).

Because of the expense of applying nitrogen as an artificial fertilizer there has been much research—and substantial achievement—in developing grass-legume mixtures for high grade pastures where these are required for small-scale intensive production. Increasing the quantity of nodulating legumes in the pasture, raises the general production of dry matter and increases the protein content of the graze. The topic of pasture improvement is further discussed in chapter 2.

Fertilizer trials in pasture-improvement experiments have shown that phosphate may also be a factor limiting production on East African soils. In the highly ferruginous soils, which are so common, phosphate may be present in substantial amounts but it may be in a form which is not readily available to the plant. It has been demonstrated (Birch 1949, 1952, 1953) that in many East African soils, with a pH below 4, much of the phosphorus is adsorbed on iron and aluminium colloids and is almost unavailable or only slowly available. At higher pH values more of the phosphate becomes base-linked, especially to calcium,

and this so-called saloid-bound phosphate is unstable and easily released into the soil solution. Although saloid-bound phosphate constitutes only a small proportion of the total phosphate it is the important form for plant nutrition.

In employing pH as a guide to the availability of soil phosphate it should be remembered that pH values vary considerably through the soil profile (Doyne, 1935). The zone of biological activity is often more acid than the lower part of the soil profile, and the maximum acidity may be at a depth of 45–60 cm below the surface. This is due to a combination of the leaching effect of drainage water, carrying calcium and other bases downwards, and the withdrawal of bases by the plant roots. These factors can reduce the pH of the subsoil to as low as 4; indeed, values below 4 are even recorded from the highly weathered red earths, many of which occur under forest and very high rainfall at high altitudes. The bases absorbed by the roots are returned to the surface in the litter and during the burn. There is therefore a higher pH in the organic matter at the surface, presumably due to the liberation of bases when the leaves decay, Bates and Baker (1960) discuss the distribution of phosphates in the soil profile of forest in Nigeria.

Potassium deficiences in the soils appear to be more local and more rare than deficiencies of nitrogen and phosporous, but responses to potassium have been obtained (Lock, 1962; Stephens, 1962).

Calcium is usually in good supply except in soils of the very high rainfall areas, for example, the Sese Islands and the highly-leached soils in the upper part of the upland forest belt. Harrop (1960) records an extremely acid topsoil, pH 2·9, in the Impenetrable Forest in south-western Uganda, the annual rainfall being from 1750 to 1880 mm.

Responses to other nutrients, including the minor and trace elements, seem not to occur widely and to be rare. Deficiencies of magnesium, sulphur, and boron are reported (Ollier and Harrop, 1959). Sometimes low yields may be caused as much by deficiencies of these elements as by deficiencies of nitrogen and phosphorus. On some highly leached soils, applications of boron triggers off healthy growth.

Bearing in mind the ancient and strongly leached character of many East African soils it is surprising that trace element deficiencies have not been recorded more frequently (Chenery, 1958, 1960; Calton and Vail, 1956; Russell and Duncan, 1956; Schutte, 1955). Why are these deficiencies not so common here as on the old land surfaces of Australia? Attempts at more intensive production may yet show trace element deficiencies to be common in East Africa. (Diekmahns, 1957; Elmer and Gosnell, 1963; Gosnell,

1964; Pinkerton, 1967; Pinkerton *et al.*, 1965; Robinson and Chenery, 1958; Schoenmaekers and Chenery, 1959; Storey and Leach, 1933; Vail and Calton, 1957.)

Some thought should also be given to the toxic effects of excessively *high* quantities of some nutrients as a limiting factor to plant growth. Saline soils provide an example of this. We may also recall the very high levels of sulphur in some reclaimed highland papyrus swamps. Manganese is another trace element which is often present in toxic amounts and patchy, unproductive soil (*lunyu*), common on overcropped areas, is caused by excess of soil manganese. Ollier and Harrop (1959) note that Uganda tea is rich in manganese and, on this account, it may be correspondingly poor in iron. This may account for the pale leaf-colour noticeable in some tea plantations in Kenya and Uganda.

Eight

The history of the vegetation

Alan Hamilton

With a section on the East African coastal—
West African forest disjunctions by R. B. Fader

Acknowledgements

I am indebted to my friend and teacher, Michael Morrison, for introducing me to the study of African ecology and prehistory. It is unfortunate that he died before starting this chapter on a subject which, among many other interests, always had a special fascination for him and was his particular field of research. I trust that this account, by his student, does justice to a meticulous and critical worker. Any shortcomings are, however, entirely my own responsibility.

I am very grateful to R. B. Faden, of the East African Herbarium, for contributing the section on East African Coastal–West African rain forest disjunctions, and to Keith Thompson for critically reading the manuscript and making many useful suggestions.

Introduction

A study of vegetation history is largely a study of change and of the causes of change. Changes can occur both in the composition and in the distribution of vegetation types. A flora may be enriched by immigration and evolution or impoverished by emigration and extinction. The physiognomy and distribution of vegetation types are affected by a wide variety of environmental factors, of which, in East Africa, the most important are climate and, in geologically recent times, human activities. Indeed, a study of the history of the vegetation is also a study of climatic change and of the history of man.

This account deals with the vegetation of East Africa during the Miocene and subsequent periods. The *Glossopteris* flora found in rocks of the Karroo System in each of the East African countries is therefore not considered. This flora predates the appearance of angiosperms in the fossil record and is hardly relevant to a study of more recent vegetation history.

It is necessary to review briefly the geomorphological background of East Africa during the Kainozoic Era. The Pleistocene is taken as extending right up to the present day and, accordingly, the Kainozoic Era is divided into the following periods: Palaeocene (65–54), Eocene (54–38), Oligocene (38–26), Miocene (26–7), Pliocene (7–2·5) and Pleistocene (2·5–0). The figures in brackets indicate the approximate time limits of each period in millions of years before the present (after Eicher, 1968).

During the early part of the Kainozoic Era East Africa had no high mountains and is believed to have consisted of a gently undulating plain. In the northern part of our area the watershed between east- and west-flowing rivers lay near the present position of the Uganda-Kenya border, with rivers from Uganda draining westwards towards the Atlantic Ocean. This tranquil period was broken in the Miocene by a phase of tectonic activity which has continued up to the present day. Down-faulting of the rift valleys is thought to have been initiated during the Lower Miocene (Gautier, 1967). This caused the drainage of Uganda rivers to be diverted into the Western Rift Valley, where very deep deposits of sediments have accumulated. Thick lacustrine and fluviatile deposits are also present in the Eastern Rift Valley (Saggerson in Russell, 1962). Rifting was succeeded by upwarping of the shoulders of the rift valleys, which, in Uganda, caused the reversal of the Kagera and other west-running rivers and the creation of shallow inland basins, such as those now occupied by Lakes Victoria and Kyoga. Associated with the rift valley movements and with the fractures that resulted from them was a very considerable amount of volcanic activity, which is still manifest today, though to a much lesser extent than in the past (Haughton, 1963).

To date, very few fossiliferous deposits have been

examined and, consequently, there are huge gaps, both temporal and geographical in the record. In East Africa macrofossils are most commonly found in association with either Miocene or Pleistocene volcanics and to my knowledge, useful assemblages, definitely assignable to the Pliocene, have not yet been discovered. Recently, the study of plant microfossils known as 'pollen analysis', has thrown much light on the vegetation during the Upper Pleistocene, since pollen grains have been found to be well preserved in upland peats and in some lake sediments.

The dating of horizons containing plant fossils used to rely solely on correlation with other deposits, often using either animal fossils or human artifacts as stratigraphic indicators. The limitations of this procedure have to some extent been overcome by the introduction of 'absolute' methods of dating. These involve measurements of the quantities of isotopes which, in the process of time, have either been formed or have been destroyed by radioactive decay. The two types of geochemical determinations so far applied to East African material involve the breakdown of C^{14} to C^{12} and the breakdown of K^{40} to A^{40}. The former transformation has a half life of about 5500 years and the C^{14} method can only be applied to carbon samples less than about 60 000 years old. (A half life is the length of time taken for half of the original mass of an unstable isotope such as C^{14} to be transformed into other isotopes, and, for any particular transformation, is a constant.) In the case of carbon, it is assumed that the atmospheric ratio of C^{14} to C^{12} in the past was the same as today, and hence it is possible to calculate the age of carbonaceous samples. The K–A transformation, with a much longer half life, can only be used on material more than about 500 000 years old and, in East Africa, has been widely used to date lava flows. Clearly there is a wide span of time not covered by either method, but it is hoped that refinement of existing techniques and the development of others will eventually bridge this gap. Both the C^{14} and the K–A methods have weaknesses which make it desirable to secure determinations from several samples at each locality. Age determinations are usually expressed as the number of years before 1950 (B.P.).

A necessary assumption in the reconstruction of past vegetation from fossil floras or other data is that the ecology of taxa was the same in the past as it is today. While for East Africa this assumption is probably generally valid for evidence relating to the Pleistocene, it becomes more doubtful with evidence relating to earlier times.

Several fields of research provide indirect evidence of the history of the vegetation. Foremost among these is the study of plant geography. The present occurrence of taxa or vegetation types may only be explicable in terms of a change from a previous pattern of distribution. Where grazing or fire is thought to be important for determining range, supporting evidence is sometimes available from exclusion experiments. Unlike fossil floras, precise dates are very rarely available for the times of vegetational change inferred from evidence of this type.

In this chapter evidence on the history of the vegetation derived from macrofossils, pollen analysis and plant geography are examined in turn. Results from these three disciplines are then compared and discussed in the light of other data. Non-botanical evidence of climatic change and of human cultural attainments are discussed at some length since, as previously pointed out, the histories of the climate and of man are closely linked with the history of the vegetation.

Evidence from macrofossils

During the upper Kainozoic, the climate of East Africa has favoured the rapid decay of dead plant material and therefore has been generally unsuitable for the preservation of fossils. Furthermore, for a long time, much of the land has been subject to erosion and, with the exception of the rift valleys, sites of non-volcanic sediment accumulation have been relatively few. Consequently the fossil record is fragmentary. Fossil plants are most often found either in beds of subaerial volcanic ash or in lake sediments trapped under or between volcanic layers, and are most commonly preserved either as impressions or as calcite casts. Fossil wood is common at many sites but usually insufficient internal structure is preserved to permit identification. Fossil leaves are also frequent but, even when well preserved, are difficult to identify, since leaves of similar morphology are produced by species of widely different ecology. Fossil fruits, seeds and flowers can be identified more reliably, but are rarely found. In some cases plants are found preserved in their original growth positions; more commonly, however, it is difficult to determine the relationship between fossil 'death' assemblages and the former vegetation.

Miocene fossils

Chaney (1933) has described a fossil flora from the Bugishu Series (sic) which in places underlies the

Elgon volcanics. This flora is therefore older than the Bukwa flora, to be mentioned presently, and may be of about the same age as the Rusinga flora. The Bugishu Series consists of lacustrine sands and other sediments laid down in a series of basins in the early Kainozoic plain. The flora consists of leaves, fruits and wood. Chaney has identified twelve species, belonging to the following genera. The most nearly related modern equivalents of some of the fossil species are given in brackets.

Acrostichum sp.
Bauhinia waylandi
Berlinia spp. (two species)
Cassia sp. (cf. *singueana*)
Dalbergia sp.
Olea sp. (cf. *africana*)
Parinari sp.
Pittosporum spp. (three species, one cf. *mannii*)
Terminalia sp.

The identifications of *Bauhinia*, *Berlinia*, *Cassia*, *Dalbergia*, *Olea* and *Parinari* are based on fossil leaves only and therefore, except in the case of *Bauhinia* with its very characteristically shaped leaves, some doubt remains as to correct determination. The inclusion of *Terminalia* in the above list is based on a report by W. N. Edwards and no fruits of this genus were seen in the material sent to Chaney for examination.

Assuming, however, that the above identifications are correct, an attempt can be made to reconstruct the type of vegetation from which the fossils were derived. By comparison with their nearest living relatives, Philips (in Chaney, 1933) considered that the fossils may have originated from a woodland or open woodland, perhaps growing under a rainfall of 850–1300 mm per annum and probably with a markedly seasonal climate.

The largest and best documented of the East African Miocene floras comes from Rusinga Island, situated at the mouth of the Kavirondo Gulf in Lake Victoria (Chesters, 1957). As with the Bugishu Series, the fossiliferous beds at Rusinga were deposited in shallow basins in the early Kainozoic plain, lacustrine and volcanic sediments being capped by a plateau lava. K–A dates have been obtained for samples from various sites in the succession and show wide divergence (15·3, 22·2, 42·0, 107·0 and 167·0 million years B.P.). This illustrates the care which must be exercised when applying this technique to Kainozoic volcanic rocks (Bishop, 1967a). On palaeozoological grounds, Bishop considered that an age of around 22 million years is of approximately the right order. Rusinga Island is the richest Miocene faunal locality in East Africa, having yielded sixty species of mammals including three species of *Proconsul* (Bishop, 1967a).

The fossil flora consists of wood, a few fragmentary leaves and a very large number of fruits and seeds. Thorny wood is very common and many of the smaller pieces bear the remains of twining stems. The fruits and seeds are mostly preserved as hollow, crystal-filled casts, but internal structures can sometimes be discerned.

Chesters has referred the majority of the fossil plant species determined from fruits and seeds to living genera. The following is an abbreviated list:

Anacardiaceae (2 species, belonging to the genera *Antrocaryon* and *Odina*)
Annonaceae (5 species)
Connaraceae (*Cnestis*)
Cuourbitaceae (3 species, one a member of the genus *Lagenaria*)
Euphorbiaceae (4 species)
Leguminosae (3 species, one close to *Pterocarpus*)
Meliaceae (*Entandrophragma*, near *E. utile*)
Menispermaceae (5 species, including members of the genera *Cissampelos*, *Stephania*, *Syntrisepalum* and *Triclizia*)
Oleaceae (1 species, close to *Schrebera*)
Rhamnaceae (4 species, including members of the genera *Berchemia* and *Zizyphus*)
Sapindaceae (4 species)
Ulmaceae (*Celtis*)
Other families identified: Burseraceae, Capparidaceae, Rutaceae.

By comparison with allied living species, Chesters concluded that at least half of the identified plants were climbers and that the remainder were trees of varying dimensions. In her opinion, the abundance of climbers suggests a forest-edge vegetation, such as would be found on the bank of a river or lake. Ecological interpretation of the fossil mollusca (van Zinderen Bakker, 1966) and mammals indicates that the forest may have been riverine, strips of gallery forest along the water courses separated from each other by more open types of vegetation.

Miocene fossil plants and animals occur at Bukwa on the north-eastern slopes of Mt Elgon (Hamilton, 1968; Walker, 1969). The fossil flora is found in beds of fine to coarse-grained calcareous ash derived from the Elgon volcano. Several K–A determinations give similar dates of about 20 million years B.P. (Brock and Macdonald, 1969). One bed of ash contains grasses and sedges preserved in their positions of growth. The presence of *Juncellus* rhizomes indicates that these monocotyledons may have been growing on the margin of an alkaline lake. Elsewhere, wood and dicotyledon leaves are common and are associated in one horizon with fruits, seeds and flowers, which are preserved as calcite casts. The detailed arrangement of the floral parts can be seen on some of the fossil flowers; one species has been tentatively identified as close to the genus *Bersama* (Melianthaceae) and another to the Sterculiaceae. It is thought that the fossil flowers, fruits, seeds and dicotyledon leaves originated from forest vegetation.

Other finds of Miocene fossil plants have been reported from Mt Elgon (Annual Reports of the Geological Survey of Uganda 1922, 1932, and 1933) and from Mt Napak in Karamoja (van Zinderen Bakker, 1966). An interesting discovery made by Wayland was the stem of a fossil *Cyathea* tree fern on Mt Elgon at an altitude of about 2150 m. Fossil wood from Elgon has been examined by Bancroft (1935), who found in some specimens anatomical features referable to the group Dipterocarpoideae (Dipterocarpaceae). Today, members of this group are characteristically tall forest trees of the South-east Asian forests and are entirely absent from Africa. This identification needs verification because, if correct, it is of great phytogeographical interest.

Pleistocene fossils

In Uganda, fossil plants are found in the Upper Pleistocene volcanics of the Western Rift Valley and adjacent areas (Osmaston, 1965). These volcanics may be classified into a northern group in the vicinity of Fort Portal and a southern group near Lake George. In both areas fossil plants occur in tuffs at 1200–1500 m on the plateau and they have also been discovered in the southern area 900–1000 m in the Rift. The rainfall is 1300–1500 mm per annum on the plateau and 600–1000 mm in the Rift. The age of the volcanism in the Fort Portal area has been dated by C^{14} at 4070 ± 120 B.P. (Osmaston, 1967).

Most of the fossils are leaves, but wood is found in and above a volcanic tuff layer in Nyakimya Swamp near Fort Portal and one species has been identified to the genus *Parinari*. A fruit and a leaf of *Parinari excelsa* were recovered from mud 1·5 m beneath the base of the tuff in this swamp. Fossil leaves in the northern group are found at the base of the tuff and are probably derived from the vegetation destroyed when the tuff was deposited. On the other hand, fossil leaves in the southern plateau group occur only within the tuff and are probably derived from successional vegetation growing on the freshly deposited volcanics.

A comparison of the fossil leaves recorded from the three areas reveals interesting differences:

Table 8.1

| | Ferns | Number of fossils of different types | | |
		Grasses	Other monocots	Dicot. trees and shrubs
Northern plateau (NP)	0	0	2	20+
Southern plateau (SP)	2	8+	8	22+
Rift (R)	0	4+	0	1

Osmaston concludes that these preliminary figures, coupled with the finding of the *Parinari* wood and the *Parinari excelsa* leaf and fruit in the swamp at Fort Portal, at once suggest that the northern plateau area was originally covered with closed forest of the same kind as still grows in the neighbouring district, that the southern plateau tuffs bore temporary vegetation of a mixed type similar to their present vegetation in the intervals between eruptions, and that the Rift Valley was covered mainly by savanna as it is today.

The following plant species were identified:

Table 8.2

Species	Occurrence	Reliability of Identification %
Phymatodes scolopendria	SP	100
Imperata cylindrica	SP+R	33
Ensete ventricosum	SP	100
Aframomum	SP	66
Parinari excelsa	NP	100
Celtis adolfi-frederici	NP	100
C. africana	NP	66
Cordia abyssinica	SP	66
Diospyros abyssinica	NP	33
Uapaca guineensis	NP	66

The fossil assemblages indicate a climate not unlike the present. The area from which the northern fossils derive is now deforested and Osmaston attributes this deforestation to man and not to climatic change. He thinks that the Kibale Forest formerly extended to the Ruwenzori and was probably continuous with the montane forests. The presence of *Celtis adolfi-frederici* and *Uapaca guineensis* suggests slightly warmer and wetter conditions than now and a study of the fossil soils associated with the volcanics supports this climatic interpretation (Bishop and Posnansky, 1960).

Evidence from pollen analysis

Pollen analysis is a very useful technique for investigating the more recent Pleistocene vegetation of East Africa. Its importance is partly due to two properties of pollen grain walls; they are very resistant to decay and they display a wide range of design which frequently permits confident identification. Many plants, particularly those which are wind pollinated, produce pollen grains in vast numbers, and some of these come to rest at sites where conditions are favourable for their preservation. By analysing the pollen content of material from such sites, a picture of past vegetation can be obtained.

In East Africa well-preserved pollen grains can usually be found in peats above an altitude of *c.* 1500 m and in sediments beneath deep lakes at all altitudes. By taking cores from these peat and lacustrine deposits, a continuous record of changes in pollen composition can be obtained and these are expressed as a pollen diagram (see Fig. 8.1). It is often possible to date horizons in these cores by C^{14}. It is normal practice in the construction of pollen diagrams to isolate those pollen types thought to be of greatest value as indicators of past vegetation and environment from those of lower indicator value. The total number of grains in the first group, and only these, constitutes the 'pollen sum' and is used as the basis in calculating the percentage of all pollen types at each level. For instance, pollen belonging to the types in the top, lefthand part of the diagram in Fig. 8.1 constitute the pollen sum. Pollen types produced by hydrophytes and swamp-edge species ('local pollen types') are excluded since they are subject to considerable fluctuations in frequency which are unrelated to major changes in either vegetation or climate. As will be seen later, it is also advisable to exclude from the pollen sum those pollen types which are very well dispersed.

Pollen diagrams from upland areas

Upland vegetation in East Africa lies in more or less clearly demarcated zones or belts whose distribution can often be related to climatic conditions. The vegetation is therefore likely to be sensitive to climatic change. This consideration, combined with the relatively large number of suitable sampling sites, has resulted in the selection of upland areas for most analytical studies of pollen in East Africa. A list of upland sites which have been sampled is given in Table 8.1. The following account begins with Coetzee's (1967) interpretation of pollen diagrams from Kenya and Tanzania. This is followed by a survey of some recent studies relevant to the interpretation of pollen diagrams from upland East Africa and the results of these studies are then applied to interpretation of pollen diagrams from Uganda. Finally, the Kenyan and Tanzanian diagrams are briefly re-examined.

Coetzee's interpretations of the Kenya and Tanzania pollen diagrams

Coetzee (1967) has given detailed interpretations of the pollen diagrams which she and van Zinderen Bakker have obtained from cores in upland Kenya and Tanzania. Changes in pollen curves are interpreted in terms of upward and downward movements of vegetation belts caused by fluctuations in temperature, which are thought to have been contemporaneous and of similar magnitudes at all sites. Changes in moisture conditions are thought to have been locally, but not regionally, important. Coetzee's interpretations of the vegetation and climatic events may be summarised as follows:

33 350–31 200 B.P. Sacred Lake lay in the lower part of the Ericaceous Belt indicating temperatures between 3·5 and 5·6°C lower than today.

31 200–26 000 B.P. A fairly dense forest of *Hagenia*, *Olea* and *Podocarpus* became established around Sacred Lake, culminating in a maximum forest expansion at 26 000 B.P. At this time, it is estimated that temperatures were between 2·0 and 4·1°C lower than today. Pollen spectra from the base of the Kaisungor core indicate that temperatures on the Cherangani Hills were depressed to the same extent as on Mt Kenya.

26 000–14 000 B.P. Both Lake Rutundu and Kaisungor Swamp were surrounded by Afroalpine grassland and Sacred Lake was again in the Ericaceous Belt. This shows that temperatures decreased from the maximum at 26 000 B.P. and remained 5·1–8·8°C

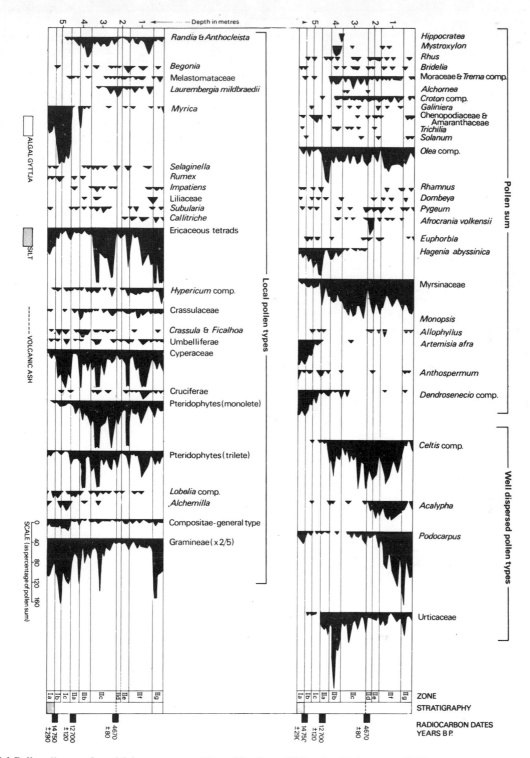

Fig. 8.1 Pollen diagram from Mahoma swamp. (From Hamilton, 1970; after Livingstone, 1967)

193

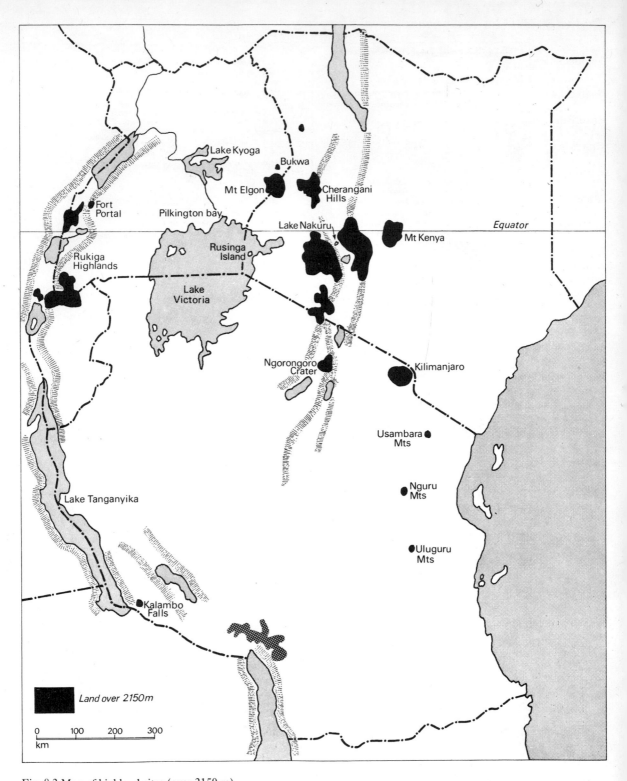

Fig. 8.2 Map of highland sites (over 2150 m).

194

Table 8.3   Upland sites which have been sampled

Sample site	Locality	Altitude (m)	Present vegetation	Max. age diagram (B.P.)	Reference or investigator
Sacred Lake	Mt Kenya	2400	Moist upland forest	33350* ± 1000	Coetzee, 1967
Lake Rutundu	„	3140	Lower ericaceous Belt	18000*	Coetzee, 1967
Crater lake	Mt Kilimanjaro	2650	Upland forest	4620* ± 50	Coetzee, 1967
Kaisungor Swamp	Cherangani hills	2900	Upper part of dry upland forest	27750* ± 600	Coetzee, 1967
Mahoma Bog	Ruwenzori	2595	Upper bamboo zone	5000*†	Osmaston, 1958
Lake Mahoma	Ruwenzori	2960	Upper bamboo zone	15500*	Livingstone, 1967
Bigo Bog	Ruwenzori	3400	Ericaceous belt	5000†	Osmaston, 1958
Bujuku Bog	Ruwenzori	3900	Afroalpine belt	3000†	Hedberg, 1954
Lake Bujuku	Ruwenzori	3920	Afroalpine belt	3000*	Livingstone, 1967
Lume Bog	Ruwenzori	3950	Upper ericaceous belt	3000†	Osmaston, 1958
Lake Kitandara	Ruwenzori	3990	Afroalpine belt	7500*	Livingstone, 1967
Lake Bunyonyi	Kigezi	1800	Cultivation	3000†	Morrison, 1965
Kankoko Swamp	Kigezi	2870	Cultivation	8000†	Morrison, 1966
Butongo Swamp	Kigezi	2200	Cultivation	7000†	Morrison, 1966
Muchoya Swamp	Kigezi	2256	Lower bamboo zone	24000*	Morrison, 1968

Symbols:*, from C^{14} dates or estimated from C^{14} dates; †, estimated from other evidence.

lower than today until 14 000 B.P. Varying degrees of local dryness are indicated by the presence of *Artemisia* pollen.

14 000–10 500 B.P. With the exception of a brief cold interval on Mt Kenya at $10\,800 \pm 100$ B.P. and two cold spells on the Cheranganis at $12\,650 \pm 100$ and *c.* 10 800 B.P., pollen diagrams indicate a continuous and gradual warming throughout this period.

10 500–8000 B.P. Very marked changes in the pollen curves are attributed to a sharp increase in temperature. *Hagenia* forest replaced Ericaceous Belt vegetation around Sacred Lake. Ericaceous Belt vegetation replaced Afroalpine vegetation both around Lake Rutundu and around Kaisungor Swamp.

After 8000 B.P. Warmer conditions were experienced at all sites. Dense humid forest became established around Sacred Lake, and Kaisungor Swamp came to lie within the Upland Forest Belt. After a slight drop, temperatures reached a maximum at 4000 B.P., when they were somewhat higher than they are today. This was marked by the presence of a humid forest with much *Ilex* around the Kilimanjaro site.

Both Coetzee (1964, 1967) and van Zinderen Bakker (1962) have correlated the more recent pollen zones recognised in these East African pollen diagrams with the pollen zones established for the postglacial period in northwest Europe. Coetzee (1967) further considers that the changes of temperature in equatorial East Africa and adjacent regions during the last 33 350 years were synchronous and parallel with those of Europe.

The above interpretation of these pollen diagrams has been hotly disputed by Livingstone (1967), who considers that far too much emphasis was placed on small fluctuations in the frequency of grass pollen. (It is noted that both Coetzee and Livingstone include grass pollen in the pollen sum.) Coetzee and van Zinderen Bakker assume that the quantity of grass pollen carried over long distances is small and, therefore, that an increase in the grass pollen curve signifies an increase in the extent of grassland near the sample site. They also assume that most of the grass pollen originates from high altitude tussock grassland and therefore that, the higher the percentage of grass pollen, the colder the climate. The second of these assumptions is questioned by Livingstone who points out that grasses are common in a wide variety of habitats in both upland and lowland East Africa.

Recent studies relevant to interpretations of pollen diagrams from upland East Africa

The controversy outlined in the preceding paragraph highlights some of the problems encountered in the interpretation of pollen diagrams from upland East Africa. Clearly, until past vegetation can be confidently reconstructed, climatic interpretation of pollen diagrams will remain in dispute. Accurate reconstruction requires knowledge of the pollen production of different types of vegetation and of the dispersal abilities of different pollen types. Following preliminary work by Hedberg (1954), Osmaston (1958) and Coetzee (1967), a more detailed study of recent pollen production and dispersal in upland East Africa has provided answers to some of the outstanding questions (Hamilton, 1972). The technique used in this investigation is the collection of surface soil, peat and sediment samples from various localities and analysis of their pollen contents. The pollen spectra are then compared with the composition of the vegetation surrounding the sample sites. In this way, a picture of pollen deposition, as relevant to diagram interpretation, has been built up. Where possible, surface samples have been collected from sites physically similar to those from which pollen diagrams have been obtained, since vegetation structure affects pollen deposition.

A surprisingly large quantity of pollen in many surface samples is of long-distance origin, that is, it is derived from plants growing at some distance from the sample sites. For example, it is found that, on average, about 25 per cent of the pollen in a surface sample from the Afroalpine and Ericaceous Belts of Mt Elgon belongs to species which grow only in the Upland Forest Belt or in other lower altitude types of vegetation. The actual percentage of long-distance pollen in these samples is much higher since some well dispersed pollen types, such as Gramineae, produced by ubiquitous taxa are not included in the figure.

By examining the quantity of long-distance pollen in surface samples, the pollen productivity of a vegetation type can be estimated. If the quantity of long-distance pollen relative to local pollen within a particular type of vegetation is high, it means that that vegetation itself produces little pollen. It is found that vegetation of the Ericaceous and Afroalpine Belts produces considerably less pollen than that of the upland forest belt and that, within this belt, the Bamboo Zone is a low pollen producer. Where the presence of a vegetation type of low pollen production is indicated in the diagram, the proportion of long-distance pollen is expected to be high.

Some pollen types are particularly well dispersed and thus are poor indicators of past vegetation. Even if abundant in a pollen diagram, their presence does not necessarily imply that their parent species were growing close to the sample site. The pollen produced by *Acalypha*, *Celtis*, *Podocarpus*, Urticaceae and Gramineae, and, to a lesser extent, by *Macaranga kilimandscharica* and *Olea* are well dispersed. The first four of these should not be included in the pollen sum during the construction of pollen diagrams and should be placed in a special category of 'very well-dispersed pollen types'.

In contrast to these well-dispersed pollen types, others are poorly dispersed. Examples are *Afrocrania volkensii*, *Dombeya*, *Ilex mitis*, *Impatiens* and *Mimulopsis* and the presence of one of these in large quantities in a pollen diagram strongly indicates that its parent species was growing near the sample site.

Pollen diagrams can only be interpreted in terms of environment if the ecology of the species concerned is known. Recent studies strongly suggest that the three environmental variables most responsible for determining the distribution of species in the highlands are temperature, availability of water and human disturbance (Hamilton, 1972).

In the absence of autecological studies it must be assumed that the total altitudinal range of a plant today indicates the range of temperature under which it grows naturally. Temperatures on the East African mountains decrease rapidly with increasing altitude (about 1°C in 160 m). Therefore, accurate vegetation reconstruction is essential before pollen diagrams can be used to estimate past temperature. Such estimations should be based only on the presence or absence of pollen types of poor dispersal ability since it is only the presence of these which strongly indicate that the parent species was growing near to the sample site.

Accurate vegetation reconstruction is not so essential for the estimation of water availability because the effects of dry or moist climatic conditions are felt over a wide area. Indeed, three of the very well-dispersed pollen types *Acalypha*, *Celtis*-type and Urticaceae, are particularly useful for assessing past moisture conditions. Species of *Acalypha* and *Celtis* do not occur at high altitude and are most common in moist lowland forests, *Acalypha* in secondary vegetation. The frequency of their pollen types in a diagram is thought to give a good indication of lowland moisture conditions. In contrast to *Acalypha* and *Celtis*, the Urticaceae are uncommon in the lowlands and are particularly characteristic of moist upland forests. The frequency of Urticaceae pollen therefore shows moisture conditions at higher altitudes. Rarity of these three pollen types in a diagram is much more significant than abundance. Abundance shows that it was wet somewhere in the area; rarity suggests a generally dry climate.

Although human disturbance of the vegetation is most marked below an altitude of 2150 m, pollen produced by species growing on disturbed land is carried up to all altitudes on the mountains. The vegetation changes induced by disturbance have some similarities to those produced by increased aridity and, on occasion, it may be difficult to determine which of these two factors is responsible for changes recorded in pollen diagrams. In surface samples, disturbance is associated with increase in the quantity of pollen belonging to plants of open vegetation, such as Chenopodiaceae, *Dodonaea viscosa* and Gramineae, at the expense of pollen of climax vegetation.

Detailed interpretation of the Lake Mahoma pollen diagram

As an example of how the type of information gathered from a study of present pollen deposition can be applied to pollen diagram interpretation, the Lake Mahoma diagram, originally published by Livingstone (1967) is first reinterpreted (Fig. 8.1). The pollen diagram has been redrawn, excluding from the pollen sum local pollen types and well-dispersed types.

Lake Mahoma is situated at an altitude of 2960 m on Ruwenzori, occupying a kettle-hole on the Mahoma moraines. Today, the lake lies in the upper part of the Bamboo Zone. Scattered trees of Ericaceae and, less commonly, of *Podocarpus milanjianus* and *Rapanea rhododendroides*, grow among the bamboo. The core, which was taken under 9·5 m of water, has an analysed length of 5·7 m. The uppermost sample lay close to the surface of the sediment and can be regarded as a surface sample. Three radiocarbon dates have been obtained and these indicate that the lowermost sample from the core is *c.* 15 500 years old.

Livingstone (1967) divided the pollen diagram into two zones, a basal Zone 1, characterised by high abundances of the pollen of a number of shrubs, herbs and small trees, and an upper Zone 2, characterised by high abundance of the pollen of a variety of closed forest trees. These Zones and the subzones described by Livingstone are shown on Fig. 8.1 and are used in the present interpretation.

The most abundant pollen types in the pollen sum of Zone 1 (c. 15 500–12 500 B.P.) are Chenopodiaceae+ Amaranthaceae, *Olea*, *Hagenia abyssinica*, *Myrsinaceae* (probably produced by *Rapanea rhododendroides*), *Artemisia afra*, *Anthospermum* and *Dendrosenecio* comp. This pollen spectrum has many similarities to those of surface samples collected between 3600 and 3950 m in the upper Ericaceous and lower Afroalpine Belts of Mt. Elgon (Hamilton, 1972) and this correspondence suggests analogous vegetation. If this is correct, it follows that *Artemisia afra*, *Anthospermum usambarensis* and *Dendrosenecio* sp. were growing close to the lake, that *Hagenia* and *Rapanea* were growing at lower altitudes than now, probably forming a *Hagenia-Rapanea* Zone and that Chenopodiaceae+Amaranthaceae and *Olea* were growing even further afield. *Artemisia* and *Anthospermum* no longer grow on Ruwenzori and are found only on the Ericaceous and Afroalpine Belt of the drier East African mountains. This reconstruction suggests temperatures some 6°C lower than now and a drier climate. The increase in *Hagenia* and decrease in *Artemisia* and *Dendrosenecio* comp. pollen towards the top of Zone 1 suggests that *Hagenia* forest was drawing closer to the lake, indicating a warming in climate.

All the pollen types included in the well-dispersed category have very low values in Zone 1, indicating the absence or rarity of their parent species on the mountain and its neighbourhood. The significance of these low values is increased by the fact that the lake appears to have been surrounded by ericaceous or Afroalpine vegetation, both of which have low pollen productivity. It is concluded from the low value of *Celtis*-type pollen that *Celtis* trees were uncommon in the lowlands around Ruwenzori and that the extent of lowland forest was very small. The low value of Urticaceae pollen probably shows that bamboo forest, which is characteristically rich in urticaceous herbs, was greatly reduced in extent, indicating a very dry climate.

The high value of *Myrica* pollen in Zone 1 is thought to indicate abundance of *M. salicifolia* trees in vegetation below the *Hagenia-Rapanea* zone. At the present time, either *Podocarpus* or bamboo forest is usually characteristic below the *Hagenia-Rapanea* zone, but the rarity of *Podocarpus* and Urticaceae pollen in Zone 1 suggests that neither of these was present. It is postulated that these were replaced during Zone 1 by a forest rich in *Myrica salicifolia* and *Olea*, implying a very dry climate.

The boundary between Zones 1 and 2 corresponds to the biggest unconformity in the pollen diagram. With the exception of the transitional stages recorded in its lower part, pollen spectra of Zone 2 (*c.* 12 600–0 B.P.) are not markedly different from those found in surface samples in and near Lake Mahoma today, and it is therefore probable that the lake was surrounded by bamboo forest for the greater part of this zone. Since bamboo forest is a relatively low pollen producer, this interpretation is supported by the relatively high value of such well-dispersed pollen types as *Celtis*-

type, *Olea* and Urticaceae. The only poorly-dispersed type with a high frequency in the pollen sum is *Afrocrania volkensii* in Sub-Zone 2d and this implies that the tree was growing near the lake. Today, *Afrocrania* is common both in the *Hagenia-Rapanea* and bamboo zones, but rarely grows at lower altitudes. The occurence of its pollen in Sub-Zone 2d therefore supports the above vegetation interpretation.

The presence of bamboo forest around Lake Mahoma indicates that temperature and moisture conditions were approximately similar to the present throughout Zone 2. *Celtis*-type, *Acalypha*, *Podocarpus* and Urticaceae all reach high values in this zone. In general this indicates much moister conditions than previously, with extensive forests in the lowlands around Ruwenzori and bamboo forest on the mountain itself. There is a delay between the rise of Urticaceae and that of *Celtis*-type and this may be due to a slower rate of establishment of *Celtis* trees, compared with Urticaceae. Possibly part of the large quantity of *Olea* pollen in Sub-Zone 2a originated from a lowland species of *Olea*, such as *O. welwitschii* growing in colonizing forest.

A maximum in *Celtis*-type pollen between Sub-Zone 2c and Sub-Zone 2f corresponds to a minimum in the grass pollen curve and is thought to indicate a period of maximum forest expansion. At the same time, a maximum in several local pollen types, such as *Begonia* and *Laurembergia*, and a minimum in Ericaceae imply slightly warmer conditions. During this period (*c*. 6000–2000 B.P.) the climate appears to have been slightly warmer and wetter than at present.

The decline in *Celtis* pollen in Sub-Zone 2g (*c*.1000–0 B.P.) is accompanied by a dramatic rise in grass pollen. The presence of certain pollen types, such as Chenopodiaceae+Amaranthaceae, *Myrica* and *Anthospermum* which are produced by plants of disturbed land, point to extensive forest clearance by man. A lower level of human disturbance, extending back to 4600 B.P., may be responsible for the rise in *Acalpha* pollen (Livingstone, 1967).

Other pollen diagrams from Ruwenzori

The above interpretation of the Lake Mahoma pollen diagram can be compared with interpretations of two other pollen diagrams obtained by Livingstone (1967) from Ruwenzori. These derive from cores in Lakes Bujuku (3920 m) and Kitandara (3990 m). The Bujuku core was taken under 13 m of water and has a total length of 4·3 m. A single C^{14} determination gives an age of 3000 B.P. for the basal sediments. The Kitandara core, taken under 9 m of water, has an analysed length of 12·5 m. Extrapolation from a single C^{14} determination gives an age of 7500 B.P. for the lowermost sediments.

Vegetation interpretation of these diagrams indicates that the lakes lay then, as now, within the Afroalpine or, in the case of Lake Kitandara, the Ericaceous Belts throughout the time periods represented. This lack of any marked change in the vegetation around the lakes supports the climatic history of the last 7500 years, as interpreted from the Lake Mahoma diagram. An Ericaceae pollen maximum and a *Dendrosenecio*-type pollen minimum at *c*. 4600 B.P. in the Kitandara diagram roughly coincide with the Ericaceae pollen minimum at Lake Mahoma, thus providing further evidence for a slight upward migration of vegetation belts on Ruwenzori at the time.

It is expected that changes in the frequencies of at least some of the very well-dispersed pollen types in one pollen diagram will be paralleled by similar changes in other pollen diagrams from the same region. Basing correlations on the available C^{14} dates, the following pollen changes are common to both the Mahoma and the Kitandara diagrams:

A rise in *Podocarpus* pollen from 7500 B.P. to the present day.

A fall in *Celtis*-type pollen after *c*. 4600 B.P.

A peak in *Acalypha* pollen between *c*. 4600 and 1000 *B.P.*

High values of Gramineae after *c*. 1000 B.P.

The Bujuku core is too short to show any but the most recent events. A very sharp rise in Gramineae matches similar rises at Lakes Mahoma and Kitandara.

It is concluded from these similarities that the vegetation changes responsible for these pollen changes were of regional and not local importance and that the environmental interpretation previously given is probably correct.

Pollen diagrams from Kigezi

Several cores were obtained by Morrison from the Rukiga Highlands of Kigezi in south-west Uganda. The ridges in this area lie at about 2450 m and are connected to the valleys by steep slopes. Many of the valleys contain swamps and lakes, with sediments very suitable for pollen analysis. A core from Muchoya swamp has been analysed in detail (Morrison, 1961 and 1968). This swamp lies at an altitude of 2256 m and is today surrounded by bamboo forest. Three C^{14} determinations indicate that the oldest sediments in the core, at a depth of 10·6 m, were deposited some 24 000 years ago. Morrison has divided the pollen diagram into four zones, which are estimated to have commenced at the following times: Zone 1, 24 000 B.P.; Zone 2, 17 000 B.P.; Zone 3, 11 000 B.P.; Zone 4, 600 B.P. The topmost sample from the core was collected about one metre below the present swamp surface and is estimated to have an age of *c.* 1200 B.P. Pollen spectra of surface samples from the swamp differ from this uppermost sample in several respects and are therefore placed in another zone, Zone 5. During Zones 1, 2 and much of 3, Muchoya was a lake, which later became overgrown by swamp vegetation.

The pollen sum of Zone 1 is characterized by high values of *Hagenia* and *Olea* and by traces of *Anthospermum* and *Stoebe*. Of the well-dispersed pollen types, *Podocarpus* occurs rather infrequently, as in the rest of the pollen diagram, and small porate grains (*Acalypha*, *Celtis*-type and Urticaceae) are uncommon. The local pollen category contains high values of *Alchemilla*, Gramineae and fern spores, and *Myrica* is also present. Vegetation interpretation of the overall spectrum suggests that the lake lay near the boundary between the *Hagenia* zone and the Ericaceous Belt. The

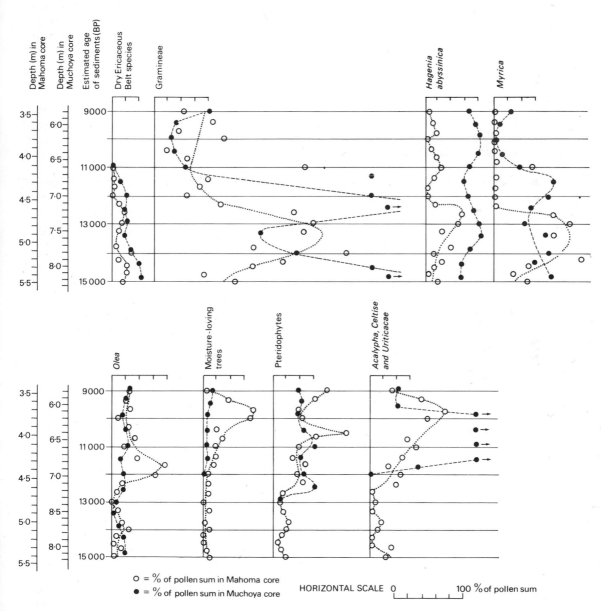

Fig. 8.3 Comparison of part of pollen diagrams from Mahoma and Muchoya swamps.

mean annual temperature is estimated to have been *c*. 6°C lower than today, or perhaps a bit warmer. Dry conditions are indicated, *inter alia*, by the rarity of small porate grains and of well-dispersed pollen types belonging to trees of moist upland forest.

In Zone 2, *Anthospermum* and *Stoebe* increase to 20 per cent and 10 per cent respectively, and there is a corresponding decrease in *Hagenia*. Small porate grains are virtually absent. In the local pollen category, *Myrica* and Gramineae are abundant, but fern spores are rare and *Alchemilla* is much less common than in Zone 1. The pollen spectra of Zone 2 are interpreted as indicating that the lake was now surrounded by dry Ericaceous Belt vegetation, in which patches of *Hagenia* forest occurred. The mean annual temperature is estimated to have been *c*. 6°C lower than today, and abundant evidence shows that the climate was very dry.

Hagenia resumes a high percentage in Zone 3 as *Anthospermum* and *Stoebe* die out. *Olea* persists and is here accompanied by the pollen of other upland forest trees, including *Afrocrania volkensii*, *Anthocleista*, *Macaranga kilimandscharica* and *Nuxia congesta*. At the base of this zone, there are sharp rises in small porate grains and pteridophyte spores, while Gramineae shows a sudden fall. *Myrica* persists in Zone 3 at about the same level as in Zone 2.

Changes in pollen spectra between Zones 2 and 3 have many similarities to those which occur between Zones 1 and 2 in the Mahoma pollen diagram and C[14] dates indicate that the events in the two areas were more or less contemporaneous. See Fig. 8.3. It is concluded that the vegetation changes responsible for these pollen variations were a result of a change in climate. Livingstone (1967) suggested that weathering of fresh moraines may have been responsible for some of the vegetation changes around Lake Mahoma, but this cannot apply in the Muchoya region, which was never glaciated.

As might be expected by comparison with the Mahoma diagram, there is clear evidence that the climate during Zone 3 at Muchoya was much wetter than during Zone 2. (Fig. 8.3). Temperature changes cannot be accurately evaluated for Zone 3 because poorly dispersed pollen types, belonging to species of restricted altitudinal range, are not sufficiently abundant to indicate that their parent species were growing near the lake. The vegetation around the lake could have been either bamboo or *Hagenia* forest.

In Zone 4, *Hagenia* pollen becomes at first rare and then absent. The pollen sum now contains higher percentages of *Macaranga kilimandscharica*, *Nuxia congesta* and *Olea*, and *Alchornea hirtella*, *Ilex mitis*

and *Neoboutonia*-type, all of which are uncommon lower in the diagram, now occur frequently. Small porate grains decrease from their high values in Zone 3, but are still common. Pollen spectra indicate that the swamp lay close to the boundary between bamboo forest and the underlying broad-leafed forest zone and temperatures were therefore a bit higher than they are today.

A decline in the incidence of some forest tree pollen types indicates slightly lower temperatures in Zone 5. Forest clearance is shown by the presence of *Anthospermum*, *Dodonaea viscosa* and *Rumex* pollen.

The pollen changes between zones 3 and 4 and between zones 4 and 5 are recorded in one or more of three other Kigezi pollen diagrams, from Butongo Swamp (2200 m), Katenga Swamp (2070 m) and Lake Bunyonyi (1980 m).

The Katenga and Bunyonyi diagrams are of special interest because they record in detail forest clearance and subsequent land-use in the Rukiga Highlands. Three stages can be recognized. During Stage 1, trees on lower slopes were felled, but land-use was not intensive and secondary forest regenerated. In Stage 2, increased pressure led to virtual elimination of lower slope forests and to the spread of such plants as *Dodonaea viscosa* and *Justicia*. Finally, in Stage 3, upper slope forests were felled. There are signs of soil deterioration from Stage 2 onwards.

Summary of upland pollen diagram interpretation

It is apparent from the interpretations of the Uganda pollen diagrams that three environmental factors, temperature, moisture availability and human disturbance, have played important roles in determining the composition and distribution of past vegetation types. Van Zinderen Bakker and Coetzee (1972) have re-examined the Kenya and Tanzania pollen diagrams in the light of the fresh data on pollen deposition (Hamilton 1972) and they now conclude that there have been big changes in moisture as well as temperature conditions on Mt. Kenya and the Cherangani Hills. Climatic changes in highland East Africa during the Upper Pleistocene as determined by pollen analysis can be summarized as follows:

A relatively warm period, from *c*. 30 000–26 000 B.P., with temperatures *c*. 2–4°C lower than now, was followed by a long cold phase, which began to terminate around 12 500 B.P. It is estimated that mean annual temperatures were *c*. 6°C lower in the upper part of this period, but may have been a bit higher before 17 000 B.P. The period between 12 500 and 10 500 B.P.

is clearly transitional and temperatures increased rapidly to values similar to the present. Since 10 500 B.P., temperatures have been fairly constant, but with a slightly warmer period centred around 4000 B.P.

Moisture changes have paralleled temperature changes in western Uganda, a cold climate being a dry climate. This appears also to have been the case on the Cherangani hills. When the Kaisungor diagram is redrawn according to the method used for the Uganda diagrams, it is found that between *c.* 17 000 and 12 000 B.P., the pollen sum contains very high values of Chenopodiaceae pollen. Grass pollen is also abundant, and there are low frequencies of all types of tree pollen, even of those which are relatively well dispersed. Taken together this indicates a very dry climate, with semi-desert conditions in the lowlands around Cherangani. Pollen spectra indicate a much moister climate after 12 000 B.P. On Mt. Kenya, pollen spectra indicate that it was somewhat drier before *c.* 12 000 B.P., but here pockets of moist upland forest containing *Macaranga kilimandscharica* were also present.

Initiation of large scale forest clearance in the lowlands around Ruwenzori has been dated at *c.* 1000 B.P. (A.D. 950). Estimates based on rates of sediment accumulation give an approximately similar date for Kigezi. This forest clearance is almost certainly linked with the immigration of ironworking, Bantu-speaking peoples into these areas. No clear evidence of forest clearance can be seen in the Kenya and Tanzania pollen diagrams.

The Lake Victoria pollen diagram

Kendall (1969) has recently published an important paper on the ecological history of the Lake Victoria basin. Numerous C^{14} dates show that a core from Pilkington Bay near Jinja extends back to 15 000 B.P. A pollen diagram obtained from this core and the interpreted vegetation history of the area can be summarised as follows:

1 Prior to 12 000 B.P., values of Gramineae pollen are very high and those of forest trees very low. Forest must have been either completely absent or of very limited extent in the lake region.

2 At 12 000 B.P., the appearance of Oleaceae, Moraceae and *Celtis* pollen and a reduction in Gramineae pollen show that forest became established and subsequently spread. A maximum in tree pollen between 8000 and 3000 B.P. corresponds with very low values of Gramineae and indicates a period of greatest forest expansion.

Climatic changes inferred by Kendall for the northern Victoria basin from pollen and other evidence are:

1 Dry from before 14 500–12 000 B.P. (at this time Lake Victoria was without an outlet).

2 Moderately wet from 12 000–10 500 B.P.

3 Moderately dry from 10 500–9500 B.P.

4 Wet from 9500–6500 B.P.

5 Slightly drier, or with a more seasonal rainfall after 6000 B.P.

The pollen diagram confirms that lowland forest was either absent or of very limited extent in Uganda before 12 000 B.P. and that there was a maximum in forest spread at *c.* 4000 B.P. These vegetation changes are clearly linked with changes in moisture availability. Temperature changes cannot be inferred from this lowland pollen diagram. The reduction in forest after 3000 B.P. may be due either to forest clearance or to a drier climate; but, most likely, to a combination of both.

Evidence from disjunct distribution

When records of the occurrence of a taxon are collected together, it is quite often found that its range is discontinuous, and that neighbouring populations are separated by a wide interval. This type of distribution is known as disjunct, and here we will consider three ways in which it can arise. First, a taxon may originate separately in different areas. Second, a taxon may spread from one to other climatically favourable areas by long-distance dispersal. Last, and of particular interest to students of past vegetation, a disjunct distribution may be due to contraction from a once continuous range, following a change in climate or in other environmental conditions.

Multiple origin of a taxon under natural conditions is certainly of rare occurrence. (A case of suspected double origin for a species is reported from Malaya by Whitmore (1969), who considers that isolated populations of *Macaranga quadricornis* in the north and south of the country may have evolved separately from the *M. triloba* complex.) In East Africa, different species of the characteristically Afroalpine subgenus *Dendrosenecio* may have evolved separately from normal types of *Senecio* growing in the upland forest (Hedberg, 1969). It is possible that the genome of *Senecio* carries a tendency towards gigantism which is realised independently on different East African mountains under the peculiar climatic conditions encountered here at high altitudes. A similar mechan-

201

ism may be present in *Lobelia* a genus considered to be normally herbaceous, but with a potential for strange giant growth forms, which are found, not only on the East African mountains, but also in many other parts of the world.

Very little is known about the long-distance dispersal abilities of plant disseminules, particularly over long periods of time. The chance of successful long-distance transport depends on a larger number of factors, including the quantity of disseminules produced, the mechanisms available for their transport and the area available for their reception. Hedberg (1969b) thinks that the importance of cyclones as agents for long-distance dispersal has been underestimated. As will be seen later, he does not consider it necessary to invoke climatic change to explain the upland forest disjunctions, an opinion which conflicts with that of Moreau (1963a and b and 1966). This disagreement between two eminent African ecologists underlies the uncertainties attached to interpretation of disjunct distributions.

Where the interval between the isolated populations is wide, where a relatively large number of taxa share the same disjunction and where a proportion of these taxa have disseminules apparently poorly adapted for long-distance dispersal, an explanation in terms of climatic change is likely. On the other hand, where relatively few taxa share the same disjunction and particularly where their disseminules are well adapted for long-distance dispersal, conclusions concerning climatic change cannot be made.

Even where it is reasonably clear that certain plants once occurred in areas where they no longer grow and that this was the result of climatic conditions different from those of today, it is extremely difficult to establish the date of separation of the now disjunct populations. Attempts to do so are based on estimates of the rates of evolution in the taxa concerned. Disjunction of two populations of the same species is likely to be due to more recent isolation that disjunction of two different species of the same genus. When comparing plant and animal disjunctions, it should be borne in mind that, if plant and animal genera and species are assumed to be comparable taxonomic units, the rate of evolution in plants in East Africa during the last twenty-five million years has been generally much slower than that of mammals.

East African coastal—West African rain forest disjunctions

Isolated lowland forests exist along the Kenyan and Tanzanian coasts from just above sea level to 760 m altitude (Moreau, 1935a). Forests of a similar type but with some upland elements occur from 760 m to 1220 m in the eastern Usambara Mountains (Moreau, *loc. cit.*) and at similar altitudes in the Nguru and Uluguru Mountains, all in Tanzania. These are about 40, 140 and 130 km from the coast respectively.

The coastal rain forests of Kenya and Tanzania are separated from the easternmost edge of the Zaïre rain forest (in the Semliki valley in western Uganda) by more than 1100 km and from the easternmost extensions of that forest type (the Kakemega forest in western Kenya and the Kigoma and Bukoba forests in western Tanzania) by 650 km in Kenya and by 850 km in Tanzania. Yet there are a number of taxa which occur in these widely separated areas and not in between.

Using principally the completed parts of the *Flora of Tropical East Africa* (1952 continuing), a typescript of the *Papilionaceae* for this flora, other regional floras and specimens in the East African Herbarium, the distributions of many clearly disjunct genera and species which occur in the coastal lowland and sub-upland rain forests have been worked out. It is possible that because of the large size of the floral areas used in the *Flora of Tropical East Africa*, apparent continuous distributions may, in fact, be quite disjunct, so that many other taxa might be included here. Also such large families as the Euphorbiaceae and Rubiaceae which have many forest species and have not yet been studied for this flora would likely furnish a number of additional examples.

Some types of distribution suggest a former connection between the coastal rain forests and the West African forests and these are summarised below:

1 Species occurring in the East African coastal forests and also in the West African (Zaïre and/or Guinea) forests or their eastern extensions.

(a) Species occurring in West Africa only, in the main Zaïre and/or Guinea forest blocks, i.e. extending no further east than the Semliki valley in Uganda; examples are *Adiantum confine, Paramaclobium caeruleum, Schefflerodendron usambarense* and *Malacantha alnifolia*.

(b) Species reaching the isolated forests further east in Uganda and/or western Kenya and/or western Tanzania. These are numerous and include *Atrophyium mammianum, Bolbitis gemmifera, Marantochloa leucantha* and *Chrysophyllum perpulchrum*.

2 Species represented by different subspecies in the East African coastal forests and the West African forests. Only two examples have been found: *Greenwayodendron suaveolens* subsp. *graveolens* occurs from West Africa to Uganda while subsp. *usambaricum* is

confined to the eastern Usambaras in Tanzania; *Pterocarpus mildbraedii* subsp. *usambarensis* confined to the eastern Usambaras is separated from subsp. *mildbraedii* (Congo-Brazzaville to Ivory Coast) by nearly the whole width of Africa.

3 African genera represented by approximately equal numbers of species in the East African coastal forests and West African forests. Examples here are *Mesogyne* (Moraceae) with one species endemic to the coastal forests of Tanzania and a second endemic to São Tomé, and *Cylicomorpha* (Caricaceae) with one subupland species occurring from Kenya to Malawi and a second in the Cameroun Republic.

4 African genera with one or two species in the East African coastal forests but with a principal distribution in the West African forests, for example, the two well-represented West African genera *Isolona* (Anonaceae) with twenty species and two in coastal forests and *Allanblackia* (Guttiferae) with seven species ranging from West Uganda to West Africa with two other species in eastern Tanzania.

Although the floristic affinities of the East African forests with the Zaire and Guinea rain forests cannot yet be computed numerically, it is apparent that the similarities are of sufficient magnitude to compel us to formulate a hypothesis for this relationship. Two explanations are possible: either large numbers of species have crossed a gap of 650–1100 km or more of inhospitable country by various means, but where the rain forest was never continuous; or conditions suitable for the growth of rain forest, i.e. higher more evenly distributed rainfall and higher temperatures than at present, existed in areas now covered by other vegetation types, permitting a corridor or connection between the coastal and West African forests.

The botanical evidence suggests that there have been at least two periods of connection between the coastal rain forests and the Zaire rain forest; a relatively recent one, which would account for the large number of species common to both, and a more ancient one which would explain the similarities at the generic level. Moreau (1966), on the basis of two levels of differentiation in bird species, also suggests the possibility of connections on two separate occasions.

To summarise, the similarities in floral composition between the coastal rain forests of Kenya and Tanzania and the West African rain forests can best be explained by a former connection between them. This would require a warmer and wetter climate with more evenly distributed rainfall that at present. From distributional evidence alone it is difficult to estimate when such climate conditions may have occurred.

The South-west Africa—North Africa arid region disjunctions

Several plant taxa show disjunctions between the arid and semi-arid regions of South-west Africa and those of North Africa. These may be classified into three categories.

1 Same species in both areas as, for example, *Sporobolus spicatus*, *Zygophyllum simplex*, *Salvadora persica* (Koch in van Zinderen Bakker, 1967); *Aizoon canariense*, *Tribulocarpus dimorphanthus* (Verdcourt, 1969, with map).

2 Closely related species pairs, one in the north and one in the south. Such as: *Heliotropium rariflorum*—*H. hereroense*, *Stapelia revoluta*—*S. prognatha*; *Kissenia capensis*—*K. spathulata* (Winter, 1966).

3 Genera in both areas, but not occurring in the interval: *Monsonia* sect. *plumosa* (Kers, 1969); *Echidnopsis*, *Erythrophysa*, *Salsola*, *Fagonia* (Winter, 1966).

Notable disjunctions between south-western Africa and north-eastern Africa (Somalia) are also found in the avifauna (Winterbottom, 1967) and in mammals (Kingdon, 1971).

On the basis of their faunal and floral similarities, several authors have suggested that the two areas were previously connected, probably on several occasions, by corridors of arid or semi-arid country. Unfortunately no detailed comparison of the floras of the two areas has yet been made, and it is difficult to assess the botanical evidence. There are, in fact, several arguments against former connection. First, in each area there are many endemic species which do not extend to the other. This may be explained by the suggestion that the corridor was never wide, nor open for long, nor indeed perhaps ever absolutely continuous, and also by the fact that many of the strict endemics have poor dispersal mechanisms (Verdcourt, 1969). Second, the areas available for production and reception of plant disseminules are large. On the other hand, many of the disjunctions have very wide intervals and, on balance, there seems to be a reasonable case in favour of former connection or connections. Possibly the most recent of these was terminated at *c.* 12 000 B.P., a date at which pollen diagrams indicate the climate, at least in some areas of East Africa, changed from very dry to comparatively wet.

Upland disjunctions

Today, upland communities in East Africa are separated from each other, often by hundreds of miles of lowland vegetation. Since numerous taxa occur in more than one upland area and not in the intervening lowlands, they display disjunct distributions within East Africa. Some also show much wider disjunction and are, for instance, next found on Mt Cameroon in West Africa or in temperate parts of the world.

Of the montane floras, the Afroalpine is the best known (Hedberg, 1957) and disjunctions occur at several taxonomic levels (Hedberg, 1969b). In some cases species occur on many or all of the high mountains without displaying any perceptible morphological differences (e.g. *Sagina afroalpina*). In other cases different mountain populations of one species show statistical differences, but their variation ranges overlap so much as to preclude taxonomic distinction (e.g. *Alchemilla johnstonii*). Hedberg (1961a) considers that the existence of a large number of pairs and groups of vicarious taxa (e.g. *Romulea congoensis*—*R. keniensis* and the *Lobelia deckenii* group) testifies that the different ecological islands of Afroalpine flora must have been effectively isolated from each other for a long time. The spread of Afroalpine plants between the mountains has probably been affected by long-distance dispersal, which may have been facilitated by Pleistocene changes of climate. These climatic changes are thought to have enlarged the ranges of Afroalpine species, but not to have caused connection.

Many of the Afroalpine taxa are not represented in the tropical African lowlands and have their nearest relatives in more distant parts of the world. Hedberg (1965) has classified Afroalpine plants into nine genetical flora elements (see Chapter 5), each of which includes species thought to have originated from a particular region.

In view of their very mixed origin, the foreign elements are thought to have been introduced by long-distance dispersal and therefore the flora is an excellent example of the effectiveness of this process over a long period of time. Geological evidence indicates that there were no high mountains in East Africa during the early Kainozoic and it is likely that the Afroalpine flora postdates the beginning of the Miocene. The fact that only a relatively small proportion of the flora appears to have originated from lowland tropical African species suggests that the local flora has been slow to adapt to new circumstances.

The taxonomy of species of the upland forest belt is less well known, but nevertheless there are indications that the proportion of species common to different highland areas is much higher than in the case of the Afroalpine belt and that the proportion of vicarious taxa is much lower (Chapman and White, 1970). Some well-known examples of widely distributed upland forest belt taxa are *Hagenia abyssinica*, *Arun-*

dinaria alpina, *Afrocrania volkensii*, *Entandrophragma excelsum* and *Aningeria adolfi-friedericii*. Subspecies of the latter are among the few vicarious taxa recorded in the upland forest flora. Other vicariads recorded are varieties of the forest-edge species *Crotalaria agatiflora*, *Millettia oblata* and *Astragalus atropilosulus* (Verdcourt, personal communication). The disjunct distributions displayed by many upland forest species has led many authors to claim previous connection or connections at times when the climate was colder and wetter than it is today. However, although a large number of species share these disjunctions, the case in favour of previous connection is by no means clear cut. The areas available for production and reception of disseminules are quite large (much larger than in the case of the Afroalpine Belt) and the intervals between different upland areas are comparatively small. Indeed, Hedberg (1969b) considers that given sufficient time most upland forest species of animals and plants should be able to pass even a fairly wide gap without depending upon continuous distribution of the whole ecosystem. However, studies on the distribution of birds (Moreau, 1966) and mammals (Coe, 1967) support previous connection between the now isolated upland forests. The avifauna above an altitude of *c.* 1500 m is apparently quite distinct from that at lower altitudes and upland forests birds are not found in lowland forest. Moreau considers that the size of the common element in different upland forests in East Africa suggests that the mountains were colonised by an already integrated community.

The possibility of former connection between the upland forests of West Africa and those of East Africa has received considerable attention. Giant Lobelias (sect. *Rhynchopetalum*), *Rumex abyssinicus*, *Crassocephalum mannii*, *Laportea alatipes* and *Agauria salicifolia* are among those plants which show a highland Cameroun—upland East Africa disjunction (Morton, 1961 and 1967). Morton is of the opinion that long-distance dispersal is insufficient to explain the disjunctions and thinks that connection occurred several times. However, this is debatable in view of the paucity of information concerning long-distance dispersal potentials of plant disseminules. As in the case of upland forest disjunctions within East Africa, studies on bird distribution favour previous connection between east and west African montane communities (Moreau, 1966). The birds of highland Cameroun fall into two categories, one that is fairly distinct from, and one which is very similar to, the East African upland types and Moreau considers that there has been more than one connection between upland ecosystems in East and West Africa. Both Morton and Moreau have calculated that the climate must have been wetter and temperatures c. 5°C lower to bring about connection.

Evidence from relict patches of vegetation

Patches of one type of vegetation are sometimes found within a much more extensive area of another. In cases where these patches are thought to be the remains of a formerly much more extensive area of the first vegetation type, they are known as 'remnants' or 'relicts'. There are, of course, many reasons why one type of vegetation comes to replace another but, in East Africa, the most important have undoubtedly been climatic change and human activities. Here we are only concerned with relicts which are thought to have been caused, either directly or indirectly, by man.

The vegetation over the greater part of East Africa has been profoundly modified by man. It is often extremely difficult to envisage what the natural vegetation would be like, but sometimes the presence of relict patches of vegetation gives us an insight into the past. Conclusions reached from a study of these relicts can sometimes be checked by experiments in which certain modifying influences, e.g. burning, are excluded from selected areas of vegetation and the effects observed.

It is useful to include in this discussion some forest remnants in the northern part of Zambia, close to the Tanzanian border (Lawton, 1963, 1972). The dominant vegetation type in this area is miombo woodland, a fire-climax community which is widespread in East Africa and which is unlikely to have covered extensive areas before man began regularly to burn the plant cover. Siszya forest lies at an altitude of 1370–1530 m, is only a few acres in extent and is surrounded by miombo. The forest is a ritual burial ground and this alone accounts for its survival. Lawton writes that the Siszya forest is probably the last remaining remnant of a tropical forest that covered part of the plateau in the past. Other data from the area, including some from fire exclusion sample plots, led Lawton to the view that much of the northern province of Zambia was once forest covered, but has been degraded to the fire climax Miombo and Chipya woodland types by repeated burning over a long period of time. It is noted that wood samples from the Kalambo archaeological site, dated at c. 50 000 B.P., include a mixture of forest species (*Cynometra* sp., probably *C. alexandri*) and woodland species (*Colophospermum mopane, Dalbergia*

sp.), suggesting that, even at that time, both forest and fire climax woodland occurred in the area.

Extensive regions of Uganda which are now covered with grassland or woodland probably once carried forest. Rabongo in Murchison Falls National Park is a *Cynometra*-dominated forest, thought to be a relict of a forest linking Budongo forest with formerly forested areas in west Acholi. Over most of this area, forest has been replaced by *Terminalia* and other woodland types which are themselves in the process of degradation to treeless grassland through the combined action of elephants and fire. The Rabongo forest has decreased greatly in area over the past four decades to reach about 280 ha in 1958. Consumption and tramping of understorey trees and shrubs by elephants combined with mortality to large trees resulting from debarking and burning permitted sufficient light penetration for the spread of *Panicum maximum* and other grasses into the forest margin. As a result, fire from surrounding grasslands is carried ever deeper into the forest, bringing about rapid attrition of the forest vegetation (Buechner and Dawkins, 1961). In this case, the basic cause of vegetation degradation is believed to be an exceptional increase in the population of elephants. This is itself due partly to evacuation of people from the area after an outbreak of sleeping sickness in 1912 and partly to concentration of elephants in the national park because of harassment in regions to which the animals formerly migrated in large numbers.

The above examples show that extensive areas of woodland and grassland in regions of relatively high rainfall may have been derived from climatic climax rain forest. In regions of lower or more seasonal rainfall, the nature of the climax vegetation is less well known, because relicts appear to be rare.

Review of the evidence

A surprising feature of the Miocene fossil plants which have been identified from East Africa, is that many are referable to present day East Africa in general. At first sight, this might suggest that the flora has changed little since that time, but the paucity of well-authenticated identifications makes this conclusion premature. The rate of evolution in the area does, however, appear to have been much slower in plants than in mammals, since few Miocene fossil mammals are referable to modern genera. A slow rate of evolution in plants is also suggested by the fact that only a few lowland African plants have succeeded in adapting themselves to the Afroalpine environment.

Many Afroalpine taxa provide the only reasonably certain examples of natural immigrants into the East African flora during the Miocene and subsequent periods. As fossil and other evidence suggests that the climate of lowland East Africa has been tropical and therefore unsuitable for these taxa for a long time it is concluded that their introduction onto the newly formed mountains was affected by long-distance dispersal.

There is some evidence that the Miocene flora of East Africa was richer in species than it is now. Modern African lowland rain forests are species poor. The total number of phanerogam species known from Zaire, Rwanda and Burundi in 1946 was 9706, which compares with over 10 000 in Borneo, a much smaller and less diversified area (Richards, 1952). There are only 377 species of orchid in Zaïre compared with an estimated 5000 in Malaysia (Richards, 1952), only 50 species of palm belonging to 15 genera in Africa compared with 1140 species in 92 genera in America and 1150 species in 107 genera in Asia and Australia (Corner, 1966). The poverty extends to the forest birds. The 600 000 square miles of Zaïre forest are said to contain only about as many birds as have been recorded in a square kilometre of rain forest in Costa Rica (Moreau, 1966). Richards considers that the environment of the African rain forest is quite as equable as that of the Malayan or South American forests and therefore that the poverty of the African forest is not due to any ecological factor operating at the present day. The cause must be sought in earlier times. Although several other explanations have been advanced (e.g. Aubreville, 1949), the most likely seems to be that the flora and fauna of the African rain forest were depauperated during a time or times of extreme climate, either during the Pleistocene (Moreau, 1966; Richards, 1952) or the Pleiocene (Cloudesley-Thompson, 1969). It must be assumed that the forests of South America and Malaysia have never been subject to such extreme conditions. Moreau links the extinction of lowland African rain forest plants with the expansion of upland forest at the expense of lowland forest at a time of cold, wet climate. However, recent evidence suggests that at least the last cold period of the Pleistocene was universally felt and hence coldness alone cannot explain the difference in floristic richness between the African and South American or Malaysian forests. Furthermore, it is thought that wet conditions are not limiting to lowland forest development. It therefore seems more likely that the forest was depauperated at a time when the climate was very dry, not very wet. For further discussion of lowland extinctions see Chapman and White (1970). The reported occurrence of Dipterocarp wood on Elgon suggests that there may have been a sizable Asian element in the Miocene flora.

The Miocene fossil floras comprise plants which, assuming similar ecology to their modern relatives, could grow at the same localities today. However, scarcity of data precludes generalisations concerning the pattern of vegetation types in East Africa at that time. It is nevertheless clear that the flora was a tropical one and that forests occurred.

Some of the taxa showing disjunct distribution between the forests of West Africa and those of the East African coast are sufficiently distinct to suggest that some time before the Pleistocene the two areas were linked by a continuous belt of moist tropical forest. This would have required a climate considerably wetter and slightly warmer than at present. Verdcourt (Personal communication) thinks that this connection, as well as possible Pleistocene connections, was through a loop of forest around Tanzania passing down the east side of Lake Tanganyika, through Kungwe, Ufipa, the mountains north of Lake Malawi, the Iringa highlands, the Ulugurus and the Usambaras.

All over the world, the Pleistocene period has been a time of climatic variability. Lands at high altitudes have experienced a succession of ice ages, the last of which began to terminate 14 000 to 10 000 years ago in north-western Europe.

Signs of ancient glaciation can be found on Elgon and other East African mountains now free of permanent ice (Nilsson, 1932). On the still glaciated mountains, Kenya, Kilimanjaro and Ruwenzori, there are extensive series of moraines, which Osmaston (1965) believes are geologically correlated sufficiently well to show that glacial events on the three mountains were contemporaneous. The glacial advances and retreats on these mountains must therefore have been determined by changes in the same environmental variables. Osmaston calculated from the positions of the moraines that if the precipitation was the same during the last extensive glaciation as it is now, the temperature must have been $c.$ 4°C lower. Alternatively, similar glacial advances could have been caused by either colder and drier, or warmer and wetter climate. According to a C^{14} date from Lake Mahoma, the retreat of the glaciers on Ruwenzori at the end of this stage commenced shortly before 14 700 ± 290 B.P. (Livingstone, 1962).

Interpretation of highland pollen diagrams supports the hypothesis that temperature changes in highland East Africa during the Upper Pleistocene have been similar to those estimated for other parts of the world

(Coetzee and van Zinderen Bakker, 1970; Hamilton, 1972). Following a slightly warmer period lasting from c. 30 000–25 000 B.P., temperatures decreased between 25 000 and 12 500 B.P. to c. 6°C lower than now. A rapid increase after 12 500 B.P. culminated in temperatures similar to today at c. 10 500 B.P. A slightly warmer period is recorded around 4000 B.P. No data are available concerning temperature fluctuations in lowland East Africa, but it seems likely that these have paralleled the highland changes.

Pollen diagrams indicate that, at both high and low altitudes, changes in moisture conditions have paralleled changes in temperature. Cold times have been dry and warm times wet. This conflicts with the well-known 'pluvial theory', according to which the glacial periods of temperate regions correspond to times in East Africa of high rainfall and/or low evaporation (pluvials) and, conversely, interglacials correspond to interpluvials (Leakey, 1965; Wayland, 1934 and 1952). The Gunz, Mindel, Riss and Wurm glaciations are correlated with the Kageran, Kamasian, Kanjeran and Gamblian pluvials.

The pluvial theory is based mainly upon interpretations of a number of raised lake sediments and river terraces, but the original climatic interpretations of these deposits have come under much criticism (Bishop, 1963, 1967b and 1969; Bishop and Posnansky, 1960; Cole, 1964; Cooke, 1963; Isaac, 1966). Many of the raised lake and river deposits are thought now to be the product of tectonic activity, rather than climatic change. In any case, many workers consider that these deposits are insufficiently well dated to permit correlation even within East Africa. Reviewing data then available, Cooke (1963) came to the conclusion that the only well-documented evidence for climatic changes in East Africa is that for the Gamblian pluvial, the two later wet phases (Makalian and Nakuran) and the intervening dry period, as shown in the Nakuru-Naivasha basin. He holds that inferences relating to deposits of pre-Gamblian date are unsatisfactory, though there is evidence favouring a general climate somewhat wetter than that of the present day in some areas.

The geological evidence for pluvial conditions in the Naivasha-Nakuru-Elmenteita basin has been re-examined by Washbourne (1967), who considers that supposed shorelines attributed to the Makalian and Nakuran wet periods are insufficiently well marked to show major standstills in the lake level. She states that the Gamblian pluvial is represented by a single shore line which lies about 180 m above the present level of Lake Nakuru and which is linked with the outlet of the basin. The lake was apparently stabilised at this level for some time. Recently detailed studies on the sediments beneath Lakes Nakuru and Elmenteita have strongly suggested that the lakes were small and alkaline before c. 12 000 B.P., that there was a period of maximum lake expansion, corresponding to the Gamblian shoreline, between 10 000 and 8 500 B.P. and that subsequently conditions have been somewhat drier (Isaac 1970; Richardson 1966 and 1972). Further evidence that the Gamblian pluvial occurred during postglacial rather than glacial times is provided by reassessment of artifacts from Gamble's cave. These are thought to be probably not older than 10 000 B.P. and might be considerably younger (Bishop, 1967).

To summarise, the pluvial theory can no longer be upheld in the light of recent evidence. This suggests that at least the latter part of the last glaciation was a time of *dry* climate in East Africa (not a pluvial) and that the subsequent postglacial period has been comparatively *wet*. It should, however, be borne in mind that there are large areas of East Africa for which no data are available and it is possible that these experienced different conditions (van Zinderen Bakker, 1969). There is, as yet, no reliable information about climatic conditions prior to 30 000 B.P., though Moreau's work on birds suggests at least one period of cold, wet climate before this date.

Pollen diagrams support the existence of a dry climate from 25 000–12 500 B.P. The vegetation belts on the mountains were depressed by 800–1000 m at this time and the dry climate encouraged the growth of plants, such as *Artemisia*, *Anthospermum*, *Stoebe* and *Myrica*, all tolerant of dry conditions, at the expense of moisture-loving taxa, such as *Alchornea Celtis* and Urticaceae. Lowland forest was greatly reduced in extent and was virtually absent from Uganda. Abundant Chenopodiaceae pollen in the Cherangani diagram indicates that semi-desert conditions prevailed around that mountain. It is possible that the ranges of some of the taxa now showing disjunct distributions between South-Western African and North African arid regions were connected, or nearly connected at this time.

With the exception of comparatively recent, but drastic, changes caused by man, the vegetation has probably not shown any great changes since 10 500 B.P. There is evidence that lowland forest was more extensive at one time during this period, and it is possible that there was then a connection between forests of West Africa and those of the East African coast. If so, subsequent forest retraction may be responsible for some of the disjunct distributions of *species* and *subspecies* of these forest floras. It will be recalled that an earlier connection has been postulated to account for disjunctions at the generic level.

The impact of man

East Africa can well be called the home of man. Evidence accumulated over the last thirty years strongly indicates that our genus, *Homo*, evolved from the ancestral australopithecine stock in the area (Tobias, 1967). For the greater part of his existence, it is unlikely that man has had any marked influence on the vegetation since, for over a million years, he lived as a hunter and gatherer and was an integral part of the ecosystem. The importance of man in the history of the vegetation is due to a number of comparatively recent cultural achievements, particularly the knowledge of how to make and use fire, animal domestication, agriculture and iron working.

In Africa, the use of fire by man may not be as ancient as in north-eastern Asia and Europe (Phillips, 1968). The oldest archaeological sites with evidence of human-made fires are Kalambo Falls in Zambia and the Cave of Hearths in the Transvaal (Cole, 1964), and a C^{14} determination from the former gives a date of $57\,300 \pm 300$ B.P. The fact that no earlier sites have been discovered is not particularly significant, since such sites are only likely to be preserved in a very few instances. Having learnt how to make a fire, it is probable that man quickly instituted the practice of setting fire to undergrowth in dry seasons, to allow greater visibility and freedom of movement and to facilitate hunting. In addition, as Hopkins (1965a) has pointed out, man enjoys the sight of a good blaze, especially at night.

The importance of burning as an ecological factor in East Africa can hardly be overemphasised. So much of the vegetation is fire climax that, in many areas, it is difficult to envisage what the natural vegetation would be like. However, in the absence of burning, many grasslands tend to revert to woodland and many woodlands to forest. It is unlikely that moist forest types can be changed to woodland or grassland by fire alone and without the aid of other factors such as forest clearance or the presence of herds of destructive large mammals such as elephants or buffaloes.

It seems likely that, before man introduced burning, much of East Africa was covered with closed woody vegetation (Aubreville, 1949). Indeed, extensive areas may have been covered by dry types of forest which, because of susceptibility to burning, have almost entirely disappeared. The accumulation of a large quantity of dead material over many years must have made occasional natural fires in these forests particularly fierce, but it is thought that these were not sufficiently frequent to have had any marked effect in determining community structure. Verdcourt and Polehill (personal communication) point out that it is unlikely that the arid northern part of Kenya and the arid 'wedge' which reaches down eastern Kenya into Tanzania, were ever covered by closed, woody vegetation.

Before man-induced burning, large mammals probably created and maintained extensive open glades within the various types of woody vegetation which covered East Africa. In this context, it is of considerable interest to note that, in Africa some 50 000 years ago, there was a big extinction of mammalian species, particularly of larger forms (Martin, 1966). Martin states that such extinctions occurred in other parts of the world during the Upper Pleistocene and that they were always associated either with the arrival of man in the region or else with a change in his culture. How man caused these extinctions is uncertain. Gillett (personal communication) suggests that they may have resulted from the discovery of the effectiveness of poison for hunting. Another possibility is that they resulted from the vegetation changes which followed the onset of widespread and regular burning. For Africa, the latter explanation receives some support from the C^{14} date from Kalambo Falls. The animals which became extinct may have been predominantly browsers in dry forests and, unlike the elephant, were not able to adapt to new conditions when their natural food sources were destroyed.

The earliest bones of domestic cattle from East Africa are found at Stone Bowl Culture sites, which, according to C^{14} dates, fall into the first millennium B.C. (Soper, personal communication) and which are situated in and near the Eastern Rift Valley in the Naivasha-Nakuru-Ngorongoro area. Artifacts belonging to this culture suggest a connection with the north-east. The impact of these pastoralists on the vegetation of East Africa as a whole was slight, since they occupied a fairly limited area.

It is not clear whether the Stone Bowl Culture people were agriculturalists as well as pastoralists. For discussion of this interesting topic see Murdock (1960) and Doggett (1970).

However, extensive agriculture as well as iron-working are thought to have been introduced into East Africa by Bantu-speaking peoples between A.D. 0 and 500 (Soper, 1969). They replaced or displaced the stone age hunters and gatherers over much of East Africa, leaving only small relict populations, such as the Ndorobo of the Mau Escarpment and the Hudza of Tanzania.

The introduction of iron implements greatly facilitated forest clearance, and it is probable that the widespread forest destruction, which is recorded in pollen diagrams from western Uganda and dated at *c*. 950 A.D., followed the arrival of the Bantu in the area. The settled way of life associated with an agricultural economy allows a big population increase and, in East Africa, this must have further hastened vegetation degradation. The main crops probably included sorghum, Eleusine millet and bananas, the two cereals having been introduced earlier from the Ethiopian region. The introduction of several crops, such as cassava, maize, sweet potatoes and groundnuts, all of which are widely grown today, postdated the Columban discovery of America.

The arrival of Europeans in the East African hinterland during the last century has been followed by a big population increase, which can be attributed mainly to the introduction of western medicine and to a reduction in the incidence of tribal warfare. This and, to a lesser extent, the new cash economies and the decimation of large mammal populations, has had marked effects on the vegetation. A summary of these is given by Shantz and Turner (1958), who base their findings on photographs of the same pieces of vegetation taken in 1919–23 and again in 1956–67. They attribute the main changes in the vegetation during this period to excessive grazing by domestic animals, particularly in drier areas, and to a great increase in the area of land under cultivation. The latter now extends into regions only marginally suitable for agriculture.

Rain forests, which once covered extensive areas of East Africa, are now largely confined to forest reserves, which have been established to help meet future timber requirements. However, even within the reserves, much of the forest is already secondary (Dale, 1952).

Overgrazing, with resulting vegetation degradation and soil erosion, has become a serious problem in many parts of East Africa. The effects of overgrazing in north-eastern Uganda have been described by Wilson (1960) who states that, only about fifty years ago, Karamoja District was covered with wooded grassland and open grass savanna, supporting large populations of wild animals. Today, the predominant plant communities are composed of succulents and thorny bushes, and large mammals are rare in most areas.

Marked changes in vegetation have resulted from schemes aimed at control of tsetse flies (see Chapter 2) the carriers of trypanosomiasis (sleeping sickness). This serious disease of man and cattle is, even today, the main factor limiting population growth in much of western Tanzania, and has been the cause of big population movements in the past. In some northern parts of the Lake Victoria shore the whole population was moved from a tsetse infected area in the early years of this century. By 1909 the mainland shore and islands were deserted and forty years later this area had been covered by dense forest.

In 1892 the country of Karagwe on the borders of Uganda and Tanzania was described by travellers as having extensive grass plains forming pasturage for handsome long-horned cattle. There was thick tree growth in the valleys and villages with banana plantations. Today this is an almost deserted bushland due to an epidemic of rinderpest which spread rapidly and drove the remaining pastoralists north to Ankole. With the removal of the human population, the country reverted to bush, and wild animals, and tsetse found their way in from Tanzania further south. A full account of various aspects of the relationship between tsetse and vegetation is given by Ford (1971).

The systems of agriculture and cattle management which are practised in East Africa today were satisfactory before the recent population increase. Land was allowed to lie under fallow long enough for soil fertility to be restored and the cattle population was low enough to avoid overgrazing. Today the situation is entirely different and, clearly, changes in the old systems are essential if the natural resources of East Africa are to be conserved.

Bibliography

ADAMS, M.E. (1967) 'A study of the ecology of *Acacia mellifera*, *A seyal* and *Balanites aegyptica* in relation to land clearing', *J. Appld. Ecol.*, **4**, 221–37.

ABRAHAM, M.F.H. (1958) 'The East African camphor forests of Mt Kenya', *E. Afr. agric. J.*, **24**, 139–41.

AFRICAN LAND DEVELOPMENT IN KENYA 1946–62. Pp. 312. Nairobi: Ministry of Agriculture, Animal Husbandry and Water Resources.

AGNEW, A.D.G. (1968) 'Observations on the changing vegetation of Tsavo National Park (East)', *E. Afri. Wildl. J.*, **61**, 75–80.

AGNEW, A.D.G. and HEDBERG, O. (1969) 'Geocarpy as an adaptation to Afroalpine solifluction soils', *J. E. Afr. Nat. Hist. Soc. and Nat. Museum*, **27**, 215–16.

ALBRECHT, F.O. (1964) 'Natural changes in grass zonation in the Red Locust outbreak centre in the Rukwa Valley, Tanganyika', *S. Afric. J. Agric. Sci.*, **71**, 123–4.

ALLT, G. (1968) 'pH and soil moisture in the substrate of giant Senecios and Lobelias on Mt Kenya', *E. Afr. Wild. J.*, **6**, 71–4.

ANDERSON, B. (1957) *A Survey of Soils in the Kongwa and Nachingwea Districts of Tanganyika*. Pp. 120. University of Reading in conjunction with Tanganyika Agricultural Cooperative.

ANDERSON, B. (1963) *Soils of Tanganyika*. Pp. 36. Bulletin No. 16 Ministry of Agriculture; Research division. Dar es Salaam: Government printer.

ANDERSON, G.D. (1963a) 'Some weakly developed soils of the eastern Serengeti plains, Tanganyika', *African Soils*, **8**, 339–47.

ANDERSON, G.D. (1963b) 'A comparison of red and yellowish-red upper-slope soils of the eastern Usambara foothills, Tanganyika', *African Soils*, **8**, 432–4.

ANDERSON, G.D. and TALBOT, L.M. (1965) 'Soil factors affecting the distribution of the grassland types and their utilisation by wild animals of the Serengeti plains, Tanganyika', *J. Ecol.*, **53**, 33–56.

ANDERSON, M.C. (1964) 'Light relations of terrestrial plant communities and their measurement', *Biol. Rev.*, **39**, 425–86.

AUBREVILLE, A. (1949) *Contribution à la Palaeohistoire des forêts de l'Afrique Tropicale*. Societé d'Editions, Paris.

ARBER, A. (1920) *Water Plants: A Study of Aquatic Angiosperms*. Pp. 436. Cambridge: University Press.

ASHTON, P.S. (1967) 'Climate versus soils in the classification of South East Asian tropical lowland vegetation', *J. Ecol.*, **55**, 67–8.

ASHTON, P.S. (1969) 'Speciation among tropical forest trees; some deductions in the light of recent evidence', *Biol. J. Linn. Soc.*, **1**, 155–96.

AUSTIN, M.P. and GREIG-SMITH, P. (1968) 'The application of quantitative methods to vegetation survey. II Some methodological problems of data from rain forest', *J. Ecol.*, **56**, 827–44.

BACKLUND, H.O. (1956) 'Aspects and succession in some grassland vegetation in the Rukwa valley; a permanent breeding area of the Red Locust', *Oikos*, Supplement 2, 1–132.

BAILEY, R.G. (1966) 'The dam fisheries of Tanzania', *E. Afr. agric. For. J.*, **22**, 1–15.

BAKER, H.G. and HARRIS, B.T. (1957) 'The pollination of *Parkia* by bats and its attendant evolutionary problems', *Evolution, Lancaster, Pa.*, **11**, 449–60.

BAKER, H.G. and HARRIS, B.T. (1958) 'Pollination in *Kigelia africana*', *J. W. Afr. Sci. Ass.*, **4**, 25–30.

BAKER, S.J.K. (1963) *The East African Environment*. Reprinted from *History of East Africa*, vol. 1, Oliver, R. and Mathew, G. (eds). Oxford: Clarendon Press.

BALL, J.B. (1965) *Working Plan for the Lake Forests. Central Forest Reserves Masaka District Uganda*. Pp. 35. Entebbe: Forest Dpt. Cyclostyled.

BALLY, P.R.O. (1954) 'Tree Euphorbias as timber trees', *J. E. Africa. nat. Hist. Soc.*, **22**, 105–6.

BALLY, P.R.O. (1969) 'Coryndon Museum Expedition to the Chyulu Hills', *J. E. Afr. Nat. Hist. Soc.*, **14**, 161–6.

BANAGE, W.B. and VISSER, S.A. (1967) 'Soil moisture and temperature levels and fluctuations in one year in a Uganda soil catena', *E. Afr. agric. For. J.*, **32**, 450–5.

BANCROFT, H. (1935) 'The Dipterocarps in Africa', *Empire For. J.*, **14**, 74–5.

BATRA, S.W. and L.K. (1967) 'The fungus gardens of insects', *Scientific American*, **217**, 112–20.

BARTLETT, H.H. 'Fire, primitive agriculture and grazing in the tropics', in *Man's Role in Changing the Face of the Earth*. W.L. Thomas (ed). Chicago: University Press.

BASINSKI, J.J. (1959) 'The Russian approach to soil classification and its recent development', *J. Soil Sci.*, **10**, 14–26.

BATES, J.A.R. (1960) 'Studies on a Nigerian forest soil I. The distribution of organic matter in the profile and in various soil fractions', *J. Soil Sci.*, **11**, 246–56.

BATES, J.A.R. and BAKER, Mrs T.C.N. (1960) 'Studies on a Nigerian forest soil II. The distribution of phosphorus in the profile and in various soil fractions', *J. Soil Sci.*, **11**, 257–65.

BATES, M. (1944) 'Observations on the distribution of diurnal mosquitoes in a tropical forest', *Ecology*, **25**, 159–70.

BATES, M. (1961) *The Forest and the Sea*. Pp. 216. New York: New American Library.

BATTISCOMBE, E. (1926) *A Descriptive Catalogue of Some of the Common Trees and Woody Plants of Kenya Colony*. Pp. 142. London: Crown Agents.

BATTISCOMBE, E. (1936) *Trees and Shrubs of Kenya Colony; A Revision and Enlargement of Battiscombe 1926*. Pp. 201. Nairobi: Government printer.

BAX, P. NAPIER and SHELDRICK, D.L.W. (1963) 'Some preliminary observations on the food of elephants in the Tsavo National Park', *E. Afr. Wild Life*, **1**, 40–53.

BAXTER, R.M., TALLING, J.F., PROSSER, M.V. and WOOD, R.B. (1965) 'Stratification in Tropical African Lakes at moderate altitude 1500–2000 m', *Limnol. Oceanogr.*, **10** (4), 510–20.

BEADLE, L.C. (1932) 'Scientific results of the Cambridge expedition to the East African lakes 1930–31. The water of some East African lakes in relation to their fauna and flora', *J. Linn. Soc. Lond.*, **38**, 157–211.

BEADLE, L.C. (1962) 'Evolution of species in the lakes of E. Africa', *Uganda J.*, **26**, 44–54.

BEADLE, L.C. (1963) 'Anaerobic life in a tropical crater lake', *Nature, Lond.*, **200**, 1223–4.

BEADLE, L.C. (1966) 'Prolonged stratification and deoxygenation in tropical lakes. Crater Lake Nkuguti, Uganda compared with Lake Bunyonyi and Lake Edward', *Limnol. Oceanogr.*, **11**, 2.

BEADLE, L.C. (1972) *Tropical Lakes and Rivers* Pp. 300, London; Longman.

BEADLE, L.C. and LIND, E.M. (1960) 'Research on the swamps of Uganda', *Uganda J.*, **24**, 84–98.

BEADLE, N.C.W. (1953) 'The edaphic factor in plant ecology with a special note on soil phosphates', *Ecology*, **34**, 426–8.

BEADLE, N.C.W. (1954 and 1962) 'Soil phosphate and the delimitation of plant communities in eastern Australia', *Ecology*, **35**, 370–5; **43**, 281–8.

BEADLE, N.C.W. (1966) 'Soil phosphate and its role in moulding segments of the Australian flora and vegetation with special reference to xeromorphy and sclerophylly', *Ecology*, **47**, 992–1007.

BEALS, E.W. (1969) 'Vegetational change along altitudinal gradients', *Science, N.Y.*, **165**, 981–5.

BEAUCHAMP, R.S.A. (1953) 'Sulphates in African Inland Waters', *Nature*, **171**, 769.

BELLIS, E. (1964) 'Soil surveys in Kenya', *African soils*, **9**, 137–43.

BERRIE, A.D. (1966) 'Fish ponds in relation to the transmission of bilharziasis in East Africa', *E. Afr. agric. For. J.*, **31**, 276–82.

BIGGER, M. (1968) 'A Check list of the flora of Kilimanjaro. College of Wildlife management, Murka, Tanzania', Cyclostyled. Pp. 54.

BIRCH, H.F. (1948) 'Soil phosphate – a review of the literature', *E. Afr. agric. J.*, **14**, 29–33.

BIRCH, H.F. (1949) 'The estimation of adsorbed and acid soluble phosphates in East African soil', *Commonw. Bur. Soils Technical Communication*, no. 46.

BIRCH, H.F. (1952) 'The relationship between phosphate response and base saturation in acid soils', *J. agric. Sci., Camb.*, **42**, 276–85.

BIRCH, H.F. (1953) 'The relationship between phosphate response and base saturation, pH, and silica content of acid soils', *E. Afr. agric. J.*, **19**, 48–9.

BIRCH, H.F. (1958) 'The effect of soil drying on humus decomposition and nitrogen availability', *Pl. Soil*, **10**, 9–31.

BIRCH, H.F. and FRIEND, M.T. (1965a) 'Humus decomposition in East African soils', *Nature, Lond.*, **138**, 500–1.

BIRCH, H.F. and FRIEND, M.T. (1956b) 'The organic matter and nitrogen status of East African soils', *J. Soil Sci.*, **7**, 156–67.

BIRCH, W.R. (1959) 'High altitude ley agronomy in Kenya. Part I. Observations on the climate and soil', *E. Afr. agric. J.*, **25**, 35–41.

BIRCH, W.R. (1963) 'Observations on the littoral and coral vegetation of the Kenya coast', *J. Ecol.*, **51**, 603–15.

BISHOP, W.W. (1963) 'The later Tertiary and Pleistocene in eastern equatorial Africa', *African Ecology and Human Evolution*. New York: Werner-Gren Foundation for Anthropological Research.

BISHOP, W.W. (1965) 'Quaternary geology and geomorphology in the Albertine rift valley, Uganda', Pp. 293–321. Reprinted from *The Geological Society of America*, special paper 84, Wright, H.E. Jr and Frey, D.G. (eds).

BISHOP, W.W. (1966) 'Stratigraphical geomorphology: a review of some East African landforms'. Pp. 139–76. Reprinted from *Essays in Geomorphology*, Drury, G.H. (ed). London: Heinemann.

BISHOP, W.W. (1967a) 'The later Tertiary in E. Africa' in *Background to Evolution on Africa*. Pp. 31–51. University of Chicago Press.

BISHOP, W.W. (1967b) 'Annotated lexicon of Quaternary stratigraphical nomenclature in E. Africa' in *Background to Evolution in Africa*. Pp. 375–95. University of Chicago Press.

BISHOP, W.W. (1969) 'Pleistocene stratigraphy in Uganda', *Geol. Survey of Uganda*, Mem. 10. Entebbe: Govt. Printer.

BISHOP, W.W. and POSNANSKY, M. (1960) 'Pleistocene environments and early man in Uganda', *Uganda J.*, **24**, 44–61.

BISHOP, W.W. and WHITE, F. (1962) 'Tertiary mammalian faunas and sediments in Karamoja and Kavirondo, East Africa', *Nature, Lond.*, **196**, 1283–7.

BLACKIE, J.R. (1964) 'Hydrology and afforestation in the Aberdares', *E. Afr. Geogr. Rev.*, **2**, 17–22.

BLOCK, W. and BANAGE, W.B. (1968) 'Population density and biomass of earthworms in some Uganda soils', *Rev. Ecol. Biol. Soc.*, **5**, 515–21.

BOALER, S.B. (1963) 'The annual cycle of stem girth increment in trees of *Pterocarpus angolensis* DC. at Keibungu, Tanganyika', *Comm. For. Rev.*, **42**, 232–6.

BOALER, S.B. (1966) 'The ecology of *Pterocarpus angolensis* DC. in Tanzania'. Pp. 128. *Ministry of Overseas Development Overseas Research Publication*, no. 12, London: HMSO, 1958.

BOGDAN, A.V. (1958) *A Revised List of Kenya Grasses with Keys for Identification*. Pp. 73. Nairobi: Govt. Printer.

BOGDAN, A.V. (1958) 'Some edaphic vegetational types at Kiboko, Kenya', *J. Ecol.*, **46**, 115–26.

BOGDAN, A.V. (1962) 'Grass pollination by bees in Kenya', *Proc. Linn. Soc. Lond.*, **173**, 57–60.

BOGDAN, A.V. and KIDNER, E.M. (1967) 'Grazing natural grassland in western Uganda', *E. Afr. agric. For. J.*, **23**, 31–4.

BOGDAN, A.V. and PRATT, D.J. (1961) *Common Acacias of Kenya*. Pp. 15. Nairobi. Govt. Printer.

BOGDAN, A.V. and KIDNER, E.M. (1967) 'Grazing natural grassland in western Uganda', *E. Afr. agric. For. J.*, **23**, 31–4.

BONIGER, J. (1956) 'Mangrove poles from Kenya coast; one of East Africa's oldest trades has a record year', *Kenya Press Office Feature*, no. 311. Nairobi: Department of Information.

BOUGHEY, A.S. (1955a) 'The nomenclature of the vegetation zones on the mountains of tropical Africa', *Webbia*, **11**, 413–23.

BOUGHEY, A.S. (1955b) 'The vegetation of the mountains of Biafra,' *Proc. Linn. Soc. Lond.*', **165**, 144–50.

BOUGHEY, A.S. (1956) 'The lowland rain forest of tropical Africa', *Proc. Trans. Rhodesia Scient. Ass.*, **44**, 1–17.

BOUGHEY, A.S. (1957) *The Origin of the African Flora*. Pp. 48. London: Oxford University Press.

BOWDEN, B.N. (1962) 'An illustrated field key to the commoner Uganda grasses', *E. Afr. agric. For. J.*, **27**, 230–51.

BOWDEN, B.N. (1964) 'The dry seasons of intertropical Africa and Madagascar', *J. trop. Geogr.*, **19**, 1–3.

BRASNETT, N.V. (1940) 'Modern trends in forestry with particular application to Uganda', *Uganda. J.*, **7**.

BRASNETT, N.V. (1944) 'The growing of *Chlorophora excelsa* in Uganda', *E. Afr. agric. J.*, **10**, 83–93.

BRAZIER, J.D. (1958) 'Nomenclature of Uganda timbers', Pp. 29. *Uganda Protectorate Forest Department Bulletin*, no. 6. Entebbe: Govt. Printer.

BREDON, R.M. and HORRELL, C.R. (1961) 'The chemical composition and nutritive value of some common grasses in Uganda. I General pattern of behaviour of grasses', *Trop. Agric., Trin.*, **38**, 297–304.

BREDON, R.M. and HORRELL, C.R. (1962) 'The chemical composition and nutritive value of some common grasses of Uganda. II The comparison of chemical composition and nutritive value of grasses throughout the year with special reference to growth', *Trop. Agric., Trin.*, **39**, 13–17.

BREDON, R.M. and WILSON, J. (1963) 'The chemical composition and nutritive values of grasses from semi-arid areas of Karamoja as related to ecology and types of soil', *E. Afr. agric. For. J.*, **29**, 134–42.

BREMNER, J.M. (1951) 'A view of recent work on soil organic matter', *J. Soil Sci.*, **2**, 67–96.

BRENAN, J.P.M. (1958) 'Leguminosae subfamily Mimosoideae' in *Flora of East Tropical Africa*. Pp. 173. London: Crown Agents. (See List of Floras.

BRENAN, J.P.M. (1967) Leguminosae subfamily Caesalpinoideae. Pp. 230. In *Flora of East Tropical Africa*. London: Crown Agents. (See List of Floras.

BRENAN, J.P.M. and GREENWAY, P.J. (1949) *Check Lists of the Forest Trees and Shrubs of the British Empire. No. 5 Tanganyika territory.* Part II, pp. 653. Oxford: Imperial Forestry Institute.

BROCK P.W.G. and MACDONALD R. (1969) 'Geological environment of the Bukwa mammalian fossil locality, E. Uganda', *Nature*, **223**, 593–6.

BROCKINGTON, N.R. (1961) 'Studies of the growth of a *Hyparrhenia* – dominated grassland in Northern Rhodesia', *J. Br. Grassld. Soc.*, **16**, 54–64.

BROOMFIELD, C. (1955) 'A study of podsolization', *J. Soil Sci.*, **6**, 284–92.

BROWN, L. (1959) *The Mystery of the Flamingoes*, London: *Country Life*.

BUCKMAN, H.O. and BRADY, N.C. (1969) *The Nature and Properties of Soils.* 7th edn. London: Macmillan.

BUECHNER, H.K. and DAWKINS, H.C. (1961) 'Vegetation changes induced by elephants and fire in Murchison Falls National park, Uganda', *Ecology*, **42**, 752–66.

BURGER, W.C. (1967) *Families of Flowering Plants in Ethiopia.* Pp. 236. Stillwater, Oklahoma: Oklahoma Univ. Press.

BURGES, A. (1960) 'The nature and distribution of humic acids', *Sci. Proc. R. Dublin Soc.*, (A), **1**, 53–8.

BURTT, B.D. (1929) 'A record of fruits and seeds dispersed by mammals and birds from the Singida District of Tanganyika Territory', *J. Ecol.*, **17**, 351–5.

BURTT, B.D. (1934) 'A botanical reconnaissance in the Virunga volcanoes of Kigezi, Ruanda, Kivu', *Kew Bull.*, **4**, 145–65.

BURTT, B.D. (1942) 'Burtt Memorial Supplement: Some East African Vegetation Communities', Jackson, C.H.N. (ed.), *J. Ecol.*, **30**, 65–146.

BURTT, B.D. (1953–57) *A Field key to the Savanna Genera and Species of Trees, Shrubs and Climbing Plants of Tanganyika Territory.* Part I. *Genera and Some Species.* 2nd edn. rev. by J.R. Welch. Part III *The Species of the More Important Genera with General Index.* Pp. 128. Rev. and ed. by P.E. Glover and C.H.N. Jackson. Dar es Salaam: Gov. printer.

BURTT, B.D. and DAVY, J. (1938) 'The classification of tropical woody vegetation', Oxford: *Imperial Forestry Institute Paper*, no. 13.

BUTT, R.A. (1965) 'Trials of species for timber planting in the savanna woodland zone of north Uganda', *E. Afr. agric. For. J.*, **31**, 54–62.

BUXTON, P.A. (1948) *Trypanosomiasis in East Africa.* London: HMSO.

BUXTON, P.A. (1955) 'The natural history of tsetse flies'. Pp. 816. *London School of Hygiene and Tropical Medicine, Memoir* 10. London: H.K. Lewis.

CALDER, E.A. (1957) 'Features of nitrate accumulation in Uganda soil', *J. Soil Sci.*, **8**, 60–71.

CALTON, W.E. (1954) 'An experimental pedological map of Tanganyika', *2nd Internat. African Soils Confer.*, **I**, 236–40. Leopoldville.

CALTON, W.E. (1959) 'Generalisations on some Tanganyika soil data', *J. Soil Sci.*, **10**, 169–76.

CALTON, W.E. (1963) 'Some data on a Tanganyika catena', *E. Afr. agric. For. J.*, **29**, 173–7.

CALTON, W.E. and VAIL, J.W., (1956), 'Micronutrient problems in Tanganyika', *Proc. 6th Internat. Congr. Soil. Sci.*

CARPENTER, G.D. HALE (1920) *A Naturalist on Lake Victoria.* Pp. 333. London: Unwin.

CARPENTER, G.D. HALE (1925) *A Naturalist in East Africa.* Pp. 187. Oxford: The Clarendon Press.

CARTER, G.S. (1956) *The Papyrus swamps of Uganda.* Pp. 25. Cambridge: Heffer.

CHAMBERLAIN, C.J. (1933) *Methods in Plant Histology.* Pp. 416. 5th ed. Chicago: University of Chicago press.

CHAMPION, A.M. (1933) 'Soil erosion in East Africa', *Geogr. J.*, **82**, 130–9.

CHANEY, R.W. (1933) 'A Tertiary flora from Uganda', *J. Geol. (Chicago)*, **41**, 702–9.

CHAPMAN, J.D. and WHITE F. (1970) *The Evergreen Forests of Malawi.* Oxford: Commonwealth Forestry Institute.

CHENERY, E.M. (1951) 'Some aspects of the aluminium cycle', *J. Soil. Sci.*, **2**, 97–109.

CHENERY, E.M. (1952) 'A digest of the chemistry of dead swamps in Kigezi district, south-west Uganda'. Pp. 1–6. *Annual report Hydrological Survey Department*, Entebbe.

CHENERY, E.M. (1954) 'Acid sulphate soils in central Africa', *Trans. 5th Internat. Congr. Soil Sci.*, **4**, 195–8. Leopoldville.

CHENERY, E.M. (1958) 'A review of the role of minor trace elements in E. African Agriculture', *Rep. of the directors of Overseas Dept. of Agric. Wye College*, Paper 6.

CHENERY, E.M. (1960) *An introduction to the soils of the Uganda protectorate.* Pp. 80. Uganda Department of Agriculture. Research Memoirs Series 1 No. 1 Kawanda: Cyclostyled.

CHESTERS, I.K.M. (1957) 'The Miocene flora of Rusinga Island, Lake Victoria, Kenya', *Palaeontographica*, **101** B, 30–71.

CHRISTY, C. (1911) *The African Rubber Industry and Funtumia elastica.* Pp. 252. London: Bale and Danielsson.

CLOUDESLEY-THOMSON J.L. (1969) *The Zoology of Tropical Africa.* Pp. 355. London: The World

Naturalist. Weidenfeld and Nicolson.

COE, M.J. (1964) 'Colonisation in the nival zone of Mt Kenya', *E. Afr. Acad. Proc.*, **2**, 137–40.

COE, M.J. (1967), 'Biogeography of the Equatorial Mountains', *Palaeoecology of Africa*, **2**, 59–73.

COE, M.J. (1967) *The Ecology of the Alpine Zone of Mt Kenya*. Pp. 136. The Hague: W. Junk.

COETZEE, J.A. (1964) 'Evidence for a considerable depression of the vegetation belts during the upper Pleistocene of the E. African mountains', *Nature*, **204**, 564–6.

COETZEE, J.A. (1967) 'Pollen analytical studies in east and southern Africa', in *Palaeoecology of Africa and of the Surrounding Islands and Antarctica*, vol. 3, 1–146. Cape Town and Amsterdam: A.A. Balkema.

COETZEE, J.A. and van Zinderen Bakker, E.M. (1970) 'Palaeoecological problems of the Quaternary of Africa', *Suid Afrikaanse Tydskrif vir Weterskap*, no. 4.

COLE, S. (1964) *The Prehistory of E. Africa*, London: Weidenfeld and Nicolson.

COOMBE, D.E. (1960) 'An analysis of the growth of *Trema guineensis*', *J. Ecol.* **48**, 219–31.

COOKE, N.B.S. (1963) 'Pleistocene mammal faunas of Africa with particular reference to Southern Africa' in *African Ecology and Human Evolution*, Howell Boulière (ed.). New York: Werner Green Foundation for Anthropological Research.

COPLEY, H. (1948) *The lakes and rivers of Kenya: A Short Guide to the Inland Waters and Their Inhabitants*. Pp. 80. London: Longman.

COPLEY, G.C., TWEEDIE, E.M. and CARROLL, E.W. (1964) 'A key and check list to Kenya Orchids. Part I'. *E. Afr. Nat. Hist. Jl.* **24**, 1–58. Part II, Ibid., **24**, 85–91.

CORNER, E.J.H. (1952) *Wayside trees of Malaya*. 2 volumes. Pp. 772. 2nd edn. Singapore: Govt. Printer.

CORNER, E.J.H. (1964) *The Life of Plants*. Pp. 315. London: Weidenfeld and Nicolson.

CORNER, E.J.H. (1966) *The Natural History of Palms*. Pp. 393. London: Weidenfeld and Nicolson.

COTTON, A.D. (1930) 'A visit to Kilimanjaro', *Kew Bull.*, 97–121.

COUSENS, J.E. (1951) 'Some notes on the composition of lowland tropical rain forest in the Rangarn Forest Reserve, Johore', *Malaya Forester*, **14**, 131–9.

COWAN, I.R. (1965) 'Transport of water in the soil-plant atmosphere system', *J. appl. Ecol.*, **2**, 221–39.

CROIZAT, L. (1952) *Manual of Phytogeography*. Pp. 587. The Hague: W. Junk.

CUNNINGHAM, R.K. (1963) 'The effect of clearing a tropical forest soil', *J. Soil. Sci.*, **14**, 334–45.

DAGG, M. (1965) 'A rational approach to the selection of crops for areas of marginal rainfall in East Africa', *E. afr. agric. For. J.*, **30**, 296–300.

DALE, I.R. (1939) 'The woody vegetation of the Coast Province of Kenya', *Imp. For. Inst. Paper* 18, pp. 28.

DALE, I.R. (1940) 'The forest type of Mt Elgon', *Jl. E. Africa Nat. Hist. Soc.*, **36**, 74–82.

DALE, I.R. (1952) 'Is East Africa drying up?' *E. Afr. agric. J.*, **17**, 90–103.

DALE, I.R. (1953) *A Descriptive List of the Introduced Trees of Uganda*. Entebbe: Govt. Printer.

DALE, I.R. (1954) 'Forest spread and climatic change in Uganda during the Christian era', *Emp. For. Rev.*, **33**, 23–9.

DALE, I.R. and GREENWAY, P.J. (1961) *Kenya Trees and Shrubs*. Pp. 653. Nairobi: Buchanan's Kenya Estates Ltd, in association with Hatchard's, London.

DARWIN, C. (1906) *The Movements and Habits of Climbing Plants*. Pp. 208. London: Murray.

DAUBENMIRE, R. (1968) 'Ecology of fire in grasslands', in *Advances in Ecological Research*. J.B. Cragg (ed.). Pp. 209–66. London and New York: Academic Press.

DAVIES, W. (1961) 'Temperate and tropical grasslands', *Proc. 8th International Grassland Congress*, Reading, England.

DAVIES, W. and SKIDMORE, C.L. (1966) *Tropical Pastures*. Pp. 215. London: Faber and Faber.

DAWE, M.T. (1906a) *Report on a Botanical Mission Through the Forest Districts of Buddu and the Western and Nile Provinces of the Uganda Protectorate*. Pp. 63. London: HMSO.

DAWE, M.T. (1906b) 'Notes on the Vegetation of Buddu and the Western and Nile Provinces of the Protectorate', *J. Linn. Soc. London*, **37**, 533–44.

DAWKINS, H.C. (1949) 'Timber planting in the *Terminalia* woodland of northern Uganda'. *Emp. For. Rev.*, **28**, 226–47.

DAWKINS, H.C. (1951) 'Graphical field keys of Uganda trees. I. Forest trees of Mengo district', *E. Afr. agric. J.*, **17**, 90–103.

DAWKINS, H.C. (1952) 'Experiments in low percentage enumerations of tropical high forest', *Emp. For. Rev.*, **31**, 131–45.

DAWKINS, H.C. (1954a) 'Timu and the vanishing forests of north-east Karamoja', *E. Afr. agric. J.*, **19**, 164–7.

DAWKINS, H.C. (1954b) 'The northern province mountains; speculations on climate and vegetational history', *Uganda J.*, **18**, 58–64.

DAWKINS, H.C. (1954c) 'The construction of commercial volume tables for tropical forest trees', *Emp. For. Rev.*, **33**, 61–70.

DAWKINS, H.C. (1956) 'Rapid detection of aberrant

girth increment of rain forest trees', *Emp. For. Rev.*, **35**, 2–8.

DAWKINS, H.C. (1958) 'The management of natural tropical high forest; with special reference to Uganda'. Pp. 155. *Institute paper*, no. 34. Oxford: Imperial Forestry Institute.

DAWKINS, H.C. (1960) *Observations on Regeneration in the Kakamega and Malaba Forests, Kakamega District, Kenya*. Pp. 5. Entebbe: Uganda Forest Department. Cyclostyled.

DAWKINS, H.C. (1964) 'The productivity of lowland tropical high forest and some comparisons with its competitors', *For. Soc. J. Oxford University*, **12**, 15–18.

DAWKINS, H.C. (1964) 'Productivity of tropical forests and their ultimate value to man', *Proc. papers International Union for the Conservation of Nature, 9th technical meeting, Nairobi*. Part II. 'Ecosystems and biological productivity', Pp. 178–82.

DAWKINS, H.C. (1966) 'Dimensional behaviour of tropical high forest trees as related to stratification.' *J. Ecol.*, **54**, 281–2.

DAWKINS, H.C. (1967) 'Wood production in tropical rain forest', *J. Ecol.*, **55**, 20–1.

DAWKINS, H.C. and PHILIP, M.S. (1962) *Working Plan for Mpanga Forest for the Period 7/7/60–30/6/65*. Pp. 31. Entebbe: Uganda Forest Department. Cyclostyled.

DEAN, G.J.W. (1967) 'Grasslands of the Rukwa Valley', *J. Appl. Ecol.*, **4**, 45–57.

DEB, B.C. (1950) 'The movement and precipitation of iron oxides in podzol soils', *J. Soil Sci.*, **1**, 112–22.

DELANY, M.J. (1964) 'A study of the ecology and breeding of small mammals in Uganda', *Proc. zool. Soc. Lond.*, **142**, 347–70.

DELANY, M.J. (1971) 'The biology of small rodents in Mayaya Forest, Uganda', *J. Zool. Lond.*, **165**, 85–129.

DELANY, M.J. (1973) *The Rodents of Uganda*, London: British Museum (Natural History).

DENNY, P. (1971) 'Zonation of aquatic macrophytes around Habukam island, Lake Bunyonyi, S.W. Uganda', *Hydrobiologia*, **12**, 249–57.

DENNY, P. (1972/3) 'Lakes of S.W. Uganda. 1. Physical and chemical studies on Lake Bunyonyi', *Freshwater Biol.*, **2**, 143–58. 2. 'The vegetation of Lake Bunyoni', *Freshwater Biol.*, **3**. In preparation.

DENNY, P. (1972) 'Sites of nutrient absorption in aquatic macrophytes'. *J. Ecol.*, **60** 819–129.

DENNY, P. and LYE, K.A. (1973) 'The *Potamogeton schweinfurthii* complex in Uganda', *Kew Bull.*, **27**.

DEUSE, P. (1959) 'Esquisse de la vegetation des tourbieres et marais tourbeux du Ruanda-Urundi', *Bull. Soc. R. Sci. Liege*, nos. 1–2, 1–7.

DIEKMAHNS, E.C. (1957) 'A boron deficiency in sisal (*Agave sisalana* Perrine)', *E. Afr. agric. J.*, **22**, 197–8.

D'HOORE, J. L. (1955) 'Tropical and subtropical black soils and clay of Africa', *African soils*, **3**, 5–15.

D'HOORE, J.L. (1960) 'The soil map of Africa south of the Sahara', *Proc. 3rd Int. Afric. Soils Conf.*, Dalaba.

DIRMHIRN, I. (1961) 'Entomological studies from a high tower in Mpanga forest, Uganda. II. Light intensity at different levels', *Trans. R. Ent. Soc. Lond.*, **113**, 270–4.

DIXON, H.H. (1964) *Transpiration and the Ascent of Sap in Plants*. Pp. 208. London: MacMillan.

DOGGETT, H. (1970) *Sorghum*. Pp. 403. London: Longman.

DOUGALL, H.W. and BOGDAN, A.V. (1958) 'Browse plants of Kenya with special reference to those occurring in south Baringo', *E. Afr. agric. J.*, **23**, 236–44.

DOUGALL, H.W., Drysdalem, V.A. and GLOVER, P.E. (1964) 'The chemical composition of Kenya browse and pasture herbage', *E. Afr. Wildl. J.*, **2**, 86– .

DOUGHTY, L.R. (1953) 'The value of fertilizers in African agriculture. Field experiments in East Africa 1947–1951', *E. Afr. agric. J.*, **19**, 30–1.

DOYNE, H.C. (1935) 'Studies in tropical soils; increase of acidity with depth', *J. agric. Sci. Camb.*, **25**, 192– .

DRING, D.M. and RAYNER R.N. (1967) 'Some Gasteromycetes from Eastern Africa', *J. E. Afr. Nat. Hist. Soc. Nat. Mus.*, **26**, 5–46.

DUDAL, R. (1963) 'Dark clay soils of tropical and subtropical regions', *Soil Sci.*, **95**, 264–70.

DYSON, W.G. (1964) 'Tree seed improvement in Kenya', *Commonw. For. Rev.*, **43**, 213–17.

DYSON, W.G. (1966) 'Wood quality assessment for tree breeding in East African *Pinus radiata* Don.,'
' *E. Afr. agric. For. J.*, **22**, 137–43.

DYSON-HUDSON, V.R. (1962) 'An Akarimojong-English check list of the trees of southern Karamoja', *Uganda J.*, **25**, 166–70.

EDELMAN, C.H. and VOORDE, P.K.J. van der (1963) 'Important characteristics of alluvial soils in the tropics', *Soil Sci.*, **95**, 258–63.

EDWARDS, D.C. (1935a) 'The reaction of Kikuyu grass herbage to management', *Emp. J. Exp. Agric.*, **8**, 101–10.

EDWARDS, D.C. (1935b) 'The grasslands of Kenya I. Areas of high moisture and low temperature', *Emp. J. exp. Agric.*, **8**, 153–9.

EDWARDS, D.C. (1940a) 'Pasture and fodder grasses of Kenya', *E. Afr. agric. J.*, **5**, 248–54.

EDWARDS, D.C. (1940b) 'A vegetation map of Kenya with special reference to grassland types', *J. Ecol.*, **28**, 377–85.

EDWARDS, D.C. (1942) 'Grass burning', *Emp. J. exp. Agric.*, **10**, 219–31.

EDWARDS, D.C. (1948) 'Food of goats in semi-arid areas', *E. Afr. agric. J.*, **13**, 221–3.

EDWARDS, D.C. (1951) 'The vegetation in relation to soil and water conservation in East Africa', Pp. 28–43, in *Management and Conservation of Vegetation in Africa. Bulletin* 41, Commonwealth Bureau of Pastures and Field Crops.

EDWARDS, D.C. (1956) 'The ecological regions of Kenya; their classification in relation to agricultural development', *Emp. J. exp. Agric.*, **24**, 89–108.

EDWARDS, D.C. and BOGDAN, A.V. (1951) *Important Grassland Plants of Kenya*. Pp. 124. Nairobi: Pitman.

EGGELING, W.J. (1934) 'Notes on the flora and fauna of a Uganda swamp', *Uganda J.*, **1**, 51–60.

EGGELING, W.J., (1935) 'The vegetation of Namanvi swamp, Uganda', *J.Ecol.*, **23**, 422–35.

EGGELING, W.J. (1938) 'The savanna and mountain forest of south Karamoja, Uganda', *Paper* 11, Pp. 14, Oxford: Imperial Forestry Institute.

EGGELING, W.J. (1940a) 'Budongo – an East African mahogany forest', *Emp. For. J.*, **19**, 179–96.

EGGELING, W.J., (1940b) *The Indigenous Trees of the Uganda Protectorate*. Pp. 491. Entebbe: Govt. Printer.

EGGELING, W.J. (1941) *An Annotated List of the Grasses of the Uganda Protectorate*. Entebbe: Gov. Printer.

EGGELING, W.J. (1947) 'Observations on the ecology of the Budongo rain forest, Uganda', *J. Ecol.*, **34**, 20-87.

EGGELING, W.J. (1955) 'The relationship between crown form and sex in *Chlorophora excelsa*', *Emp. For. Rev.*, **34**, 294.

EGGELING, W.J. and DALE, I.R. (1947) *Notes on the Forests of Uganda and Their Products*. Entebbe: Govt. Printer.

EGGELING, W.J. and DALE, I.R. (1948) 'A review of some vegetation studies in Uganda', *Uganda J.*, **12**, 39– .

EGGELING, W.J. and DALE, I.R. (1951) *The Indigenous Trees of the Uganda Protectorate*. Pp. 491. 2nd edn. Entebbe: Govt. Printer.

EGGELING, W.J., and HARRIS, C.M. (1939) *Fifteen Uganda Timbers*. Pp. 120. Oxford: Clarendon Press.

EICHER, D.L. (1968) *Geologic Time*, Pp. 150. New Jersey: Prentice Hall.

ELLIS, B.S. (1952) 'Genesis of a tropical red soil', *J. Soil Sci.*, **3**, 52–62.

ELMER, J.L. and GOSNELL, J.M. (1963) 'The role of boron and rainfall n the incidence of wattle die-back in East Africa', *E. Afr. agric. For. J.*, **29**, 31–8.

EMERSON, W.W. (1959) 'The structure of soil crumbs', *J. Soil Sci.*, **10**, 235–44.

EVANS, A.C. and MITCHELL, H.W. (1962) 'Soil fertility studies in Tanganyika. I. Improvement to crop and grass production on a leached sandy soil in Bukoba', *E. Afr. agric. For. J.*, **27**, 189–96.

EVANS, A.C. (1963) 'Soil fertility studies in Tanganyika II Continued application of fertilizer on the red and red-brown loams of the Nachingwea series', *E. Afr. agric. For. J.*, **27**, 228–30.

EVANS, G.C. (1939) 'Ecological studies on the rain forest of southern Nigeria. II. The atmospheric environmental conditions', *J. Ecol.*, **27**, 436–82.

EVANS, G.C. (1956) 'An area survey method of investigating the distribution of light intensity in woodlands with particular reference to sunflecks', *J. Ecol.*, **44**, 391–428.

EVANS, G.C. (1966) 'Temperature gradients in tropical rain forest', *J. Ecol.*, **54**, 20–1.

EVANS, G.C. (1969) 'The spectral composition of light in the field I. Its measurement and ecological importance', *J. Ecol.*, **57**, 109–25.

EVANS, G.C., WHITMORE, T.C. and WONG, Y.K. (1960) 'The distribution of light reaching the ground vegetation in a tropical rain forest', *J. Ecol.*, **48**, 193–204.

FEDEROV, A.A. (1966) 'The structure of the tropical rain forest and speciation in the tropics', *J. Ecol.*, **54**, 1–11.

FANE, R. (1957) *Fifty-two Kenya Trees and How to Recognise Them*. Pp. 19. Nairobi: East African Standard Ltd.

FANSHAWE, D.D. (1966) 'The Doun Palm – *Hyphaene thebaica* (Del.) Mart.', *E. Afr. agric. For. J.*, **22**, 108–16.

FIELD C.R. (1968) 'The food plants of some wild ungulates in Uganda', *E. Afric. agric. For. J.*, **33**, 159–62.

FIELD, C.R. and LAWS, R.M. (1970) 'The distribution of the larger herbivores in the Q.E. National Park, Uganda', *J. Appl. Ecol.*, **7**, 273–94.

FIELD, C.R. and ROSS, I.C. (1973) *Ibid* 2. 'Nutrition of elephant and giraffe', *E.Afr. Wildl. J.*, In preparation.

FIENNES. T.W. (1940) *The Ecology of the Grasses of Lango, Uganda, East Africa*. Pp. 22. Entebbe: Govt. Printer.

FISH, G.R. (1952) *E. Afric. Fisheries research Organization Rept.*, 1951.

FISH, G.R. (1953) *E. Afric. Fisheries Research Organization Rept.*, 1952.

FISH, G.R. (1954) *E. Afric. Fisheries Research Organization Rept.*, 1953.

FISH, G.R. (1955a) 'The oxygen content of the water in dams in Ankole, Uganda', *E. Afric. agric. For. J.*, **20**, 178–82.

FISH, G.R. (1955b) 'The food of *Tilapia* in E. Africa', *Ug. J.*, **19**, 1, 85–9.

FISH. G.R. (1956) 'Chemical factors limiting the growth of phytoplankton in Lake Victoria', *E. Afr. agric. J.*, **21**, 152–8.

FISHLOCK, C.W.L. and HANCOCK, G.L.R. (1933) 'Notes on the flora and fauna of Ruwenzori with special reference to the Bujuku valley', *J. E. Africa Nat. Hist. Soc.*, **44**, 205–29.

FISKE, W.F. (1927) *A History of Sleeping Sickness and Reclamation in Uganda*. Pp. 37. Entebbe: Govt. Printer.

FORD, J. 'The Ankole pilot scheme of tsetse reclamation'. *Paper 1. The development of the Ankole tsetse belts and bibliography*. Pp. 40. Kampala: E. Afr. Trypanosomiasis Research Organization. Cyclostyled.

FORD, J. (1971) *The role of the trypanosomiases in African Ecology. A study of the Tsetse-fly problem*—Oxford: Clarendon Press.

FORD, J. and HALL, R de Z. (1947) 'The history of Karagwe, Bukoba district', *Tanganyika Notes and Records*, **24**, 3–27.

FORD, J. and CLIFFORD, H.R. (1968) 'Changes in the distribution of cattle and of bovine trypanosomiasis associated with the spread of Tsetse flies (Glossina) in S.W. Uganda', *J. Appl. Ecol.*, **5**, 301–37.

FREEMAN, R.W.T. (1940) 'The forestry problem of eastern province of Uganda', *Emp. For, J.*, **19**.

FREISE, F. (1936) 'Das Binnenkline von Urwalden in subtropischen Brasilien', *Peterm. Mitt.*, 301–4, 346–8.

FRIES, R.E. and FRIES, T.C.E. (1948) 'Phytogeographical researches on Mt Kenya and Mt Aberdare, British East Africa', *K. svenska Vetensk-Akad, Handl.*, **25**, 5–83.

GANF, C.G. (1972) 'Lake George, Uganda. Studies on a tropical freshwater ecosystem.' In *Productivity Problems of Freshwaters. I.B.P., PF section, UNESCO*. Polish Scientific Publishers, Krakow.

GARDNER, H.M. (1932) 'Conifers of Kenya'. Pp. 281–4 in *Report of the Royal Horticultural Society Conifer Conference*.

GARNHAM, P.C.C., HARPER, J.O. and HIGHTON, P.B. (1946) 'The mosquitoes of Kaimosi forest, Kenya colony with special reference to yellow fever', *Bull. Ent. Res.*, **36**, 473–4.

GATES, D.M. (1962) *Energy Exchanges in the Biosphere*. Pp. 151. New York: Harper and Row.

GATES, D.M. (1968) 'Leaf temperature of desert plants', *Science*, **159** (3818), 994.

GAUTIER, A. (1967) 'New observations on the later Tertiary and early Quarternary in the W. Rift', *Background to evolution in Africa*, Bishop and Clark (eds.). Univ. of Chicago Press.

GEIGER, R. (1965) *The Climate Near the Ground*. Pp. 611. Cambridge, USA: Harvard University press.

GIBB, Sir A and partners (1955) *Water resources survey of Uganda 1954–55 applied to irrigation and swamp reclamation*. Pp. 134. Entebbe: Govt. Printer.

GILBERT, V.C. (1970) *Plants of Mt Kilimanjaro*. Pp. 117. US National Park Service, Washington, DC.

GILCHRIST, B. (1952) In *Report of Central African rail Link-Development Survey* 1–2. Overseas Consultants Inc and Sir Alexander Gibb and Partners. London: Colonial Office.

GILLET, A.C. (1951) *Northern Frontier Kenya, District Tour. Pp.* 14. Nairobi: Kenya Forest Department. Cyclostyled.

GILLETT, J.B. (1961) 'The history of the botanical exploration of the area of The Flora of Tropical E. Africa', *Comptes rendues de la IVme reunion de l'AETEAT*, Lisbon, 205–9.

GILLETT, J.B. (1967) 'The identification of Aloes in East Africa', *J. E. Afr. Nat. Hist. Soc.*, **26**, 65–73.

GILLETT, J.B. and MCDONALD, P.G. (1970) *A Numbered Check List of Trees and Shrubs and Noteworthy Lianes Indigenous to Kenya*. Pp. 67. Nairobi: Govt. Printer.

GILLMAN, C. (1927) 'South-west Tanganyika territory', *Geogr. J.*, **69**, 97–131.

GILLMAN, C. (1936) 'East African vegetation types', *J. Ecol.*, **24**, 502–5.

GILLMAN, C. (1947) 'Soil reconnaissance survey through part of Tanganyika', *J. Ecol.*, **35**, 192–265.

GILLMAN, C. (1949) 'A vegetation types map of Tanganyika territory', *Geogr. Rev.*, **29**, 7–37.

GLASGOW, J.P. (1963) *The Distribution and Abundance of Tsetse*, Pp. 241. Oxford and New York: Pergamon Press.

GLOVER, J. (1963) 'The elephant problem in Tsavo', *E.A. Wild Life J.*, **1**, 30–9.

GLOVER, J. (1948) 'Water demands by maize and sorghum', *E. Afr. agric. J.*, **13**, 171.

GLOVER, J. and KENWORTHY, J.M. (1957) 'Effect of altitude on temperature and dew point of the air', *Ann. Report EAFFRO*, 18–19.

GLOVER, J. and Gwynne, M.D. (1962) 'Light rainfall and plant survival in East Africa. I. Maize', *J. Ecol.*, **50**, 111–18.

GLOVER, J., ROBINSON, R. and HENDERSON, J. (1954) 'Provisional maps of the reliability of annual rainfall in East Africa', *Q. Jl. R. meterol. Soc.*, **80**, 602–9.

GLOVER, P.E. (1939) 'A preliminary report on the comparative ages of some important East African trees in relation to their habitats', *S.Afr. J. Sci.*, **36**, 316–27.

GLOVER, P.E. (1950) 'Rain water penetration in British Somaliland soils', *E. Afr. agric. J.*, **16**, 26–33.

GLOVER, P.E. (1950–52) 'The root system of some British Somaliland plants. I.' *E. Afr. agric. J.*, **16**, 98–112 (1950); II. *Ibid.*, **16**, 154–73 (1951); III. *Ibid.*, **16**, 205–17 (1951); *Ibid.*, **17**, 38–50 (1952).

GLOVER, P.E. (1968) 'The role of fire and other influences on the savanna habitat, with suggestions for further research', *E. Afr. Wildl. J.*, **6**, 131–137.

GLOVER, P.E. (1969) *Report on an ecological survey of the proposed Shimba Hill National Reserve. Kenya National Parks, Nairobi.*

GLOVER, P.E., GLOVER, J., and GWYNNE, M.D. (1962) 'Light rainfall and plant survival in East Africa. II. Dry grassland vegetation', *J. Ecol.*, **50**, 199–206.

GLOVER, P.E., Trump. E.C. and WATERIDGE, L.E.D. (1964) 'Termitaria and vegetation patterns on the Loita plains of Kenya', *J. Ecol.*, **52**, 367–77.

GOMA, L.K.H. (1961a) 'The fluctuations of swamp water and the breeding of mosquitoes in Uganda', *J. ent. Soc. Sth Afr.*, **24**, 224–6.

GOMA, L.H.K. (1961b) 'The influence of man's activities on swamp breeding mosquitoes in Uganda', *J. ent. Soc. Sth Afr.*, **24**, 231–47.

GOOD, R. (1953) *The Geography of the Flowering Plants.* 2nd edn. Pp. 452. London: Longman.

GOSNELL, J.M. (1964) 'A sulphur deficiency in wattle (*Acacia mearnsii* de Wild = *A. molissima* Willd.)' *E. Afr. agric. For. J.*, **30**, 1–7.

GRAHAM, R.M. (1929) 'Notes on the mangrove swamps of Kenya', *J. E. Afric. Uganda nat. Hist. Soc.*, **38**, 157–64.

GRANT, D.K.S. (1938) 'Mangrove woods of Tanganyika territory, their silviculture and dependent industries', *Tanganyika Notes and Records*, **5**, 5–16.

GREENLAND, D.J. (1958) 'Nitrate fluctuations in tropical soils', *J. agric. Sci., Camb.*, **50**, 82–92.

GREENLAND, D.J. and CHANCELLOR, R.J. (1958) 'West Nile district. Some soil and vegetation types', *Uganda J.*, **22**, 64–73.

GREENLAND, D.J. and NYE, P.H. (1959) 'Increases in the carbon and nitrogen contents of tropical soils under natural fallow', *J. Soil Sci.*, **10**, 284–99.

GREENLAND, D.J., OADES, J.M. and SHERWIN, T.W. (1968) 'Electron microscope observations on iron oxides in some red soils', *J. Soil Sci.*, **19**, 123–6.

GREENWAY, P.J. (1933) 'The vegetation of Mpwapwa', *J. Ecol.*, **21**, 28–43.

GREENWAY, P.J. (1943) *Second Draft Report on Vegetation Classification for the Approval of the Vegetation Committee.* Pp. Nairobi: Pasture research conference. Cyclostyled.

GREENWAY, P.J. (1955) 'Ecological observations on an extinct East African volcanic mountain', *J. Ecol.*, **43**, 544–63.

GREENWAY, P.J. (1965) 'The vegetation and flora of Mt Kilimanjaro', *Tanganyika Notes Rec.*, **64**, 97–108.

GREENWAY, P.J. (1969) 'A check list of plants collected in Tsavo National Park', *J. E. Afric. Nat. Hist. Soc. N. Museum*, **27**, 169–209.

GREENWAY, P.J. and VESEY-FITZGERALD, D.F. (1969) 'The vegetation of Lake Manyara national park', *J. Ecol.*, **57**, 127–49.

GREIG-SMITH, P. (1964) *Quantitative Plant Ecology.* Pp. 246. 2nd edn. London: Butterworth.

GREIG-SMITH, P. and CHADWICK, M.J. (1965) 'Data on pattern within plant communities. III. Acacia-Capparis semi-desert scrub in the Sudan', *J. Ecol.*, **53**, 465–74.

GRIFFITH, A.L. (1951) 'East African enumerations I. The Rondo plateau, south Tanganyika', *Emp. For. Rev.*, **30**, 179–82.

GRIFFITH, A.L. (1952) 'East African enumerations; *Pterocarpus angolensis* in mixed woodland, Tanganyika', *Emp. For. Rev.*, **31**, 146–9.

GRIFFITH, G. ap. (1949a) 'A note on the nitrate content of soil under *Pennisetum purpureum*', *E. Afr. agric. J.*, **14**.

GRIFFITH, G. ap. (1949b) 'Provisional Account of the Soils of Uganda', Technical Communication, 46. Pp. 16. Farnham Royal: Commonwealth Agricultural Bureau.

GRIFFITH, G. ap and MANNING, H.L. (1949) 'A note on nitrate accumulation in Uganda soils', *Trop. Agric. Trin.*, **26**, 108– .

GRIFFITH, G. ap and MANNING, H.L. (1950) 'Nitrate accumulation in Uganda soils', *Nature, Lond.*, **165**, 571.

GRIFFITHS, J.F. (1958) 'Climatic zones of East Africa', *E.Afr. agric. J.*, **23**, 179–81.

GRIFFITHS, J.F. (1962) 'The climate of E. Africa' in *The Natural Resources of E. Africa*, Pp. 144. E.W. Russell (ed) Nairobi: D.A. Dawkins Ltd in Association with E.A. Literature Bureau.

GRIFFITHS, J.F. (1967) in *Thompson and Sansome: Climate in Nairobi, City and Region*, W.W.T. Morgan, (ed), Oxford University Press.

GRIM, R.E. (1962) 'Clay mineralogy', Science, 135, 890–8.

GRONBLAD, R., SCOTT, A.M. and CROASDALE, H. (1964) 'Desmids from Uganda and Lake Victoria collected by Dr E.M. Lind', *Acta bot. fenn.*, **66**, 1–57.

218

GROOME, J.S. (1955) 'Muninga (*Pterocarpus angolensis* D.C.) in the western province of Tanganyika. I. Description, distribution and silvicultural characters', *E. Afr. agric. J.*, 21, 130–7; 'II. Growth and form statistics', *Ibid.*, 21, 189–200, 'III. Yields, yield control and management', *Ibid.*, 21, 248–54.

GROOME, J.S., LEES, H.M.N., and WIGG, L.T. (1957) 'A summary of information on *Pterocarpus angolensis*', *Forestry Abstracts*, 18, leading article series No. 25.

GRUBB, P.J. (1970) 'Interpretation of the "Massenerhebung" effect on tropical mountains', *Nature*, 229, 44–5.

GRUBB, P.J., LLOYD, J.R., PENNINGTON, T.D. and WHITMORE, T.C. (1963) 'A comparison of montane and lowland forest in Ecuador I. Forest structure, physiognomy and floristics', *J. Ecol.*, 51, 567–601.

GRUBB, P.J. and WHITMORE, T.C. (1966) 'A comparison of montane and lowland forest in Ecuador. II. The climate and its effects on the distribution and physiognomy of forests', *J. Ecol.*, 54, 303–33.

GUILLOTEAU, J. (1958) 'The problem of bush fires and burns in land development and soil conservation in Africa south of the Sahara', *African Soils*, 4, 64–102.

GWYNNE, M.D., TAERUM, R. and OPILE, W.R. (1966) *Record of research EAFFRO*, 89–100.

GWYNNE, M.D. and BELL, R.H.V. (1968) 'Selection of vegetation components by grazing ungulates in Serengeti National Park', *Nature*, 220, 390–3.

HAARER, A.E. (1951) 'Grasses and grazing problems of East Africa', *Wild Crops*, 3, 8–10.

HADDOW, A.J. (1945a) 'On the mosquitoes of Bwamba county, Uganda. I. Description of Bwamba with special reference to mosquito ecology', *Proc. zool. Soc. Lond.*, 115, 1–13.

HADDOW, A.J. (1945b) 'The mosquitoes of Bwamba county, Uganda. II. Biting activity with special reference to the influence of microclimate', *Bull. ent. Res.*, 36, 33–73.

HADDOW, A.J. (1952) 'Field and laboratory studies on an African monkey *Cercopithecus ascanius schmidti* Matschie', *Proc. zool. Soc. Lond.*, 122, 297–394.

HADDOW, A.J. and CORBET, P.S. (1961) 'Entomological studies from a high tower in Mpanga forest, Uganda. II. Observations on certain environmental factors at different levels', *Trans. R. ent. Soc. Lond.*, 113, 257–69.

HADDOW, A.J., CORBET, P.S. and GILLETT, J.B. (1961) 'Entomological studies from a high tower in Mpanga forest, Uganda. I. Introduction', *Trans. R. ent. Soc. Lond.*, 113, 249–53.

HADDOW, A.J., GILLETT, J.B. and HIGHTON, R.B. (1947) 'The mosquitoes of Bwamba county, Uganda. V. The vertical distribution and biting cycle of mosquitoes in rain forest with further observations on microclimate'. *Bull. ent. Res.*, 37, 301–30.

HALLSWORTH, E.G. (1965) 'An examination of some factors affecting movement of clay in an artificial soil', *J. Soil Sci.*, 14, 360–71.

HAMILTON, A.C. (1968) 'Some plant fossils from Bukwa', *Ug. J.*, 32, 157–64.

HAMILTON, A.C. (1969) 'The vegetation of S.W. Kigezi', *Ug. J.*, 34 (1), 175–99.

HAMILTON, A.C. (1970) 'The interpretation of pollen diagrams from Highland E. Africa'. *Palaeoecology of Africa*, 7, 45–149.

HANCOCK, W.J. (1970) 'Oxford Univ. Expedition to the Cherangani Hills', 18, (3), 55–86. Oxford Univ. Exploration Club Bulletin, 18, (3), 55–86.

HANNA, L. W. (1971) 'Effect of water availability on tea yields in Uganda', *J. Appld. Ecol.*, 8, 791–813.

HARDY, F. (1946) 'Seasonal fluctuations of soil moisture and nitrate in a humid tropical climate', *Trop. Afric. Trin.*, 23, 40–9.

HARKER, K.W. (1959) 'An *Acacia* weed of Uganda grasslands', *Trop. Agric. Trin.*, 36, 45–51.

HARKER, K.W. (1962) 'Grassland development in the fertile crescent of Uganda', *J. Br. Grassld. Soc.*, 17, 188–93.

HARKER, K.W. and NAPPER, D.M. (1960) *An Illustrated Guide to the Grasses of Uganda*. Pp. 63, 123 plates. Entebbe: Govt. Printer.

HARRINGTON, G.N. 'Fire ecology of the savanna grasslands of Ankole, Uganda. 1, Herbs, 2, Shrubs.' *J. Appl. Ecol.* In preparation.

HARRINGTON, G.N. and ROSS, I.C. (1973) 'The savanna ecology of Kidepo Valley National Park, Uganda. 1. The effect of burning and browsing on the vegetation.' *E. Afri. Wildl. J.* In preparation.

HARRIS, S.A. (1963) 'On the classification of latosols and tropical brown earths in high rainfall areas', *Soil. Sci.*, 96, 210–15.

HARRIS, W.V. (1932) 'Native beekeeping in Tanganyika territory', *Trop. Agric. Trin.*, 9, 231–5.

HARRIS, W.V. (1940a) 'Native honey production for export', *E. Afr. agric. J.*, 6, 14–16.

HARRIS, W.V. (1940b) 'Termites in East Africa. III Field key and distribution, by territories', *E. Afr. agric. J.*, 6, 201–5.

HARRIS, W.V. (1948) 'Insects which alter the landscape', *Uganda J.*, 12, 57–9.

HARRIS, W.V. (1961) *Termites: Their Recognition and Control*. Pp. 187. London: Longman.

HARRISON, J.L. (1962) 'The distribution and feeding habits among animals in a tropical rain forest', *J. Anim. Ecol.*, 31, 53–63.

HARROP, J.F. (1960) *The Soils of the Western Province*

of Uganda. Pp. 79. Memoirs of the Research Division. Series 1, No. 6, Kawanda. Uganda: Department of Agriculture. Cyclostyled.

HAUGHTON, S.H. (1963) *The Stratigraphic History of Africa South of the Sahara.* Pp. 365. Edinburgh and London: Oliver and Boyd.

HAUGHTON-SHEPPARD, P.C. (1958) 'A note on African blackwood *Dalbergia melanoxylon*', *Commonw. For. Rev.*, 37, 327–30.

HAUMAN, L. (1933) 'Esquisse de la végétation des hautes altitudes sur le Ruwenzori', *Bull. Acad. r. Belg.*, 19.

HAUMAN, L. (1955) 'The Region Afroalpine" in phytogeographic centre Africaine', *Webbia*, 11, 467–9.

HEADY, H.F. (1960) *Range Management in East Africa.* Pp. 125. Nairobi. Govt. Printer.

HEADY, H.F. (1966) 'Influence of grazing in *Themeda* grassland', *J. Ecol.*, 54, 705–27.

HEDBERG, O. (1951) 'Vegetation belts of the East African mountains', *Svensk. bot. Tidskr.*, 45, 140–202.

HEDBERG, O. (1954) 'A pollen analytical reconnaissance in tropical East Africa', *Oikos*, 5, 137–66.

HEDBERG, O. (1955) 'Altitudinal zonation of the vegetation of the East African mountains', *Proc. Linn. Soc. Lond.*, session 165, 134–50.

HEDBERG, O. (1959) 'An "Open-air hothouse" on Mt Elgon, tropical East Africa', *Svensk. bot. Tidskr.*, 53, 160–6.

HEDBERG, O. (1957) 'Afroalpine vascular plants', *Symb. Bot. Upsaliensis*, 15, 1–411.

HEDBERG, O. (1961a) 'The phytogeographical position of the Afroalpine flora', Pp. 914–19. *Recent Advances in Botany.* Toronto: University Press.

HEDBERG, O. (1961b) 'Intercontinental crosses in *Arabis alpina*', *Caryologia*, 15, 253–60.

HEDBERG, O. (1964a) 'Features of Afroalpine plant ecology', *Acta phytogeogr. suec.*, 19, 1–144.

HEDBERG, O. (1964b) 'Etudes ecologiques de la flore afroalpine', *Bull. Soc. Roy. Bot. Belgique*, 97, 5–18.

HEDBERG, O. (1965) 'Afroalpine floral elements', *Webbia*, 19, 519–29.

HEDBERG, O. (1969) 'Evolution and speciation in a tropical high mountain flora', *Biol. J. Linn. Soc. London.*, 1, 135–48.

HEDBERG, O. (1969a) 'Growth rate of the East African giant Senecios', *Nature*, 222, 163–4.

HEDBERG, O. (1969b) 'Taxonomic and ecological studies on the Afroalpine flora of Mt. Kenya', *Hochgebirgsforschung*, 1, 171–94, Innsbruck, Munchen.

HEDBERG, O. and HEDBERG, I. (1968) 'Conservation of vegetation in Africa, south of the Sahara', *Acta phytogeogr. suec.*, 54, 1–320.

HEMMING, C.F. and TRAPNELL, C.G. (1957) 'Soils of Turkana', *J. Soil Sci.*, 8, 167–83.

HENDERSON, J.P. (1949) 'Some aspects of climate in Uganda, with special reference to rainfall', *Uganda J.*, 13, 154–70.

HENNINGS, R.O. (1961) 'Grazing management in the pastoral areas of Kenya', *J. Afr. Admin.*, 13, 131–203.

HERIZ-SMITH, S. (1962a) *Wild Flowers of the Nairobi National Park*, Pp. 56. East Afr. Nat. Hist. Soc.

HERIZ-SMITH, S. (1962b) 'Flowering plants of the Ngong Hills', *J. E. Afr. Nat. Hist. Soc.*, 24, 26–8.

HESSE, P. R. (1955) 'A chemical and physical study of the soils of termite mounds in East Africa', *J. Ecol.*, 43, 449–61.

HESSE, P.R. (1957) 'Fungus combs in termite mounds', *E. Afr. Agric. J.*, 23, 104–8.

HESSE, P.R. (1958a) 'Sulphur and nitrogen changes in forest soils of East Africa', *Pl. Soil*, 9, 86–96.

HESSE, P.R. (198b) 'The distribution of sulphur in muds and water and vegetation of L. Victoria', *Hydrobiologica*, 11, 29–37.

HICKMAN, G.M. and DICKINS, W.H. (1960) *The Lands and People of East Africa.* Pp. 232. London: Longman.

HILL, M.F. (1964) *The History of the Magadi Soda Company.* Pp. 199. Birmingham: Kynoch Press for the Magadi Soda Company.

HOPKINS, B. (1960) 'Rainfall interception by a tropical forest in Uganda', *E. Afr. agric. J.*, 25, 255–8.

HOPKINS, B. (1965a) *Forest and Savanna.* Pp. 100. London: Heinemann.

HOPKINS, B. (1965b) 'Vegetation of the Olokemeji forest reserve, Nigeria. III. Microclimates with special reference to their seasonal changes', *J. Ecol.*, 53, 125–38.

HOPKINS, B. (1967) 'A comparison between productivity of forest and savanna in Africa', *J. Ecol.*, 55, 19–20.

HORA, F. and GREENWAY, P.J. (1940) *Check Lists of the Forest Trees and Shrubs of the British Empire. No. 5 Tanganyika Territory Part I.* Pp. 312. Oxford: Imperial Forestry Institute.

HORNBY, H.E. (1949) *Animal Trypanosomiasis in East Africa.* Pp.   . London: Colonial Office.

HORNBY, H. E. and R. M. (1943) 'A contribution to the study of the vegetation of Mpwapwa', *Tanganyika Notes and Records*, 15, 25–48.

HOSEGOOD, P.H. (1963) 'The root distribution of Kikuyu grass and wattle trees', *E. Afr. agric. For. J.*, 29, 60–1.

HOSEGOOD, P.H. and HOWLAND, P. (1966) 'A preliminary study of the root distribution of some exotic tree crops evaluated by a rapid sampling method', *E. Afr. agric. For. J.*, **22**, 16–18.

HUBBARD, C.E. (1926) *East African Pasture Plants. I. East African Grasses.* Pp. 55. London: Crown Agents.

HUBBARD, C.E. (1927) *East African Pasture Plants. II. East African Grasses.* Pp. 55. London: Crown Agents.

HUGHES, J.F. (1949) 'Forests and water supply in East Africa', *Emp. For. Rev.*, **28**, 314–23.

HUGHES, J.F. (1961) 'Diagnostic sampling technique in tropical high forest', *Commonw. For. Rev.*, **40**, 350–61.

HUGHES, J.F. (1965) 'A long-term plan for conversion to monocyclic working in the central forest reserves of south Mengo district, Buganda province, Uganda', *E. Afr. agric. For. J.*, **31**, 91–9.

HUNT, K. (1956) 'Kenya's forest department clothes bare hills', *Emp. For. Rev.*, **35**, 313–6.

HUNT, K. (1962) 'Charcoal exports revive Lamu's economy; Persian Gulf trade switches from *boritis*', *Kenya News Press office feature* 593. Nairobi: Kenya Government Information Service.

HUNT, M.A. and WESTENBERG, H.J.W. (1964) 'Oxidation ponds in Kenya', *J. Inst. Sew. Purif.*, 230–7.

HURCH, C.R. and PEREIRA, H.C. (1953) 'Field moisture balance in the Shimba hills, Kenya', *E. Afr. agric. J.*, **18**, 139–45.

HUTCHINS, D.E. (1909) *Report of the forests of British East Africa.* London: HMSO.

HUTCHINSON, G.E. (1957) *Limnology: Geography, Physics, Chemistry.* Vol. 1, pp. 1015. New York: Wiley.

HUTCHINSON, G.E. (1967) *Limnology: Introduction to lake biology and the limnoplankton.* Vol. 2, pp. 1115. New York: Wiley.

HUXLEY, J. (1960) *Conservation of Wild Life and Natural Habitats in Central and East Africa.* Pp. 112. Paris: UNESCO.

HUXLEY, P.A. (1960) 'Meteorological data for Makerere University College Farm, Kabanyolo, Uganda', *Meteorological Bulletin* 1. Kampala: Faculty of Agriculture, Makerere University College, Cyclostyled.

HUXLEY, P.A. (1963a) 'Meteorological data for Makerere University College Farm, Kabanyole, Uganda', *Meteorological bulletin* 2. Kampala: Faculty of Agriculture, Makerere University College. Cyclostyled.

HUXLEY, P.A. (1966) 'A preliminary study of the effect of differences in local climate on the early growth of some crop plants in the southern region of Uganda', *J. Appl. Ecol.*, **3**, 251–60.

HUXLEY, P.A. and BEADLE, M. (1964) 'A local climatic study in typical dissected topography in the southern region of Uganda', *Meteorol. Mag.*, **93**, 321–33.

ISAAC, G.L. (1966) 'The geological history of the Olorgesailie area', *Actes de Ve Congres Panafricain de Prehistoire de l'etude du Quaternaire.* Museo Arqueologico de Tenerife, Canary Islands.

ISAAC, G.L. (1960) 'Preliminary report on the work of the Univ. of California Archaeological Research Group in Kenya (1967–70)'. Unpublished.

ISAAC, P.C.G. (1960) 'The use of algae for sewage treatment in oxidation ponds', *J. Inst. Sew. Purif.*, pl. 14, 376–93.

ISAAC, F.M. (1968) 'Marine Botany of the Kenya Coast. IV. Angiosperms', *Jl. E. Afr. Nat. Hist. Soc.*, **27**, 29–47.

ISAAC, W.E. (1967) 'Marine botany of the Kenya coast. I. A first list of Kenya marine algae', *J. E. Africa nat, Hist. Soc.*, **26**, 75–83.

ISAAC, W.E. (1968) 'Marine Botany of the Kenya Coast. II. A second list of Kenya marine algae', *Jl. E. Afr. Nat. Hist. Soc.*, **27**, 1–6.

ISAAC, W.E. (1971) 'Marine botany of the Kenya Coast. V. A third list of Kenya marine algae', *Jl. E. Afr. Nat. Hist. Soc.*, **28**, 2–23.

ISAAC, W.E. and ISAAC, F.M. (1968) 'Marine botany of the Kenya Coast. III. General account of the environment, flora and vegetation', *J. E. Afr. Nat. Hist. Soc.*, **27**, 7–28.

IRVINE, F.R. (1957) 'Indigenous African methods of beekeeping', *Bee World*, **38**, 113–27.

JACKSON, J.K. (1956) 'The vegetation of the Imatong mountains', *J. Ecol.*, **44**, 341–74.

JACKSON, G. (1964) 'Grasslands of tropical Africa'. Appendix B, 98–111. in *An Introduction to Tropical Grassland Husbandry*, R.J. McIlroy. London: Oxford University Press.

JACKSON, G. and GARTLAN, J.S. (1965) 'The flora and fauna of Lolui island, Lake Victoria; a study of vegetation, men and monkeys', *J. Ecol.*, **53**, 573–97.

JAMESON, J.D. and STEPHENS, D. (1970) *Agriculture in Uganda.* Pp. 432. London: Oxford University Press.

JEFFERS, J.N.R. and BOALER, S.B. (1966) 'Ecology of a miombo site, Lupa north forest reserve, Tanzania. I. Weather, plant and growth 1962–1964', *J. Ecol.*, **54**, 447–63.

JENKINS, P.M. (1936) 'Reports on the Percy Sladen expedition to some rift valley lakes in Kenya in 1929. II. Summary of the ecological results with special reference to the alkaline lakes', *Ann. Mag. Nat. Hist.*, **18**, 134–63.

JENNY, H., GESSEL, S.P. and BINGHAM, F.T. (1949) 'Comparative study of decomposition rates of organic matter in temperate and tropical regions', *Soil Sci.*, **68**, 419–32.

JOHNSON, D.H. and NORTH, H.T. (1960) 'Forestry research in East Africa'. Pp. 56–137. In *Tropical Meteorology in Africa* D.T. Bargman (ed.). Nairobi: Munitalp Foundation. 56–137.

JOHNSTONE, H. (1902) *The Uganda Protectorate*. 2 vols. Pp. 1018. London: Hutchinson.

JONES, E.W. (1952) 'African Hepatics', *Trans. Brit. Bryol. Soc.*

JONES, E.W. (1957) Report on *Chlorophora*. Pp. 108. London: HMSO.

KEAY, R.W.J. (1955) 'Montane vegetation and flora, in the British Cameroons', *Proc. Linn. Soc.*, Session 165, 1952–53, Pt. 2, 140–3.

KEAY, R.W.J. (1959) *Vegetation Map of Africa South of the Tropic of Cancer*. Explanatory notes, pp. 5–24. London: Oxford University Press.

KEEN, B, and DOTHIE, D.W. (1953) 'Crop responses to fertilisers and manures in East Africa', *E. Afr. agric. J.*, **19**, 19–57.

KELLOGG, C.E. (ED.) (1949) 'Preliminary suggestions for the classification and nomenclature of the great soil groups in tropical and equatorial regions', *Commonwealth Bur. of Soil Sci. Tech. Comm.* 46, 76–85.

KENDAL, R.L. (1969) 'An ecological history of the Lake Victoria Basin', *Ecol. Monogr.*, **39**, 121–76.

KENDALL, R.L. and LIVINGSTONE, D.A. (1967) 'Paleoecological studies on the East African Plateau, Congrès panafricain de prehistoire, Dakar.'

KENWORTHY, J.M. (1966) 'Temperature conditions in the tropical highland climates of East Africa', *E. Afr. Geogr. Rev.*, **4**, 1–11.

KENWORTHY, J.M. and GLOVER, J. (1958) 'The reliability of the main rains in Kenya', *E. Afr. agric. J.*, **23**, 267–71.

KERFOOT, O. (1961) 'Tea root system', *Tea research institute pamphlet* 19.

KERFOOT, O. (1962a) 'Root systems of forest trees, shade trees and tea bushes', *E. Afr. agric. For. J.*, **27** (special issue), 24.

KERFOOT, O. (1962b) '*Juniperus procera* Endl. in Africa and Arabia', *E. Afr. agric. For. J.*, **26**.

KERFOOT, O. (1963) 'The root systems of tropical forest trees', *Commonw. For. Rev.*, **42**, 19–25.

KERFOOT, O. (1963) 'The vegetation of the south-west Mau forest', *E. Afr. agric. For. J.*, **29**, 295–318.

KERFOOT, O. (1964) 'A first check list of the vascular plants of Mbeya Range, S. Highland region of Tanzania', *Tanganyika Notes and Records*, **62**, 1–17.

KERFOOT, O. (1965) 'The vegetation of an experimental catchment in the semi-arid ranchland of Uganda', *E. Afr. agric. For. J.*, **30**, 227–45.

KERS, L.E. (1969) 'Studies in Cleome II', *Svensk. Bot. Tidskr.*, **63**, 1–48.

KERSHAW, K.A. (1968) 'Classification and ordination of Nigerian savanna vegetation', *J. Ecol.*, **56**, 467–82.

KERSHAW, K.A. (1964) *Quantitative and Dynamic Ecology*. Pp. 183. London: Arnold.

KINGDON, J. (1971). *East African Mammals. An atlas of evolution in Africa. Vol.* 1. Academic Press, London and New York.

KINGSTON, B. (1967) *Working Plan for the Mgahinga Central Forest Reserve, Kigezi District, Uganda. Period* 1967–1977. Pp. 19. Entebbe: Uganda Forest Department. Cyclostyled.

KRAUSKOPF, K.B. (1967) *Introduction to Geochemistry.* Pp. 721. New York: McGraw-Hill.

KUBIENA, W.L. (1958) 'The classification of soils'. *J. Soil Sci.* **9**, 9–19.

LAMERTON, J.F. (1962) 'Manda valleys in Tanganyika', *J. Ecol.*, **50**, 771–3.

LAMPREY, H.F. (1963) 'Ecological separation of the large mammal species in the Tarangire game reserve, Tanzania', *E. Afr. Wild Life J.*, **1**, 63–92.

LANG BROWN, J.R. and HARROP, J.F. (1962) 'The ecology and soils of Kibale grasslands, Uganda', *E. Afr. agric. For. J.*, **27**, 264–72.

LANGDALE-BROWN, I. (1959a) 'The vegetation of the Eastern province of Uganda', *Memoirs of the research division*, series 2, no. 1, pp. 154. Kawanda, Kampala: Department of Agriculture, Uganda. Cyclostyled.

LANGDALE-BROWN, I. (1959b) 'The vegetation of Buganda', *Memoirs of the research division*, series 2, no. 2, pp. 90. Kawanda, Kampala: Department of Agriculture, Uganda. Cyclostyled.

LANGDALE-BROWN, I. (1960a) 'The vegetation of the West Nile, Acholi and Lango districts of the northern province of Uganda', *Memoirs of the research division*, series 2, no. 3, pp. 106. Kawanda, Kampala: Department of Agriculture, Uganda. Cyclostyled.

LANGDALE-BROWN. I. (1960b) 'The vegetation of the Western Province of Uganda', *Memoirs of the research division*, series 2, no. 4, Kawanda, Kampala: Department of Agriculture, Uganda. Cyclostyled.

LANGDALE-BROWN, I. (1960c) 'The vegetation of Uganda, excluding Karamoja', *Memoirs of the research division*, series 2, no. 6, pp. 45. Kawanda, Kampala: Department of Agriculture, Uganda. Cyclostyled.

LANGDALE-BROWN, I. OSMASTON, H.A. and WILSON, J.G. (1964) *The Vegetation of Uganda and Its Bearing on Land Use*. Pp. 159. Entebbe: Govt. Printer.

LANGLANDS, B.W. (1962) 'Concepts of the Nile',

Uganda J., 1–22.

LAWSON, G.W., JENIK, J. and ARMSTRONG-MENSAH, K.O. (1968) 'A study of a vegetation catena in Guinea savanna at Mole game reserve, Ghana', *J. Ecol.*, **56**, 505–22.

LAWS, R.M. (1968) 'Interactions between elephant and hippopotamus and their environments', *E. Afr. agric. For. J.*

LAWTON, R.M. (1963) 'Palaeoecological and ecological studies in the Northern Province of North Rhodesia', Kirkia 3, 46–77.

LAWTON, R.M. (1972) 'A vegetation survey of Northern Zambia', *Palaeoecology of Africa, the surrounding islands and Antarctica*, **6**, 253–6.

LAUER, W. (1952) 'Humide und Aride Jahreszeiten in Afrika und Sudamerika und ihre Berzeihung zu Vegetations gurteln', in *Studien zur Klima und Vegetationskunde der Tropen*. Bonner Geogr. Abhandl. 8.

LEAKEY, L.S.B. (1965) *Olduvai Gorge* 1951–61. Vol. 1. Cambridge: University Press.

LEBRUN, J. (1935) 'Les essences forestieres des regions montagneuses du Congo orientale', *Publs. Inst. Agron. Congo Belge*, ser sci. 1.

LEBRUN, J. (1937) 'Observations sur les epiphytes de la foret equatoriale congolaise', *Annls. Soc. Scient. Brux.*, ser. 2, 57.

LEBRUN, J. (1937) 'Exploration du Parc National Albert; mission J. Lebrun, Brussels, 1937–38', vol. 1. *La vegetation de la plaine alluviale au sud du lac Edouard.*

LEBRUN, J. (1942) *La vegetation du Nyiragongo.* Pp. 121. Aspects de végétation des parcs nationaux du Congo Belge. Ser 1. Parc national Albert. Bruxelles: Institut des parcs nationaux du Congo Belge.

LEBRUN, J. (1960a) 'Sur un method de delimitation des horizons et etages de vegetation des montagnes du Congo Orientale', *Bull. Jard. Bot. Bruxeller*, **20**, 75–95.

LEBRUN, J. (1960b) 'Sur les horizons et etages de vegetation de divers volcans du massif des Virung (Kivu, Congo)', *Bull. Jard. bot. Bruxelles*, **30**, 255–77.

LEE, K.E. and WOOD, T.G. (1971) *Termites and Soils.* Pp. 252. London: Academic Press.

LEEPER (1956) The classification of soils. *J. Soil Sci.*, **7**, 59–67.

LEGGAT, G.J. (1954) 'A Uganda softwood scheme', *Emp. For. Rev.*, **33**, 345–51.

LEGGAT, G.J. (1957) *Working Plan for the Kitomi-Kashoya Central Forest Reserve, Ankole District, Western Province.* Period 1957–1961 Extended to 1967. Entebbe: Uganda Forest Department. Cyclostyled.

LEGGAT, G.J. (1961) *Working Plan for the Semliki Central Forest Reserve, Toro District, Uganda.* First revision; Period 1961–1971. Pp. 17. Entebbe: Uganda Forest Department. Cyclostyled.

LEGGAT, G.J. and BEATON, A. (1961) *Working Plan for the Ruwenzori Central Forest Reserve.* First Revision. Period 1961–1971. Pp. 16. Entebbe: Uganda Forest Department. Cyclostyled.

LEGGAT, G.J. and OSMASTON, H.A. (1961) *Working Plan for the Impenetrable Forest Reserve, Kigezi District.* Period 1961–1971. Pp. 16. Entebbe: Uganda Forest Department. Cyclostyled.

LEMARE, P.H. (1959) 'Soil fertility studies in three areas of Tanganyika', *Emp. J. exp. Agric.*, **27**, 197–222.

LEUCHARS, D. (1965) 'The planning and practice of trials of exotic species', *E. Afr. agric. For. J.*, **31**, 83–90.

LEVRING, T. and FISH, G.R. (1956) 'The penetration of light in some tropical East African water', *Oikos*, **7**, 98–109.

LEWIS, E.A. (1953) 'Land use and tsetse control', *E. Afr. agric. J.*, **18**, 160–8.

LIND, E.M. (1956a) 'Studies in Uganda swamps', *Uganda J.*, **20**, 166–76.

LIND, E.M. (1956b) 'The natural vegetation of Buganda', *Uganda J.*, **20**, 13–16.

LIND, E.M. (1965) 'The phytoplankton of some Kenya waters', *J. E. Africa Nat. Hist. Soc.*, **25**, 76–91.

LIND, E.M. (1967) 'Some East African Desmids', *Nova Hedwigia*, **13**, 361–87.

LIND, E.M. (1968) 'Notes on the distribution of phytoplankton in some Kenya waters', *Br. Phyc. Bull.*, **3**, (3), 481–93.

LIND, E.M. and TALLANTIRE, A.C. (1962) *Some Common Flowering Plants of Uganda.* Pp. 257. London: Oxford University Press.

LIND, E.M. and VISSER, S.A. (1936) 'A study of a swamp at the north end of Lake Victoria', *J. Ecol.*, **50**, 599–613.

LIND, E.M. (1971) 'Some Desmids from Uganda'. *Nova Hedwigia*, **22**, 535–84.

LIVINGSTONE, D.A. (1962) 'Age of deglaciation in the Ruwenzori Range, Uganda', *Nature*, **194**, 589–60.

LIVINGSTONE, D.A. (1965) 'Sedimentation and the history of water level changes in Lake Tanganyika', *Limnol. Oceanogr.*, **10**, 607–10.

LIVINGSTONE, D.A. (1967) 'Postglacial vegetation of the Ruwenzori mountains in equatorial Africa', *Ecol. Monogr.*, **37**, 25–52.

LOCK, G.W. (1962) *Sisal: Twenty-five Years Sisal Research.* Pp. 355. London: Longman.

LOCK, J.M. (1967) 'Vegetation in relation to grazing

and soils in Queen Elizabeth National Park, Uganda'. Pp. 254. Ph.D. thesis, Cambridge.

LOCK, J.M. (1970) 'The grasses of the Queen Elizabeth National Park', *Ug. J.*, **34**, 49–63.

LOGIE, J.P.W. and DYSON, W.G. (1962) *Forestry in Kenya, a Historical Account*. (Reprint 1970). Nairobi: Govt. Printer.

LOVE, D. (1963) 'Dispersal and survival of plants'. Pp. 189–205. Reprinted from *North Atlantic Biota and Their History*. London: Pergamon Press.

LOVERIDGE, J.P. (1968) 'Plant ecological investigations in the Nyamagasani valley, Ruwenzori mountains, Uganda', *Webbia*, **6**, 153–68.

LOWE, R. (1956) 'The breeding behaviour of *Tilapia* species in natural waters', *Behaviour*, **9**, 140–63.

LOWE-MCCONNEL, R. H. (ed.) (1969) *Speciation in Tropical Environments*. Pp. 246. London: Academic Press.

LUMSDEN, W.H.R. (1951) 'The night resting habits of monkeys in a small area on the edge of Semliki forest, Uganda', *J. Anim. Ecol*, **20**, 11–30.

MACMILLAN, H.F. (1943) *Tropical Planting and Gardening*. Pp. 560. London: MacMillan.

MACNAE, W. (1963) 'Mangrove swamps in South Africa', *J. Ecol.*, **51**, 1–25.

MACNAE, W. and KALK, M. (1962) 'The ecology of the mangrove swamps at Inhaoa island, Mocambique', *J. Ecol.*, **50**, 19–34.

MAIGNIEN, R. (1966) *Review of Research on Laterite*. Pp. 148. Paris: UNESCO.

MAKIN, J. (1969a) 'A soil catena and associated laterite levels in western Kenya', *E. Afr. agric. For. J.*, **34**, 485–92.

MAKIN, J. (1969b) 'Soil formation in the Turkana desert', *E. Afr. agric. For. J.*, **34**, 493–6.

MANIL, G. (1959) 'General considerations on the problem of soil classification', *J. Soil Sci.*, **10**, 5–13.

MANNING, H.L. (1956) 'The statistical assessment of rainfall probability and its application in Uganda agriculture', *Proc. Roy. Soc. A.*, **193**, 120–45.

MARSHALL, B. and BREDON, R.M. (1963) 'The chemical composition and nutritive value of elephant grass *Pennisetum purpureum*', *Trop. Agric., Trin.*, **40**, 63–6.

MARSHALL, B. and BREDON, R.M. (1967) 'The nutritive value of *Themeda triandra*', *E. Afr. agric. For. J.*, **22**, 375–9.

MARTIN, A.E. and REEVE, R. (1960) 'Chemical studies of podzolic alluvial horizons. IV. The flocculation of humus by aluminium', *J. Soil Sci.*, **11**, 369–81. 'V. Flocculation of humus and ferric and ferrous iron by nickel', *Ibid.*, **11**, 382–93.

MARTIN, P.S. (1966) 'Africa and Pleistocene overkill', *Nature*, **212**, 339–42.

MARTIN, W.S. (1940) 'Soil erosion problems in Uganda'. Pp. 73–87. In *Agriculture in Uganda*, J.D. Tothill (ed.). London: Oxford University Press.

MARTIN, W.S. and GRIFFITH, G. (1940) 'Uganda Soils'. Pp. 59–87. In *Agriculture in Uganda*, J.D. Tothill (ed.). London: Oxford University Press.

MARTIN, W.S. (1944) 'Grass covers in their relation to soil structure', *Emp. J. exp. Agric.*, **12**, 21–32.

MASEFIELD, G.B. (1948) 'Grass burning; some Uganda experience', *E. Afr. agric. J.*, **13**, 135–8.

MASON, B. (1958) *Principles of Geochemistry*. 2nd edn. Pp. 310. New York: Wiley.

MCCARTHY, J. (1961) 'Growth conditions and regeneration of Abura (*Mitragyna stipulosa*)', *Emp. For. Rev.*, **40**, 124–33.

MCCARTHY, J. (1962a) 'The form and development of knee roots in *Mitragyna stipulosa*', *Phytomorphology*, **12**, 20–30.

MCCARTHY, J. (1962b) 'The colonisation of a swamp forest clearing, with special reference to *Mitragyna stipulosa*', *E. Afr. agric. For. J.* **28**, 22–8.

MCCULLOCH, J.S.G. (1965) 'Tables for the rapid computation of the Penman estimate of evaporation', *E. Afr. agric. For. J.*, **33**, 286–95.

MCFARLANE, M.J. (1969) 'Lateritisation and landscape development in Kyagwe, Uganda', *Q. J. Geol. Soc. London*, **126**, 501–39.

MCILROY, R.J. (1964) *An Introduction to Tropical Grassland Husbandry*. Pp. 128. London: Oxford University Press.

MCLEAN, B.J. (1971) 'Land use and ecological problems' in *Studies in East African Geography and Development*. S.H. Ominde (ed.). London: Heinemann.

MEEL, L. VON (1954) *Exploration hydrobiologique du lac Tanganyika* 1946–1947. *Resultats scientifiques*, *IV* (1) *Le phytoplankton. A. Texte, B. Atlas*. Pp. 681 and 76 plates. Institut Royal des Sciences Naturelles de Belgique, Bruxelles.

MEIKLEJOHN, J. (1953) 'The microbiological aspects of soil nitrification', *E. Afr. agric. J.*, **19**, 54–

MEIKLEJOHN, J. (1954) 'Notes on nitrogen fixing bacteria from East African soils', *Trans. 5th Internat. Congr. Soil Sci.*, **3**, 123–5.

MEIKLEJOHN, J. (1955) 'Effect of bush burning on the microflora of a Kenya upland soil', *J. Soil Sci.*, **6**, 111–8.

MEIKLEJOHN, J. (1957) 'Numbers of bacteria and actinomycetes in a Kenya soil', *J. Soil Sci.*, **8**, 240–7.

MEIKLEJOHN, J. (1962) 'Microbiology of the Nitrogen cycle in some Ghana soils', *Emp. J. exp. Agric.*, **30**, 115–26.

MEIKLEJOHN, J. (1968) 'Numbers of nitrifying bacteria in some Rhodesian soils under natural grass and improved pastures', *J. appl. Ecol.*, **5**, 291–300.

MELVILLE, R. (1954) 'The Podocarps of East Africa', *Kew Bull.*, **4**, 563–74.

MELVILLE, R. (1958) 'Gymnosporae', in the *Flora of Tropical East Africa*. London: Crown Agents.

MICHELMORE, A.P.G. (1939) 'Observations on tropical African grasslands', *J. Ecol.*, **27**, 282–312.

MILBRAED, J. (1933) 'Ein botanischer Ausflug in das grasland des Kamerungebirges', *Kolon, Rundschau*, **25**, 139–47.

MILLBANK, J.W. (1956) 'Nitrogen in tropical soils', *E. Afr. agric. J.*, **22**, 73–5.

MILLOT, G. (1970) *Geology of Clays: Weathering, Sedimentology, Geochemistry*. Berlin: Springer Verlag.

MILLS, W.R. (1953) 'Nitrate accumulation in Uganda soils', *E. Afr. agric. J.*, **19**, 53–4.

MILNE, G. (1935) 'Some suggested units of classification and mapping, particularly for East African soils', *Soil Research*, **4**, 183–98.

MILNE, G. (1936) 'A provisional soil map of East Africa. Kenya, Uganda, Tanganyika and Zanzibar with explanatory memoir'. Pp. 34. *Amani Memoirs*. London: Crown Agents.

MILNE, G. (1937a) 'Notes on soil conditions and two East African vegetation types', *J. Ecol.*, **25**, 254–8.

MILNE, G. (1937b–38) 'Essays in applied pedology. I. Soil type and soil management in relation to plantation agriculture in East Usambara', *E. Afr. agric. J.*, **3**, 7–20 (1937). 'II. Some factors in soil mechanics', *Ibid*, **3**, 350–61 (1938), 'III. Bukoba, high and low fertility on a laterized soil', *Ibid.*, **4**, 13–24 (1938).

MILNE, G. (1940a) 'Soil and vegetation', *E. Afr. agric. J.*, **5**, 294–8.

MILNE, G. (1940b) 'Soil conservation – the research side', *E. Afr. agric. J.*, **6**, 26–37.

MILNE, G. (1944) 'Soils in relation to native population in West Usambaras', *Geography*, **29**, 107–13.

MILNE, G. (1947) 'A soil reconnaissance journey through parts of Tanganyika territory, December 1935 to February 1936'. C. Gillman (ed.). *J. Ecol.*, **35**, 192–265.

MILNE, G. and CALTON, W.E. (1942) 'Mechanical composition of East African soils', *E. Afr. agric. J.*, **8**, 202–8.

MILNE, G. and CALTON, W.E. (1944) 'Soil salinity related to the clearing of natural vegetation', *E. Afr. agric. J.*, **10**, 7–11.

MILNER, C. and HUGHES, E. (1968) *Methods for the Measurement of the Primary Production of Grassland*. Pp. 70. IBP Handbook, no. 6. Oxford and Edinburgh: Blackwell Scientific Publications.

MODHA, K.L. (1970) 'Shallow soils and their vegetation in the region of Nairobi, Kenya', *J. E. Afr. Nat. Hist. Soc.*, **28** (1), 1–6.

MOGGRIDGE, J.Y. (1950) 'The relations of the coastal tsetses of Kenya to the plant communities', *Bull. Ent. Res.*, **40**, 43–7.

MOOMAW, J.C. (1960) *A Study of the Plant Ecology of the Coast Region of Kenya Colony*. Pp. 52. Nairobi: Govt. Printer.

MOORE, A.W. (1960) 'The influence of annual burning on a soil in the derived savannah zone of Nigeria', *Trans. 7th Int. Cong. Soil Sci.*, **4**, 257–64.

MONTEITH, J.L. (1972) 'Solar radiation and productivity in tropical ecosystems', *J. Appld. Ecol.*, **9**, 3.

MOORE, A.W. (1963) 'Occurrence of non-symbiotic nitrogen fixing organisms in Nigerian soils', *Pl. Soil*, **19**, 385–95.

MOREAU, R.E. (1933) 'Pleistocene climatic changes and the distribution of life in East Africa', *J. Ecol.*, **21**, 415–35.

MOREAU, R.E. (1935a) 'A synecological study of Usambara, Tanganyika territory'. *J. Ecol.*. **21**, 1–43.

MOREAU, R.E. (1935b) 'Some eco-climatic data for closed evergreen forest in tropical Africa', *J. Linn. Soc.* (*Zool.*), **39**, 285–

MOREAU, R.E. (1936) 'A contribution to the ornithology of Mt Kilimanjaro and Mt Meru', *Proc. zool. Soc. Lond.*, part 4, 843–91.

MOREAU, R.E. (1938) 'Climatic classification from the stand-point of East African biology', *J. Ecol.*, **26**, 467–96.

MOREAU, R.E. (1963a) 'The distribution of tropical African birds as an indicator of past climatic changes' in *African Ecology and Human Evolution*, no. 36. New York: Viking Fund Publications.

MOREAU, R.E. (1966) *The Bird Faunas of Africa and Its Islands*. Pp. 424. New York: Academic Press.

MOREAU, R.E. (1972) *The Palaearctic-African Bird Migration Systems*. Academic Press. Pp. 386.

MOREAU, R.E. and MOREAU, W.M. (1943) 'An introduction to the epiphytic orchids of East Africa', *J. E. Africa Nat. Hist. Soc.*, **17**, 1–32.

MORGAN, W.T.W. (1967) *Nairobi city and region*, Oxford University Press.

MORGAN, W.T.W. (1970) *East Africa its people and resources*, pp. 312. Oxford University Press.

MORRISON, C.G.T., HOYLE, A.C. and HOPE-SIMPSON, J.F. (1948) 'Tropical soil vegetation catenas and mosaics', *J. Ecol.*, **36**, 1–84.

MORRISON, J. (1966) 'Productivity of grass and grass-legume swards in the Kenya highlands', *E. Afr. agric. For. J.*, **22**, 25–30.

MORRISON, M.E.S. (1961) 'Pollen analysis in Uganda', *Nature*, **190**, 483–6.

MORRISON, M.E.S. (1966) 'Low latitude vegetational

history with special reference to Africa'. Pp. 142–8. *Roy. Meter. Soc. Proc. Internat. Symp. World.*, climate 8000 to 0 B.C.

MORRISON, M.E.S. (1968) 'Vegetation and climate in the uplands of South-west Uganda during the later Pleistocene period. I. Muchoya Swamp, Kigizi district', *J. Ecol.*, **156**, 363–84.

MORTON, J.K. (1961) 'The upland floras of West Africa, their composition, distribution and significance in relation to climatic changes', *Proc. 4th Conf. AETFAT, Lisbon*, 391–409.

MORTON, J.K. (1967) 'The flora of the Lorna Mountains and Tingi Hills (Sierra Leone)'. *Palaeoecology of East Africa*, **2**, 67–9.

MOSS, B. (1969) 'Limitations of algal growth in some central African waters', *Limnol. Oceanogr.*, **14**, 591–601.

MOSS, R.P. (ed.) (1968) *The Soil Resources of Tropical Africa*. Pp. 226, Cambridge: University Press.

MUIR, A., ANDERSON, B. and STEPHEN, I. (1957) 'Characteristics of some Tanganyika soils', *J. Soil Sci.*, **8**, 1–17.

MULLER, D. (1967) 'Dry matter production in tropical rain forest', *J. Ecol.*, Pp. 20.

MUNRO, P.E. (1966a) 'Inhibition of nitrate-oxidisers by roots of grasses', *J. Appl. Ecol.*, **3**, 227–9.

MUNRO, P.E. (1966b) 'Inhibiting of nitrifyers by grass root extracts', *J. Appl. Ecol.*, **3**, 231–9.

MURDOCK, G.P. (1960) *Africa: Its Peoples and Their Culture and History*. Pp. 456. New York: McGraw-Hill.

MURPHY, P.W. (1953) 'The biology of forest soils with special reference to the mesofauna or meiofauna', *J. Soil Sci.*, **4**, 155–93.

MUTERE, F.A. (1965) 'The biology of the African fruit bat, *Eidolon helvum*'. Kampala: Ph.D thesis, University of East Africa.

NAKAMIRA, K. (1969) 'Equatorial westerlies over East Africa and their climatological significance', *Japanese Progress in Climatology*, 9–27.

NAPPER, D.M. (1963–71) 'Cyperaceae of East Africa I', *J.E. Afr. nat. Hist. Soc.* **24**, 1–18 (1963). 'II', *Ibid.*, (1964) **24**, 23–46. 'III' (1965) *Ibid.*, **25**, 1–27. 'IV' (1966) *Ibid.*, **26**, 1–17. 'V' (1971) *Ibid.*, **28**, 1–24.

NAPPER, D.M. (1965) 'Grasses of Tanganyika with keys for identification'. Pp. 146. *Bulletin* 18, Ministry of Agriculture, Forests and Wildlife, Tanzania. Dar es Salaam: Govt. Printer.

NAVEH, Z. (1966a) 'Selection of promising pasture plants for northern Tanzania. III. The early performance of drought resistant *Glycine javanica* ecotypes', *E. Afr. agric. For. J.*, **22**, 103–7.

NAVEH, Z. (1966b) 'The determination of range condition and trend on East African range lands',

E. Afr. agric. For. J., **22**, 159–62.

NAVEH, Z. and ANDERSON, G.D. (1966a) 'The introduction and selection of promising pasture plants for the Arusha and Kilimanjaro regions of north Tanzania. I. Introduction – problems of pasture research and development', *E. Afr. agric. For. J.*, **22**, 41–4.

NAVEH, Z. and ANDERSON, G.D. (1966b) 'Selection of promising pasture plants for northern Tanzania. II. A comparison of drought resistant selections of Rhodes grass (*Chloris gayana* Kunth) and Buffel grass (*Cenchrus ciliaris* L.)', *E. Afr. agric. For. J.*, 96–102.

NEWBOULD, P.J. (1967) *Methods for Estimating the Primary Production of Forests*. Pp. 62. IBP Handbook No. 2, Oxford and Edinburgh: Blackwells Scientific Publications.

NICHOLSON, J.W. (1936) 'The influence of forests on climate and water supply in Kenya', *E. Afr. agric. J.*, **2**, 48–53.

NILSSON, E. (1932) 'Quaternary glaciations and pluvial lakes in British East Africa', Ph.D. thesis, Stockholm.

NORRIS, D.O. (1962) 'The biology of nitrogen fixations'. Pp. 113–129. In *A Review of Nitrogen in the Tropics with Special Reference to Pastures. A Symposium*. Bulletin 46. Farnham Royal, England: Commonwealth Agricultural Bureau.

NTENGA, G. (1967) *Economics of beekeeping in miombo woodlands of Tanzania*. Pp. 4. Tabora: Ministry of Agriculture, Forests and Wildlife. Forest Division, Beekeeping Section. Cyclostyled.

NYE, P.H. (1954–55) 'Some soil forming processes in the humid tropics. I. A field study of a catena in the West African forest', *J. Soil Sci.*, **5**, 7–21 (1954). 'II. The development of the upper slope member of the catena', *Ibid.*, **6**, 51–62 (1955). 'III, Laboratory studies on the development of a typical catena over granitic gneiss', *Ibid.*, **6**, 63–72 (1955). 'IV. The action of the soil fauna', *Ibid.*, **6**, 73–83 (1955).

NYE, P.H. (1961) 'Organic matter and nutrient cycles under moist tropical forest', *Pl. Soil*, **13**, 333–46.

NYE, P.H. (1963) 'Soil analysis and the assessment of fertility in tropical soils', *J. Sci. Fd. Agric.*, **14**, 277–80.

NYE. P.H. (1968) 'Processes in the root environment', *J. Soil Sci.*, **19**, 205–15.

NYE, P.H. and GREENLAND, D.J. (1960) *The Soil Under Shifting Cultivation*. Pp. 156. Technical communication **51**, Farnham Royal, England: Commonwealth Agricultural Bureau.

OLLIER, C.D. (1959a) *The Soils of the Northern Province of Uganda*. Memoirs Research Division, series I, no. 3. Kawanda, Kampala: Department of

Agriculture, Uganda. Cyclostyled.

OLLIER, C.D. (1959b) 'A two-cycle theory of tropical pedology,' *J. Soil Sci.*, **10**, 137–48.

OLLIER, C.D. and HARROP, J.E. (1959) *The Soils of the Eastern Province of Uganda*. Memoirs Research Division, series 1, no. 2. Kawanda, Kampala: Department of Agriculture, Uganda. Cyclostyled.

OLPHEN, H. van (1963) *An Introduction to Clay Colloidal Chemistry for Clay Technologists, Geologists and Soil Scientists*. Pp. 301. New York: Interscience Publishers.

OSMASTON, H.A. (1951) 'The termite and its uses for food', *Uganda J.*, **15**, 80–3.

OSMASTON, H.A. (1953) 'Kalinzu forest fruit bats', *J. E. Africa nat. Hist. Soc.*, **22**, 74–5.

OSMASTON, H.A. (1956a) 'Determination of age-girth and similar relationships in tropical forestry', *Emp. For. Rev.*, **35**, 193–7.

OSMASTON, H.A. (1956b) *Working Plan for the Budongo, Siba and Kitogo Forest Reserves*. 2nd revision. Period 1955–1964. Pp. Entebbe: Uganda Forest Department. Cyclostyled.

OSMASTON, H.A. (1959a) *Working Plan for the Kibale and Itwara Forests*. First revision. Period 1959–1965. Entebbe: Uganda Forest Department. Cyclostyled.

OSMASTON, H.A. (1959b) *Working Plan for the Bugoma Forest*. First revision. Period 1960–1970. Pp. 33. Entebbe: Uganda Forest Department. Cyclostyled.

OSMASTON, H.A. (1960) *Working Plan for the Kalinzu Forest*. Period 1960–1970. Pp. 25. Entebbe: Uganda Forest Department.

OSMASTON, H.A. (1965) 'Pollen and seed dispersal in *Chlorophora excelsa* and other Moraceae and in *Parkia filicoidea* with special reference to the role of fruit bat *Eidolon helvum*', *Commonw. For. Rev.*, **44**, 96–103.

OSMASTON, H.A. (1965) 'The past and present climate and vegetation of Ruwenzori and its neighbourhood'. D. Phil. thesis, Oxford.

OSMASTON, H.A. (1966) 'Pollen analysis in the study of the past vegetation and climate of Ruwenzori and its neighbourhood' *Palaeoecology of Africa* 1, 48–50.

OSMASTON, H.A. (1967) 'Plant fossils in volcanic tuffs near the Ruwenzori', *Palaeoecology of Africa*, **2**, 25–6.

OSMASTON, H.A. (1971) 'The vegetation of Murchison Falls and Queen Elizabeth National Parks', in *Uganda National Parks Handbook*. Longman (Uganda) for the Trustees.

PARRY, M.S. (1953–54) 'Tree planting in Tanganyika. I. Methods of planting', *E. Afr. agric. J.*, **18**, 102–15 (1953). 'II. Species for the highlands', *Ibid.*, **19**, 89–102 (1953). 'III. Species for dry areas', *Ibid.*, **19**,
154–60. (1954).

PARRY, M.S. (1966) 'Recent progress in the development of miombo woodland in Tanganyika', *E. Afr. agric. For. J.*, **31**, 307–16.

PEARSALL, W.H. (1950) *Mountains and Moorlands*. Pp. 312, London: Collins.

PEARSALL, W.H. (1956) *Report on an Ecological Survey of the Serengeti National Park, Tanganyika*. Pp. 64. London: The Fauna Preservation Society.

PENDLETON, R.L. and SHARASUVINA (1946) 'Analysis of some Siamese laterites', *Soil Sci.*, **62**, 423–40.

PENMAN, H.L. (1948) 'Natural evaporation from open water, bare soil and grass', *Proc. R. Soc.*, **193**, A, 120–45.

PENMAN, H.L. (1950) 'The dependence of transpiration on weather and soil conditions', *J. Soil Sci.*, **1**, 74–89.

PENMAN, H.L. (1956) 'Estimating evaporation'. *Trans. Am. Geophys. Un.*, **37**, 43–50.

PENMAN, H.L. (1963) *Vegetation and Hydrology*. Pp. 124. Technical Communication 53. Farnham Royal, England: Commonweath Bureau of Soils.

PEREIRA, H.C. (1953) 'Interception of rainfall by Cypress plantations', *E. Afr. agric. J.*, **18**, 73–6.

PEREIRA, H.C. (1954) 'The physical importance of forest cover in the East African highlands', *E. Afr. agric. For. J.*, **19**, 233–6.

PEREIRA, H.C. (1953) 'The assessment of structure in tropical soils', *J. Agric. Sci. Camb.*, **45**, 401–10.

PEREIRA, H.C. (1956) 'A rainfall test for structure of tropical soils', *J. Soil Sci.*, **7**, 68–74.

PEREIRA, H.C. (1957) 'The seasonal assessment of water needs in the irrigation of coffee', *E. Afr. agric. J.*, **22**, 188–93.

PEREIRA, H.C. (1962) 'Hydrological effects of changes in land use in some East African catchment areas', *E. Afr. Agric. For. J.*, **27**, 1–129.

PEREIRA, H.C., CHENERY, E.M. and MILLS, W.R. (1954) 'The transient effects of grasses in the structure of tropical soils', *Emp. J. exp. Agric.*, **22**, 148–60.

PEREIRA, H.C., DAGG, M. and HOSEGOOD, P.H. (1962) 'Hydrological effects of changes of land use in some East African catchment areas', *E. Afr. Agric. For. J.*, **27**, 1–131 Special communication.

PEREIRA, H.C. and HOSEGOOD, P.H. (1962a) 'Comparative water use of softwood plantations and bamboo forest', *J. Soil Sci.*, **13**, 299–313.

PEREIRA, H.C. and HOSEGOOD, P.H. (1962b) 'Soil moisture effects of *kudzu* as a clove orchard cover crop', *E. Afr. agric. For. J.*, **27**, 225–9.

PERRIN, R.M.S. (1965) In *Experimental Pedology*. Hallsworth and Crawford (eds). London: Butterworth.

PHILIP, M.S. (1964) *Working Plan for Budongo Central Forest Reserve, Including Budongo, Siba and Kitigo Forests*. 3rd revision. Pp. 130. Entebbe: Govt. Printer.

PHILIP, M.S. (1965) 'The management of tropical high forest with special reference to the introduction of monocyclic felling in Uganda', *E. Afr. agric. For. J.*, 31, 100–8.

PHILIPS, J.F.V. (1929) 'Some important vegetation communities in the Central province of Tanganyika', *S. Afr. J. Sci.*, 26, 332–72.

PHILIPS, J.F.V. (1930a) 'Fire: its influence on biotic communities and physical factors in South and East Africa', *S. Afr. J. Sci.* 27, 352–67.

PHILIPS, J.F.V. (1930b) 'Vegetation communities of Central province of Tanganyika', *J. Ecol.*, 18, 193–234.

PHILIPS, J.F.V. (1931) 'A sketch of the floral regions of Tanganyika territory', *Trans. R. Soc. S. Afr.*, 19, 363–72.

PHILIPS, J.F.V. (1959) *Agriculture and Ecology in Africa*. Pp. 412. London: Faber and Faber.

PHILIPS, J.F.V. (1968) 'The influence of fire in Trans-Saharan Africa', *Acta Phytogeogr. Suec.*, 54, 13–20.

PIELOU, E.C. (1952) 'Notes on the vegetation of Rukwa Rift valley, Tanganyika', *J. Ecol.*, 40, 383–92.

PIERS, F. (1968) *Orchids of East Africa*. Pp. 304. Lehre: J. Cramer.

PINKERTON, J. (1967) 'Copper deficiency of wheat in the Rift valley, Kenya', *J. Soil Sci.*, 18, 18–26.

PINKERTON, J., BARRETT, M.W. and GUTHRIE, E.J. (1965) 'A note on copper deficiency in the Njoro area, Kenya', *E. Afr. agric. For. J.*, 30, 257–62.

PITMAN, C.R.S. (1934) 'The Mabira forest', *Uganda J.*, 1, 7–16.

PITT-SCHENKEL, C.J.W. (1938) 'Some important communities of warm temperate rain forest at Magambo, West Usambara', *J. Ecol.*, 26, 50–75.

PLUMTRE, R.A. *Uganda Timber Users Handbook*. Pp. 16. Bulletin no. 9. Entebbe: Information and Visual Aids Centre.

POLE EVANS, I.B. (1948) *Roadside Observations on the Vegetation of East and Central Africa*. Pp. 305. Botanical Survey Memoir 22. Department of Agriculture, South Africa. Pretoria: Govt. Printer.

POLHILL, P. (1962) 'Common perennial lilies of Kenya with ephemeral flowering shoots', *J. E. Afr. Nat. Hist. Soc.* 24, 1–25.

POORE, M.E.D. (1968) 'Studies in Malayan rain forest. I. The forest on triassic sediments in Jengka Forest Reserve', *J. Ecol.*, 56, 143–96.

POULTNEY, R.G. (1959) 'Preliminary investigations on the effect of fertilizers applied to natural grassland', *E. Afr. agric. J.*, 25, 47–9.

PRATT, D.J. (1964) 'Bush control studies in the drier areas of Kenya. II. Preliminary work with the Holt machines', *Emp. J. Expt. Agric.*, 32, 18–24.

PRATT, D.J. (1964) 'An evaluation of the Holt IXa "Bushbreaker" in *Tarchonanthus–Acacia* thicket', *J. Appl. Ecol.*, 3, 97–115.

PRATT, D.J. (1966) 'Control of *Disperma* in semi-desert dwarf shrub grassland', *J. Appl. Ecol.*, 3, 277–91.

PRATT, D.J., GREENWAY, P.J. and GWYNNE, M.D. (1966) 'A classification of East African rangeland with an appendix on terminology', *J. appl. Ecol.*, 3, 369–82.

PRESCOTT, J.A. (1950) 'A climatic index for the leaching factor in soil formation', *J. Soil Sci.*, 1, 9–19.

PRESCOTT, J.A. and PENDLETON, R.L. (1952) *Laterite and Lateritic Soils*. Pp. 51. Technical Communication 47. Farnham Royal, England: Commonwealth Bureau of Soils.

PUDDEN, H.H.C. (1957) *Exotic Forest Trees in the Kenya Highlands*. Pp. 34. Nairobi: Govt. Printer.

PURSEGLOVE, J.W. (1968) *Tropical crops; Dicotyledons I*. Pp. 332. *II*. Pp. 387. Longman.

RADWANSKI, S.A. (1960) *The Soils and Land Use of Buganda*. Memoirs of the Research Division, series 1, no. 4. Pp. 134. Kawanda, Kampala: Department of Agriculture, Uganda. Cyclostyled.

RADWANSKI, S.A. and OLLIER, C.D. (1959) 'A study of an East African catena', *J. Soil Sci.*, 10, 149–68.

RATTRAY, J.M. (1961) *The Grass Cover of Africa*. Pp. 168. Rome: Food and Agricultural Organization of the United Nations.

REA, R.J.A. (1935) 'Forest type vegetation in Tanganyika territory', *Emp. For. J.*, 14, 202–8.

RENSBERG, H.J. van (1947) 'The role of pasture development in soil conservation in Tanganyika territory', *E. Afr. agric. J.*, 13.

RENSBERG, H.J. VAN (1954) 'Run-off and soils erosion tests, Mpwapwa, Central Tanganyika', *E. Afr. agric. J.*, 20, 228–31.

REYNOLDS, G.W. (1950) *The Aloes of South Africa*. PP. 520. Johannesburg: The Aloes of South Africa Book Fund.

REYNOLDS, G.W. (1966) *The Aloes of Tropical Africa and Madagascar*. Pp. 537. Mbabane, Swaziland: The Trustees of the Aloes Book Fund.

RICH, F. (1932) 'Report of the Percy Sladen expedition to the Rift Valley lakes of Kenya in 1929. IV. Phytoplankton from the Rift Valley lakes of Kenya', *Ann. Mag. Nat. Hist.* X 233–62.

RICH, F. (1933) 'Scientific results of the Cambridge expedition to the East African lakes, The Algae.' *J. Linn. Soc. (Zool)*, 38, 249–75.

RICHARDS, P.W. (1952) *The Tropical Rain Forest.* 1st edn. Pp. 450. Cambridge: University Press.

RICHARDS, P.W. (1962) 'Plant life and tropical climate'. Pp. 67–75. In *Biometrology.* London: Pergamon Press.

RICHARDS, P.W. (1963a) 'What the tropics can contribute to ecology', *J. Ecol.*, 51, 231–41.

RICHARDS, P.W. (1963b) 'Soil conditions in some Bornian lowland plant communities'. Reprinted from *Symposium on Ecological Research in Humid Tropics Vegetation.* Kuching Sarawak.

RICHARDS, P.W. (1963c) 'Ecological notes on West African vegetation. III. The upland forests of Cameroon Mountain', *J. Ecol.*, 51, 529–54.

RICHARDS, P.W. (1969) 'Speciation in the tropical forest and the concept of the niche', *Biol. J. Linn. Soc.*, 1, 149–53.

RICHARDSON, J.L. (1966) 'Changes in level of Lake Naivasha, Kenya, during post glacial times', *Nature*, 209, 290–1.

RICHARDSON, J.L. (1968) 'Diatoms and lake typology in East and Central Africa', *Int. Rev. ges. Hydrobiol.*, 53, 238–99.

RICHARDSON, J.L. (1972) 'Palaeolimnological records from Rift Lakes in Central Kenya', *Palaeoecology of Africa*, 6, 131–6.

RICHARDSON, J.L. and RICHARDSON, A.E. (1972) 'History of an African Rift Lake and its climatic implications', *Ecological monographs*, 42, 499–534, Fall.

RIJKS, D.A. (1964) *Tables for the Calculation of Evaporation from Open Water or Plants for use in Uganda.* Namulonge, Kampala: Cotton Research Station. Cyclostyled.

RIJKS, D.A. (1968) 'Agrometerology in Uganda – a review of methods', *Expl. Agric.*, 4, 263–74.

RIJKS, D.A. and OWEN, W.G. (1965) *Hydrometrical Records from Areas of Potential Agricultural Development in Uganda.* Ministry of Mineral and Water Resources. Uganda Govt.

RIQUIER, J. (1955) 'Effects of krilium on the structure of two types of tropical soils', *African Soils*, 3, 238–49.

ROBERTSON, A.G. and BERNACCA, J.P. (1957) 'Game elimination as a tsetse control measure in Uganda', *E. Afr. Agric. J.*, 23, 254–61.

ROBINSON, J.B.D. (1958) 'Some chemical characteristics of termite soil in Kenya coffee fields', *J. Soil Sci.*, 9, 58–65.

ROBINSON, J.B.D. (1967) 'The effect of exotic softwood crops in the chemical fertility of a tropical soil', *E. Afr. agric. For. J.*, 23, 175–89.

ROBINSON, J.B.D. (1968) 'A simple, available-soil-nitrogen index. II. Field crop evaluation', *J. Soil Sci.*, 19, 280–90.

ROBINSON, J.B.D. and CHENERY, E.M. (1958) 'Magnesium deficiency in coffee with special reference to mulching', *Emp. J. exp. Agric.*, 26, 259–73.

ROBINSON, J.B.D. and GACOKA, P. (1962) 'Evidence of upwards movement of nitrate during the dry season in the Kikuyu red loam coffee soil', *J. Soil Sci.*, 13, 133–40.

ROBYNS, W. (1946) 'Sur l'existence du *Juniperus procera* Hochst. au Congo Belge', *Bull. Jard. bot. Etat Brux.*, 18, 125–31.

ROBYNS, W. (1948) *Les Territoires Biogeographiques du Parc National Albert.* Pp. 51. Brussels: Institut des Parcs Nationaux du Congo Belge.

RODIN, L.E. and BAZILEVICH, N.I. (1968) *Production and Mineral Cycling in Terrestrial Vegetation.* Pp. 288. Edinburgh and London: Oliver and Boyd.

RODRIGUES, G. (1954) 'Fixed ammonia in tropical soils', *J. Soil Sci.*, 5, 264–74.

ROSE, C.W. (1961) 'Rainfall and soil structure', *Soil Sci.*, 91, 49–54.

ROSE, C.W. (1962) 'Some effects of rainfall, radiant drying, and soil factors on infiltration under rainfall into soils', *J. Soil Sci.*, 13, 286–98.

ROSE, D.A. (1968) 'Water movement in dry soils. I. Physical factors affecting sorption of water by dry soil', *J. Soil Sci*, 19, 81–93.

ROSS, R. (1955a) 'Some aspects of the vegetation of the sub-alpine zone on Ruwenzori', *Proc. Linn. Soc. London*, session 165, part 2, 136–40.

ROSS, R. (1955b) 'The algae of the East African Great Lakes', *Proc. Internat. Ass. Theoretical and Appl. Limnol.*, 12, 320–6.

ROSS, I.C., FIELD, C.R. and HARRINGTON, G.N. (1973) *Ibid.* 3. 'Animal populations and park management recommendations', *E. Afr. Wildl. J.* In preparation.

ROWELL, T.E. (1966) 'Forest living baboons in Uganda', *J. Zool.*, 149, 344–64.

RUSSELL, E.W. (1960) 'The level of humic carbon and nitrogen in tropical soil', *African soils*, 5, 101–5.

RUSSELL, E.W. (1961) *Soil Conditions and Plant Growth.* 9th edn. Pp. 688. London: Longman.

RUSSELL, E.W. (ed.) (1962) *The Natural Resources of East Africa.* Pp. 144. Nairobi: East African Literature Bureau.

RUSSELL, F.C. and DUNCAN, D.L. (1956) *Minerals in Pasture.* Technical Communication 15. Commonwealth Bureau of Animal Nutrition.

RUSSELL, T.A. (1965) 'The *Raphia* palms of West Africa', *Kew Bull.*, 19, 173–96.

RUTTER, A.J. (1967) 'Evaporation in forests', *Endeavour*, 26, 39–43.

SAGGERSON, E.P. (1962) 'The geology of E. Africa' in

Russell (1962).

SALT, G. (1951) 'The Shira plateau of Kilimanjaro', *Geogr. J.*, 117, 150–64.

SALT, G. (1954) 'A contribution to the ecology of upper Kilimanjaro', *J. Ecol.*, 42, 375–423.

SALTER, P.J. and WILLIAMS, J.B. (1965) 'The influence of texture on the moisture characteristics of soils. I. A critical comparison of techniques for determining the available water capacity and moisture characteristic curve of a soil', *J. Soil Sci.*, 16, 1–6. II. 'Available-water capacity and moisture release characteristics' *Ibid.*, 16, 310–17.

SALTER, P.J., BERRY, G. and WILLIAMS, J.B. (1966) 'The influence of texture on the moisture characteristics of soils. III. Quantitative relationships between particle size, composition and available-water capacity', *J. Soil Sci.*, 17, 93–8.

SANFORD, W.F. (1968) 'Distribution of epiphytic orchids in semi-deciduous forest in southern Nigeria', *J. Ecol.*, 56, 697–705.

SANSOM, H.W. (1954) 'The climate of East Africa, based on Thornthwaite's classification', *E. Afr. Meteorol. Dept. Mem.*, 3, 1–49.

SAUNDER, D.H. (1956) 'Determination of available phosphate in tropical soils by extraction with sodium hydroxide', *Soil Sci.*, 82, 457–63.

SAYERS, G.F. (1930) *Handbook of Tanganyika*. Pp. 636. London: Macmillan.

SCHALLER, G.B. (1963) *The Mountain Gorilla*. Pp. 431. Chicago: University Press.

SCHOENMAEKERS, J. and CHENERY, E.M. (1959) 'Magnesium deficiency in tea bushes in East Africa', *E. Afr. agric. J.*, 25, 25–7.

SCHOFIELD, R.K. (1955) 'Can a precise meaning be given to available soil phosphate?' *Soils Fert.*, 18, 373–5.

SCHOLANDER, P.F., LOVE, W.E. and KANWISKER, J.W. (1955) 'The rise of sap in tall grapevines', *Pl. Physiol. Lancaster*, 30, 93–104.

SCHOLANDER, P.F., RUDD, B. and LEIVESTAD, H. (1957) 'The rise of sap in a tropical liana', *Pl. Physiol. Lancaster*, 32, 1–6.

SCHULZ, J.P. (1960) *Ecological Studies on Rain Forest in Northern Suriname. The vegetation of Suriname. 2.* Verh. K. ned. Akad. Wet. 13(1) 1–267.

SCHUTTE, K.H. (1955) 'A survey of plant minor element deficiencies in Africa', *African soils*, 3, 284–97.

Scientific Council for Africa South of the Sahara (1965) Specialist meeting on phytogeography. Yangambi 28 July–8 August 1956 CCTA CSA. Publ. no. 22.

SCOTT, J.D. (1934) 'The ecology of certain plant communities of the Central province of Tanganyika territory', *J. Ecol.*, 22, 177–229.

SCOTT, R.M. (1962) 'Exchangeable bases of mature, well-drained soils in relation to rainfall in East Africa', *J. Soil Sci.*, 13, 1–9.

SCULTHORPE, C.D. (1967) *The Biology of Aquatic Vascular Plants.* Pp. 610. London: Arnold.

SHANTZ, H.L. and MARBUT, C.F. (1923) *The Vegetation and Soils of Africa*, Pp. 263. New York: Amer. Geogr. Soc. and Nat. Res. Council.

SHANTZ, H.L. and TURNER, B.L. (1958) 'Photographic documentation of vegetation in Africa over a third of a century'. Report 169. University of Arizona.

SHMUELI, E. (1960) 'Chilling and frost damage in banana leaves', *Bull. Res. Council of Israel*, 8 D, 225–38.

SIMMONDS, N.W. 1962 *The Evolution of the Bananas.* Pp. 170. London: Longman.

SIMPSON, J.R. (1960) 'The mechanism of surface nitrate accumulation on a bare fallow soil in Uganda', *J. Soil Sci.*, 11, 45–60.

SIMPSON, J.R. (1961) 'The effects of several agricultural treatments on the nitrogen status of a red earth in Uganda', *E. Afr. agric. For. J.*, 26, 158–63.

SINGER, R. and MORELLO, J.H. (1960) 'Ectotrophic mycorrhiza and forest communities', *Ecology*, 41, 549–51.

SINGH, S. (1954) 'A study of the black cotton soils with special reference to their colouration', *J. Soil Sci.*, 5, 289–99.

SINGH, S. (1956) 'The formation of dark-coloured clay-organic complexes in black soils', *J. Soil Sci.*, 7, 43–58.

SLAYTER, R.O. (1967) *Plant Water Relationships.* London and New York: Academic Press.

SMALL, W. (1937) 'The unsuitability of certain virgin soils to the growth of grain crops', *E. Afr. agric. J.*, 2, 348.

SMITH, F.G. (1956) *Beeswax.* Pp. 15. Dar es Salaam: Govt. Printer.

SMITH, F.G. (1958) 'Beekeeping observations in Tanganyika', *Bee World*, 39, 29–36.

SMITH, F.G. (1960) *Beekeeping in the Tropics.* Pp. 265. London: Longman.

SMITH, F.G. (1966) 'Beekeeping as a forest industry', *E. Afr. agric. For. J.*, 31, 350–5.

SMITH, J. (1949) *Distribution of Species in the Sudan in Relation to Rainfall and Soil Texture.* Bulletin 4. Ministry of Agriculture Republic of Sudan.

SNOWDEN, J.D. (1933) 'A study in altitudinal zonation in south Kigezi and on Mounts Muhavura and Mahinga, Uganda', *J. Ecol.*, 21, 7–26.

SNOWDEN, J.D. (1953) *The Grass Communities and Mountain Vegetation of Uganda.* Pp. 94, London: Crown Agents.

SOMEREN VAN and BALLY P.R.O. (1939) 'Chyulu hill forest, Kenya', *J. Kenya and Uganda Nat. Hist. Soc.*, 14, 1–14 and 161–6.

SOPER, R.C. (1969) 'Radio carbon dating of "Dimpled-base ware" in W. Kenya', *Azania*, 4, 148–52.

SOUTHERN, H.N. and HOOK, O. (1963) 'A note on small mammals in East African forests', *J. Mammal*, 44, 126–9.

SPINAGE C.A. and GUINESS F.E. 'Tree survival in the absence of elephants in Akagere National Park, Ruanda', *J. Appld. Ecol.*, 8, 723–

STAPLES, R.R. (1936) 'Vegetation types and water supplies', *E. Afr. agric. J.*, 1, 453–8.

STAPLES, R.R. (1942) 'Combating soil erosion in the Central province of Tanganyika territory. I', *E. Afr. agric. J.*, 7, 156–65; 'II' *Ibid.*, 7, 190–5.

STAPLES, R.R., HORNBY, H.E. and HORNBY, R.M. (1942) 'A study of the comparative effects of goats and cattle on a mixed grass-bush pasture', *E. Afr. agric.. J.*, 8, 62–70.

STEENIS J. VAN (1935) 'On the origin of the Malaysian mountain flora. II. Altitudinal zones, general consideration and renewed statement of the problem', *Bull. Jard. Bot. Buitenz.*, (Sec 3) 13, 289–417.

STEPHENS, D. (1962) 'Upward movement of nitrate in a bare soil in Uganda', *J. Soil. Sci.*, 13, 52–9.

STEPHEN, I., BELLIS, E. and MUIR, A. (1956) 'Gilgai phenomena in tropical black clays of Kenya', *J. Soil Sci*, 7, 1–8.

STEPHENS, C.G. (1950) 'Comparative morphology and general relationship of certain Australian, N. American and European soils', *J. Soil Sci.*, 1, 123–49.

STEPHENS, J.E.M. (1955) 'Tree growth in a seasonally dry swamp in eastern Uganda', *E. Afr. agric. J.*, 20, 232–

STONEMAN, J. (1966) 'Development of fish farming in Uganda', *E. Afr. agric. For J.*, 31, 441–4.

STOREY, H.H. and LEACH, R. (1933) 'A sulphur deficiency disease of the tea bush', *Ann. appl. Biol.*, 20, 23–

STUART-SMITH, A.M. (1962a) *Working Plan for the Nyio Bamboo Forest Reserve, West Nile District, Uganda.* Period 1962–1972. Pp. 7. Entebbe: Uganda Forest Department. Cyclostyled.

STUART-SMITH, A.M. (1962b) *Working Plan for the Zoka Central Forest Reserve, Madi District, Uganda.* Period 1962–1972. Pp. 11. Entebbe: Uganda Forest Department. Cyclostyled.

SWABY, R.J. (1950) 'The influence of humus on soil aggregation', *J. Soil Sci.*, 1, 182–94.

SWYNNERTON, C.F.M. (1918) 'Some factors in the replacement of ancient East African forest by wooded pastureland', *S. Afr. J. Sci.*, 14, 122–

SYMOENS, J.J. (1968) *La Mineralisation des Eaux Naturelles.* Exploration hydrobiologique du bassin du lac Bangwerte et du Luapula. Vol 2, pt 1, pp. 199. Bruxelles.

SYNNOTT, T.J. (1971) 'Annotated list of the perennial woody vegetation of the West Ankole forests', *Ug. J.*, 35, 1–21.

TALBOT, L.M. *et al.* (1961) 'The possibility of using wild animals for animal production on East African rangelands, etc.', *Rept. L. Manyara Conf. Tanganyika.*, Feb. 1961.

TACK, C.H. (1969) *Uganda Timbers.* A list of the more common Uganda timbers and their properties. Pp. 138. Entebbe: Govt. Printer.

TACKHOLM, V. and MOHAMMED DRAR (1950) *Flora of Egypt.* Vol. 2, 1st edn. Pp. 547. Cairo: Fouad I University Press.

TALLING, J.F. (1957a) 'Diurnal changes of stratification and photosynthesis in some tropical African waters', *Proc. R. Soc.*, 147 B, 57–83.

TALLING, J.F. (1957b) 'Some observations on the stratification of Lake Victoria', *Limnol. Oceanogr.*, 3, 213–21.

TALLING, J.F. (1963) 'Origin of stratification in an African rift lake', *Limnol. Oceanogr.*, 8, 68–78.

TALLING, J.F. (1964) 'The annual cycle of stratification and primary production in Lake Victoria', *Verh. int. Verein. theor. angew. Limnol.*, 15, 384–5.

TALLING, J.F. (1965) 'The photosynthetic activity of phytoplankton in East African lakes', *Int. Rev. geo. Hydrobiol.*, 50, 1–32.

TALLING, J.F. (1966) 'The annual cycle of stratification and phytoplankton growth in Lake Victoria', *Int. Revue ges. Hydrobiol. Hydrogr.*, 51, 569–645.

TALLING, J.F. and TALLING, I.B. (1966) 'The chemical composition of African lake waters', *Int. Rev. ges. Hydrobiol. Hydrogr.*, 50, 421–63.

TARDIEU-BLOT, M.L. (1953) *Les Pteridophytes de l'Afrique Intertropicale Francaise.* Pp. 241. Memoires de l'Institut Francais d'Afrique Noire, no. 28.

TARDIEU-BLOT, M.L. (1964) *Flore du Cameroun.* Pp. 372. Paris: Museum National d'Histoire Naturelle.

TAYLOR, J. and THOMPSON, A. (1954) 'Notes on Sphagna from Uganda', *Kew Bull.*, 4, 517–21.

TAYLOR, W.R., (1960) *Marine Algae of the Eastern Tropical and Subtropical Coasts of the Americas.* Pp. 870. Ann Arbor: University of Michigan Press.

TEMPLE, P.H. (1964a) 'Lake Victoria levels', *Proc. E. Afr. Acad.*, 2, 50–8.

TEMPLE, P.H. (1964b) 'Evidence of lake-level changes from the northern shoreline of Lake Victoria, Uganda'. Pp. 31–56. From *Geographers and the Tropics*, Steel, R.W. and Prothero, R.M. (eds.).

TEMPLE, P.H. (1966) 'Evidence of changes in the level

231

of Lake Victoria and their significance'. Pp. 364. Ph.D. thesis: London.

THERON, J.J. (1951) 'The influence of plants on the mineralisation of nitrogen and the maintenance of organic matter in the soil', *J. Agric. Sci. Camb.*, **41**, 289–96.

THOMAS, A.S. (1940) 'Grasses as indicator plants in Uganda', *E. Afr. agric. J.*, **6**, 19–22.

THOMAS, A.S. (1941) 'The vegetation of the Sese islands, Uganda. An illustration of edaphic factors in tropical ecology', *J. Ecol.*, **29**, 330–53.

THOMAS, A.S. (1942a) 'Distribution of *Chlorophora excelsa* in Uganda', *Emp. For. Rev.*, **21**, 42–3.

THOMAS, A.S. (1942b) 'Lowland tropical podzols and Sphagnum', *Nature*, **149**, 195.

THOMAS, A.S. (1943) 'The vegetation of Karamoja district, Uganda', *J. Ecol.*, **31**, 149–77.

THOMAS, A.S. (1945–46) 'The vegetation of some hillsides in Uganda: illustrations of human influence in tropical ecology. Part I', *J. Ecol.*, **33**, 10–43 (1945). 'Part II', *Ibid.*, **33**, 153–72 (1946).

THOMAS, D.B. and PRATT, D.J. (1967) 'Bush control studies in the drier areas of Kenya. IV. Effects of controlled burning on secondary thicket in upland *Acacia* woodland', *J. Appl. Ecol.*, **4**, 325–35.

THOMAS, H.B. and SCOTT, R. (1953) *Uganda*. Pp.559. London: Oxford University Press.

THOMAS, I. (1963) 'The Brathay exploration group's expedition to Uganda 1962 III The flat roofed houses of the Sebei at Benet', *Uganda J.*, **27**, 115–22.

THOMPSON, B.W. (1965) *The Climate of Africa*. Pp. 132. Nairobi and New York: Oxford University Press.

THORNTHWAITE, C.W. (1948) 'An approach towards a rational classification of climate', *Geogr. Rev.*, **38**, 55–94.

TOBIAS, P.V. (1967) 'Fossil Hominids and Oldovai Gorge', *Paleoecology of E. Africa*, **2**, 49–51.

TOTHILL, J.G. (1940) *Agriculture in Uganda*. Pp. 551. London: Oxford University Press.

TRACEY, J.G. (1969) 'Edaphic differentiation of some forest types in eastern Australia. I. Soil physical factors', *J. Ecol.*, **57**, 805–16.

TRAPNELL, J.G., BRUNT, M.A. and BIRCH, W.R. (1970) *Map of Vegetation – Land Use Survey of S.W. Kenya*. Sheets 1–3, Survey of Kenya, Nairobi.

TRAPNELL, C.G. and GRIFFITHS, J.F. (1960) 'The rainfall-altitude relation and its ecological significance in Kenya', *E. Afr. agric. J.*, **25**, 207–13.

TROLL, C. (1959) 'Die tropische Gebirge', *Bonner geogr. Abhandl*, **25**.

TROLL, C. and WEIN, K. (1949) 'Der Lewisgletscher am Mt Kenya', *Geogr. Ann.*, **31**, 257–74.

TROUP, R.S. (1922a) *Report on Forestry in Kenya Colony*. London: Crown Agents.

TROUP, R.S. (1922b) *Report on Forestry in Uganda*. Pp. 39. London: Crown Agents.

TURNER, B.J. (1966) 'The composition, pattern and survival of savanna woodland in Bunyoro, Uganda'. Pp. 274. Ph.D. thesis: London.

TURNER, B.J. (1967) 'Ecological problems of cattle ranching on *Combretum* savanna woodland in Uganda', *E. Afr. geogr. Rev.*, **5**, 9–19.

TWEEDIE, E.M. (1965) 'Periodic flowering of some Acanthaceae on Mt Elgon', *J. E. Africa Nat. Hist. Soc.*, **25**, 92–4.

VAGELER, W.E. (1933) *An Introduction to Tropical Soils*. Pp. 240. H. Green (trans.). London: Macmillan.

VAIL, J.W. and CALTON, W.E. (1957) 'Die back of wattle—a boron deficiency', *E. Afr. agric. J.*, **23**, 100–3.

VAN ZINDEREN BAKKER, E.M. (1962) 'A late-glacial and post-glacial correlation between East Africa and Europe. *Nature, London*, **194**, 201–3.

VAN ZINDEREN BAKKER, E.M. (1964) A pollen diagram from equatorial Africa, Cherangani, Kenya. Geol. Mijnbouw, **43**, 123–8.

VAN ZINDEREN BAKKER, E.M. (1966) *Palaeoecology of Africa and of the surrounding islands and Antarctica*. vol. 1. A re-issue of *Palynology in Africa*. Reports 1–8 (1950–1936) pp. 270. Cape Town and Amsterdam: A. A. Balkema.

VAN ZINDEREN BAKKER, E.M. (1967) *Palaeoecology of Africa and of the surrounding islands and Antarctica*. vol. 2 covering the years 1964–1965. Cape Town and Amsterdam: A. A. Balkema.

VAN ZINDEREN BAKKER, E.M. (1969) 'Reconstruction of Quaternary climates'. *Palaeoecology of Africa*, **4**, 183–9.

VAN ZINDEREN BAKKER, E.M. (1972) A re-appraisal of late-Quaternary climatic evidence from Tropical Africa. *Palaeoecology of Africa*, **7**, 151–81.

VERDCOURT, B. (1962) 'The vegetation of the Nairobi Royal National Park', Pp. 38–49. In *The Wild Flowers of the Nairobi Royal National Park*. S. Heriz-Smith. Nairobi: E. Africa nat. Hist Soc.

VERDCOURT, B. (1963) 'The Miocene non-marine Mollusca of Rusinga Island, L. Victoria and other localities in Kenya', *Palaeontographica*, **121**, 1–37.

VERDCOURT, B. (1969) 'The arid corridors between the North East and South West areas of Africa', *Palaeoecology of Africa*, **4**, 140–44.

VESEY-FITZGERALD, D.F. (1963) 'Central African grasslands', *J. Ecol.*, **51**, 243–73.

VESEY-FITZGERALD, D.F. (1965) 'The utilization of natural pastures by wild animals in the Rukwa

Valley, Tanganyika', *E. Afr. Wildl. J.* **3**, 38–48.

VESEY-FITZGERALD, D.F. (1970) 'Origin and distribution of valley grasslands in E. Africa', *J. Ecol.*, **58**, 51–75.

VILLIERS, J.M. de (1964) 'Present soil-forming factors and processes in tropical and subtropical regions', *Soil Sci.*, **99**, 50–7.

VISSER, S.A. (1961) 'Chemical composition of rain water in Kampala, Uganda and its relation to meteorological and topographical conditions', *J. Geophysic. Res.*, **66**, 3759–65.

VISSER, S.A. and MIDDLETON, D. (1969) 'Investigation into the influence of the moisture content of the environment on the occurrence of soil microorganisms in the tropics', *Rev. Ecol. Biol. Soc.*, **2**, 99–113.

VOLKENS, G. (1897) *Der Kilimandscharo*. Berlin.

VOS, A. de and JONES, T. (1968) 'Proceedings of the symposium on wildlife management and land use'. Special issue *E. Afr. agric. For. J.*, **33**, 1–297.

WALKER, A. (1968) 'Lower Miocene fossil site of Bukwa, Sebei', *Ug. J.*, **32**, 149–56.

WALKER, A. (1969) 'Lower Miocene fossils from Mt Elgon, Uganda', *Nature*, **223**, 591–3.

WALKER, B. and SCOTT, G.D. (1968) 'Grazing experiments at Ukiriguru, Tanzania I. Comparisons of rotational and continuous grazing on natural pastures hardpan soils', *E. Afr. agric. For. J.*, **34**, 224–34. 'III. A comparison of three stocking rates on the productivity and botanical composition of natural pastures on hardpan soils', *Ibid.*, **34**, 245–55.

WALTER, H. (1963) 'Productivity of vegetation in arid countries; the savanna problem and the bush encroachment after overgrazing'. Pp. 17–20 in *The Impact of Man on the Tropical Environment*. 9th technical meeting. International Union for the Conservation of Nature and Natural Resources. Nairobi.

WALTER, H. (1964) *Die vegetation der Erde*. Vol 1, pp. 592. Stuttgart: Gustav Fischer.

WALTER, H. (1971) *Ecology of Tropical and subtropical Vegetation*. Pp. 539. Translation of *Vegetation der Erde I*. Edinburgh: Oliver and Boyd.

WALTER, H. and STEINER, M. (1936) 'Die okologie des Ostafrikanischen Mangroven', *Z. Bot.*, **30**, 65–193.

WAMBEKE, A.R. VAN (1962) 'Criteria for classifying tropical soils by age', *J. Soil Sci.*, **13**, 124–32.

WARMING, E. (1925) *Ecology of Plants; An Introduction to the Study of Plant Communities*. Pp. 422. London: Oxford University Press.

WASHBOURNE, C.K. (1967) 'Lake levels and quaternary climates in the E. Rift Valley of Kenya', *Nature*, **216**, 672–3.

WATKINS, G. (1960) *Trees and Shrubs for Planting in Tanganyika*. Pp. 158. Dar es Salaam; Govt. Printer.

WATSON, J.P. (1967) 'A termite mound in an Iron Age burial ground in Rhodesia', *J. Ecol.*, **55**, 663–9.

WATSON, J.P. (1969) 'Water movement in two termite mounds in Rhodesia', *J. Ecol.*, **57**, 441–51.

WATT, A.W.M. (1956) *Working Plan for Echuya Central Forest Reserve, Kigezi District, Uganda*. Period 1956–60. Pp. 15. Entebbe: Uganda Forest Department. Cyclostyled.

WATT, J.M. and BREYER-BRANDWYK, M.G. (1962) *Medicinal and Poisonous Plants of Southern and Eastern Africa*. 2nd edn. Pp. 1457. London: Livingstone.

WAYLAND, E.J. (1934) 'Rifts, rains and early man in Uganda', *J. Roy. Anthrop. Inst.*, **64**, 333–52.

WAYLAND, E.J. (1952) 'The study of past climates in tropical Africa'. *Proc. Pan African Congress on pre-history* 1947. L.S.B. Leakey (ed.). Oxford: Blackwell.

WAYLAND, E.J. and BRASNETT, N.V. (1938) *Soil Erosion and Water Supplies in Uganda*. Pp. 89. Geol. Survey, Uganda. Memoir No. 4. Entebbe: Govt. Printer.

WEATHERBY, J.M. (1962) 'Intertribal warfare on Mt Elgon in the nineteenth and twentieth centuries', *Uganda J.*, **26**, 200–12.

WEBB, L.J. (1969) 'Edaphic differentiation of some forest types in eastern Australia. II. Soil chemical factors', *J. Ecol.*, **57**, 817–30.

WEBER, N.A. (1959) 'Isothermal conditions in tropical soils', *Ecology*, **40**, 153–4.

WEBSTER, G. (1961) *Working Plan for South Mengo Forests*. First revision. Period 1961–1971. Pp. 43. Entebbe: Uganda Forest Department. Cyclostyled.

WELCH, J.R. (1960) 'Observations on deciduous woodland in the Eastern province of Tanganyika', *J. Ecol.*, **48**, 557–73.

WEST, G.S. (1907) 'Report on the freshwater algae, including phytoplankton of the 3rd Tanganyika expedition conducted by Dr. W.A. Cunnington, 1904–1905', *J. Linn. Soc. (Bot)*., **38**, 81–197.

WESELAAR, R. (1961) 'Nitrate distribution in tropical soils. I. Possible causes of nitrate accumulation near the surface after a long dry period', *Pl. Soil*, **15**, 110–20. 'II. Extent of capillary accumulation of nitrate during a long dry period', *Ibid.*, **15**, 121–32.

WHITMORE, T.C. (1969) 'First thoughts on species evolution in Malayan Macaranga', *Biol. J. Linn. Soc.* **1**, 223–31.

WHYTE, R.O. (1962) 'The myth of tropical grasslands'. *Trop. Agric.*, **39**, 1–11.

WILD, A. (1950) 'The retention of phosphate by soil', *J. Soil Sci.*, **1**, 221–38.

WILD, H. (1961) *Harmful Aquatic Plants of Africa and*

Madagascar. Kirkea 2, Pp. 68. Salisbury.

WILKINSON, W. (1940) 'Termites in East Africa; biology and control', *E. Afr. agric. J.*, 6, 67–72.

WILKINSON, W. (1965) 'The principles of termite control in forestry', *E. Afr. agric. For. J.*, 31, 212–17.

WILLAN, R.L. (1965) 'Natural regeneration of high forest in Tanganyika', *E. Afr. agric. For. J.*, 31, 43–53.

WILLIAMS, R.O. (1949) *The Useful and Ornamental Plants of Zanzibar and Pemba*. Pp. 497. Zanzibar: St Ann's Press, Timperley, Altrincham, England.

WILLIS, J.C. (1966) *A Dictionary of the Flowering Plants and Ferns*. 7th edn. Pp. 1214. Revised by H.K. Airy Shaw. Cambridge: University Press.

WILSON, J.G. (1960) *The Soils of Karamoja District, Northern Province, Uganda*. Pp. 70. Memoirs Research Division, series 1, no. 5 Kawanda, Kampala: Uganda Department of Agriculture. Cyclostyled.

WILSON, J.G. (1962) *The Vegetation of Karamoja District, Northern Province, Uganda*. Pp. 159. Memoirs Research Division, series 2, no. 5. Kawanda, Kampala: Uganda Department of Agriculture. Cyclostyled.

WIMBUSH, S.H. (1957a) 'Rainfall interception by cypress and bamboo', *E. Afr. agric. J.*, 13, 123–5.

WIMBUSH, S.H. (1947b) 'The African alpine bamboo', *E. Afr. agric. J.*, 13, 56–60.

WIMBUSH, S.H. (1950) *Catalogue of Kenya Timbers*. Pp. 74. Nairobi: Govt. Printer.

WINTER, B. de (1966) 'Remarks on the distribution of some desert plants in Africa', *Palaeoecology of Africa*, 1, 188–9.

WINTERBOTTOM, J.M. (1967) 'Climatological implications of avifaunal resemblances between S.W. Africa and Somaliland', *Palaeoecology of Africa*, 2, 77–9.

WOOD, G.H.S. (1960) *A Study of the Plant Ecology of Busoga District, Uganda Protectorate*. Pp. 69. Impr. For. Inst. paper 35.

WOOD, G.H.S. and Chenery, E.M. (1955) 'Regeneration of *Chlorophora excelsa* (Mvule) in Uganda in relation to soil-root conditions', *E. Afr. agric. J.*, 21, 34–41.

WOODHEAD, T. (1968a) *Studies of Potential Evaporation in Kenya*. Water Development Department, Kenya.

WOODHEAD, T. (1968b) *Studies of Potential Evaporation in Tanzania*. EASCO with Water Development Department, Tanzania.

WOODHEAD, T. (1970) 'A Classification of East African Rangeland. II. The Water Balance as a Guide to Site Potential', *J. Appld. Ecol.*, 7, 647–52.

WORTHINGTON, S. and E.B. (1933) *Inland Waters of Africa*. Pp. 259. London: MacMillan.

WURTZ, A.G. (1961) *Report to the Government of Uganda on an Experimental Fish Culture Project in Uganda, 1959–1960*. Pp. 32. Report no. 1387. Rome: Food and Agricultural Organisation of the United Nations. Cyclostyled.

ZEUNER, F.E. (1950) 'Frost soils on Mt Kenya and the relation of frost soils to aeolian deposits', *J. Soil Sci.*, 1, 20–30.

ZAHLBRUCKNER, A. and HAUMAN, L. (1936) 'Les lichens des hautes altitudes au Ruwenzori', *Mem. Inst. r. Colonial Belge*, 5, 1–31.

Floras

AGNEW A.D.Q. *et al Flora of Upland Kenya* In preparation. Oxford University Press.

ALSTON, A.H.G. (1959) *The Ferns and Fern Allies of West Tropical Africa*; a supplement to the 2nd ed. of the *Flora of West Tropical Africa*. Pp. 89. London: Crown Agents.

ANDREWS, F.W. (1950–56) *Flowering plants of the Sudan* Arbroath, Scotland. T. Buncle and Co.

EXELL, A.W. and LAUNERT, E., Edits (1960 in continuation) *Flora Zambesiaca* London, HMSO.

HUBBARD, C.E. and MILNE-REDHEAD, E. (eds.). (1952 in continuation) *Flora of East Tropical Africa* London. Crown Agents.

KEAY, R.W.J. and HEPPER (1954, 1958) *Flora of West Tropical Africa* 2nd Edition, London Crown Agents.

LIND, E.M. and TALLANTIRE, A.C. (1972) *Some Common Flowering Plants of Uganda*, 2nd edition, revised Tallantire London, Oxford University Press.

WHITE, F. (1962) *A Forest Flora of Northern Rhodesia*. Pp. 455, London, Oxford University Press.

Index of plant names

General Index

255